T0338625

A Guide to
Mathematical Methods
for Physicists

Advanced Topics and Applications

Advanced Textbooks in Physics

ISSN: 2059-7711

The *Advanced Textbooks in Physics* series explores key topics in physics for MSc or PhD students.

Written by senior academics and lecturers recognised for their teaching skills, they offer concise, theoretical overviews of modern concepts in the physical sciences. Each textbook comprises of 200–300 pages, meaning content is specialised, focussed and relevant.

Their lively style, focused scope and pedagogical material make them ideal learning tools at a very affordable price.

Published

A Guide to Mathematical Methods for Physicists: Advanced Topics and Applications
by Michela Petrini, Gianfranco Pradisi & Alberto Zaffaroni

Trapped Charged Particles: A Graduate Textbook with Problems and Solutions
edited by Martina Knoop, Niels Madsen & Richard C Thompson

Studying Distant Galaxies: A Handbook of Methods and Analyses
by François Hammer, Mathieu Puech, Hector Flores & Myriam Rodrigues

An Introduction to Particle Dark Matter
by Stefano Profumo

Quantum States and Scattering in Semiconductor Nanostructures
by Camille Ndebeka-Bandou, Francesca Carosella & Gérald Bastard

Forthcoming

An Introduction to String Theory and D-Brane Dynamics: With Problems and Solutions (Third Edition)
by Richard J Szabo

Advanced Textbooks in Physics

A Guide to Mathematical Methods for Physicists

Advanced Topics and Applications

Michela Petrini
Sorbonne Université, Paris, France

Gianfranco Pradisi
University of Rome Tor Vergata, Italy

Alberto Zaffaroni
University of Milano-Bicocca, Italy

World Scientific

NEW JERSEY · LONDON · SINGAPORE · BEIJING · SHANGHAI · HONG KONG · TAIPEI · CHENNAI · TOKYO

Published by

World Scientific Publishing Europe Ltd.

57 Shelton Street, Covent Garden, London WC2H 9HE

Head office: 5 Toh Tuck Link, Singapore 596224

USA office: 27 Warren Street, Suite 401-402, Hackensack, NJ 07601

Library of Congress Cataloging-in-Publication Data

Names: Petrini, Michela, author. | Pradisi, Gianfranco, 1961–   author. |
    Zaffaroni, Alberto, author.
Title: A guide to mathematical methods for physicists : advanced topics and applications /
    by Michela Petrini (Sorbonne Université, Paris, France), Gianfranco Pradisi
    (University of Rome Tor Vergata, Italy), Alberto Zaffaroni (University of Milano-Bicocca, Italy).
Description: New Jersey : World Scientific, 2018. | Series: Advanced textbooks in physics |
    Includes bibliographical references and index.
Identifiers: LCCN 2018012718 | ISBN 9781786345486 (hc : alk. paper)
Subjects: LCSH: Mathematical physics--Textbooks. | Physics--Textbooks. |
    Functional analysis--Textbooks.
Classification: LCC QC20 .P43495 2018 | DDC 530.15--dc23
LC record available at https://lccn.loc.gov/2018012718

British Library Cataloguing-in-Publication Data
A catalogue record for this book is available from the British Library.

For any available supplementary material, please visit
https://www.worldscientific.com/worldscibooks/10.1142/Q0162#t=suppl

Desk Editors: V. Vishnu Mohan/Jennifer Brough/Koe Shi Ying

Typeset by Stallion Press
Email: enquiries@stallionpress.com

Printed in Singapore

Preface

This book provides a self-contained and rigorous presentation of the main mathematical tools behind concrete physical problems in classical and quantum physics. It covers advanced topics that are usually treated in MSc programmes in physics: conformal mapping, asymptotic analysis, methods for solving integral and differential equations, among which Green functions, and an introduction to the mathematical methods of quantum mechanics. The book originates from lectures given at the École Normale Supérieure in Paris, the University of Milano-Bicocca and the University of Rome Tor Vergata, Italy. It complements the companion volume of the same authors *A Guide to Mathematical Methods for Physicists* Petrini *et al.* (2017) covering more elementary topics like complex functions, distributions and Fourier analysis.

The different arguments are organised in three main sections: Complex Analysis, Differential Equations and Hilbert Spaces, covering most of the standard mathematical tools in modern physics. These topics can be treated at an extremely elementary level or at a very mathematically advanced one. The aim of this book and of its companion is to keep a rigorous mathematical level, providing at the same time examples of the general theory rather than proving complicated theorems. For all the topics that involve advanced functional analysis, like the theory of linear operators in Hilbert spaces, we chose to emphasise the physicist point of view.

There are about 150 exercises with solutions, divided among the nine chapters. Together with the examples, they are a central part of the book and illustrate the general theory. The exercises range from very simple and basic, which the readers are invited to solve by themselves, to more difficult or theoretical ones, denoted with a star. The latter sometimes provide proofs of theorems given in the main text.

A selection of arguments in this book is appropriate for a one semester course in MSc programmes in Physics. The knowledge of real and complex calculus and Hilbert-space theory at the level of undergraduate studies is assumed. All necessary prerequisites can be found in the companion book *A Guide to Mathematical Methods for Physicists* Petrini *et al.* (2017), which also serves to establish notations and conventions. In order to be as self-contained as possible, we tried to limit the number of explicit references to the previous volume and we have recalled and summarised the relevant facts in the text or in Appendix A.

It is impossible to cover in a single book all mathematical methods used in physics with a satisfactory level of analysis. For this reason, some fundamental topics, like probability, group theory and differential geometry, are missing both in this book and in its companion. We have instead covered in details the mathematics of quantum physics, including a dedicated final chapter, using a physicist language and solving concrete problems, so that the book can be used in parallel to a (moderately) advanced course in quantum mechanics.

Michela Petrini is a professor in physics at Sorbonne Université in Paris. Gianfranco Pradisi is a professor at the University of Rome Tor Vergata, Italy. Alberto Zaffaroni is a professor at the University of Milano-Bicocca. They all work in high energy physics and mathematical physics, in particular string theory.

Contents

Part III. Hilbert Spaces

Part IV. Appendices

PART I
Complex Analysis

Introduction

Complex analysis is very rich and interesting. A central role is played by holomorphic functions. These are functions of a complex variable, $f : \mathbb{C} \to \mathbb{C}$, that are differentiable in complex sense. Differentiability on $\mathbb{C}$ is stronger than in the real sense and has important consequences for the behaviour of holomorphic functions. For instance, holomorphic functions are infinitely differentiable and analytic, and do not admit local extrema. The basic properties of holomorphic functions are discussed in the first volume of this book (see Petrini *et al.*, 2017) and are briefly summarised in Appendix A. In this volume we will discuss some extensions and applications of complex analysis that are relevant for physics. Typical applications are conformal mapping, integral transforms and saddle-point methods.

The idea of conformal mapping is to consider holomorphic functions as maps from the complex plane to itself. As holomorphic functions are analytic, these maps have very specific features. For instance, a bijective holomorphic function f preserves angles, namely it maps two lines intersecting in a point z_0 with a given angle to two curves whose tangents intersect in $f(z_0)$ with the same angle. Such a map f is called conformal. Conformal transformations can be used to map domains of the complex plane to simpler ones. In particular, any domain $\Omega \subset \mathbb{C}$ can be mapped to the unit disk by an appropriately chosen conformal map. This is very useful in physics to solve problems in two dimensions. As we will see in Section 1.3, one can find solutions of the Laplace equation on complicated domains of the plane by mapping them to the corresponding solution on the unit disk.

A tool to solve differential equations is provided by integral transforms. An example is the Laplace transform

$$F(s) = \int_0^{+\infty} dt\, e^{-st} f(t),$$

where f is a function of one real variable and s is a complex number. The Laplace transform can be seen as a holomorphic generalisation of the Fourier transform[1] and, as we will see, is very useful to solve Cauchy problems for linear differential equations (see Chapter 4), namely finding a solution of a differential equation that also satisfies given initial conditions.

Another useful application of complex analysis is given by the saddle-point method, which allows to estimate the behaviour of integrals of the type

$$f(x) = \int_\gamma dz\, e^{xh(z)} g(z),$$

with γ a curve on the complex plane and $h(z)$ and $g(z)$ holomorphic functions, for large values of the real parameter x. A similar analysis can be also performed for integrals of Laplace type or generalisations thereof. This is the realm of asymptotic analysis, which we discuss in Chapter 3. As we will see in Chapter 7, asymptotic expansions can also be applied to the study of the local behaviour of the solutions of differential equations. Linear differential equations with non-constant coefficients can have singular points and it is interesting to see how the solutions behave in a neighbourhood of a given point. Depending on whether a point is a singularity or not for the equation, and on the nature of the singularity, the solution can be expressed as a power series, a convergent series or an asymptotic one.

---

[1]See Petrini *et al.* (2017) and Appendix A for a short summary.

1

Mapping Properties of Holomorphic Functions

Holomorphic functions can be seen as maps from the complex plane to itself. In this chapter, we will analyse their mapping properties, namely how they transform sets of the plane $\mathbb{C}$. The main question we want to answer is the following: given an open set Ω in $\mathbb{C}$ and a holomorphic function f, what can we say about the shape and the properties of the set $f(\Omega)$? Important theorems of complex analysis, like the open mapping theorem and the Riemann mapping theorem, deal with such question. Mapping a domain Ω to a simpler one $f(\Omega)$ is a useful method for solving many problems of mathematical physics in two dimensions. We will see how these ideas apply to the solution of problems in electrostatics and fluid mechanics.

1.1. Local Behaviour of Holomorphic Functions

We start by analysing the local behaviour of a holomorphic function, namely how it transforms the neighbourhood of a point. Since holomorphic functions are analytic, their local behaviour is determined by their Taylor expansion.

Consider a function f holomorphic in a domain Ω.[1] In an open neighbourhood of any point $z_0 \in \Omega$, f can be expanded in a convergent Taylor series

$$f(z) = f(z_0) + \sum_{n=1}^{\infty} a_n (z - z_0)^n, \qquad a_n = \frac{1}{n!} f^{(n)}(z_0), \qquad (1.1)$$

where $f^{(n)}$ denotes the nth derivative of f. Assume that f is not identically constant and let m be the first integer such that $a_m \neq 0$. We then have

$$f(z) - f(z_0) = a_m (z - z_0)^m + \cdots, \qquad (1.2)$$

where the dots refer to higher powers of $(z - z_0)$. In a sufficiently small neighbourhood of z_0, the function $f(z)$ can be approximated with high precision by the leading term $a_m (z - z_0)^m$. This approximation is sufficient to study the local properties of the map f. These depend on the value of the exponent m. We need to distinguish two cases, namely when the exponent m is equal to one or bigger.

Consider first the case where $m = 1$. This means that $a_1 = f'(z_0) \neq 0$. Then there exists an open neighbourhood U of z_0 such that

(i) f maps U into $f(U)$ in a one-to-one way;

(ii) $f(U)$ is an open set;

(iii) f has a holomorphic inverse $f^{-1} : f(U) \to U$;

(iv) f preserves angles.

The last statement means that two lines L and L' that intersect in z_0 at an angle θ are mapped into two curves $f(L)$ and $f(L')$ whose tangents in $f(z_0)$ form the same angle θ, as in Figure 1.1. A map that preserves angles is called a *conformal map*. A map that sends open sets into open sets is called an *open map*.

Proof. By performing a translation in the planes z and $f(z)$, we can always take $z_0 = 0$ and $f(z_0) = 0$. Consider the open ball $U = B(0, \epsilon)$ centred in the origin. If ϵ is sufficiently small, f can be approximated by the leading term of the expansion (1.1), $\tilde{f}(z) = a_1 z$, where $a_1 = f'(0)$. A complex number $z = \rho e^{i\phi}$ in U is mapped into $\tilde{f}(z) = \rho |a_1| e^{i(\phi + \beta)}$, where $a_1 = |a_1| e^{i\beta}$. Since $0 \leq \rho < \epsilon$ and

---

[1] In this context, a domain in $\mathbb{C}$ is a connected open subset of the plane $\mathbb{C}$.

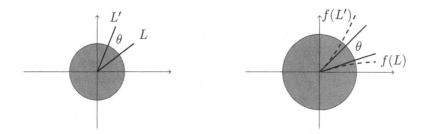

Fig. 1.1. A conformal map preserves angles.

$0 \leq \phi \leq 2\pi$, $\tilde{f}(U)$ is an open ball of radius $|a_1|\epsilon$. Therefore $\tilde{f}(U)$ is an open set and $\tilde{f}$ is obviously bijective in U. Since $a_1 \neq 0$, $\tilde{f}$ is invertible with a holomorphic inverse $\tilde{f}^{-1}(w) = w/a_1$. Moreover, $\tilde{f}$ maps the half-line $L = \{\arg(z) = \phi\}$ of angle ϕ into the half-line $\tilde{f}(L) = \{\arg(z) = \phi + \beta\}$ of angle $\phi + \beta$. Thus $\tilde{f}$ preserves the angle between two lines. We conclude that $\tilde{f}$ satisfies properties (i)-(iv). The function f has the same local behaviour of its linearisation $\tilde{f}$, and therefore it also satisfies (i)-(iv). Notice however that f does not necessarily map lines into lines; the images of two lines L and L' are two curves $f(L)$ and $f(L')$ whose tangents at the origin make the same angle as the original lines, as in Figure 1.1. $\qquad\square$

The inverse function f^{-1} that we construct in this way is holomorphic. Then, it can be expanded in Taylor series in the neighbourhood of the point $w_0 = f(z_0)$,

$$f^{-1}(w) = \sum_{k=0}^{\infty} b_k(w - w_0)^k. \tag{1.3}$$

The coefficients of the Taylor expansion are given by the *Lagrange Formula*

$$b_0 = z_0, \qquad b_n = \frac{1}{n!}\frac{d^{n-1}}{dz^{n-1}}\left(\frac{z - z_0}{f(z) - f(z_0)}\right)^n\bigg|_{z=z_0}, \qquad n \geq 1. \tag{1.4}$$

Proof. Consider the point $w = f(z)$, where z is sufficiently close to z_0. Since f is bijective in a neighbourhood U of z_0, the function $g(\zeta) = f(\zeta) - w$ has only a simple zero in U located at the point $\zeta = z$. The function $\zeta f'(\zeta)/(f(\zeta) - w)$ has then a simple pole in $\zeta = z$ with residue equal to z. By using the residue theorem we can write

$$f^{-1}(w) = z = \frac{1}{2\pi i} \int_\gamma \frac{\zeta f'(\zeta)}{f(\zeta) - w} d\zeta, \tag{1.5}$$

where γ is a closed, simple curve in U encircling z_0 and z. Since $f^{-1}(w)$ is holomorphic, we can derive (1.5) and then integrate by parts, obtaining

$$\frac{d}{dw} f^{-1}(w) = \frac{1}{2\pi i} \int_\gamma \frac{d\zeta}{f(\zeta) - w}. \tag{1.6}$$

We can always choose a curve γ such that, when ζ runs over γ, $\left| \frac{w - w_0}{f(\zeta) - w_0} \right| < 1$. Then

$$\frac{1}{f(\zeta) - w} = \frac{1}{f(\zeta) - w_0} \sum_{n=0}^{\infty} \left(\frac{w - w_0}{f(\zeta) - w_0} \right)^n. \tag{1.7}$$

By inserting this expression in (1.6) and integrating term by term we find

$$\frac{d}{dw} f^{-1}(w) = \sum_{n=0}^{\infty} \left[\frac{1}{2\pi i} \int_\gamma d\zeta \frac{1}{(f(\zeta) - w_0)^{n+1}} \right] (w - w_0)^n. \tag{1.8}$$

Comparing with the derivative of (1.3), $\frac{d}{dw} f^{-1}(w) = \sum_{n=1}^{\infty} n b_n (w - w_0)^{n-1}$, we have, for $n \geq 1$,

$$b_n = \frac{1}{n} \frac{1}{2\pi i} \int_\gamma \frac{d\zeta}{(f(\zeta) - w_0)^n} = \frac{1}{n} \frac{1}{2\pi i} \int_\gamma \frac{(\zeta - z_0)^n}{(f(\zeta) - w_0)^n} \frac{d\zeta}{(\zeta - z_0)^n}. \tag{1.9}$$

Then (1.4) follows from the Cauchy formula for the derivatives (see (A.4)).   □

Example 1.1. As a simple example of the Lagrange formula, consider the inverse function of $f(z) = 1/(z + 1)$. The coefficients of the Taylor series of f^{-1} around $w_0 = f(0) = 1$ are easily calculated from (1.4) and read $b_0 = 0$, $b_n = (-1)^n$ for $n \geq 1$. They obviously agree with the expansion of $f^{-1}(w) = \frac{1-w}{w}$ around $w_0 = 1$. In this case we do not really need (1.4), since f^{-1} can be computed directly. Lagrange formula is useful in cases where the inverse function is not analytically known.

Consider now $m > 1$. In this case, there exists an open neighbourhood U of z_0 such that
  (i) f is m-to-one and it is not invertible;
 (ii) $f(U)$ is an open set;
(iii) f magnifies angles by a factor m.

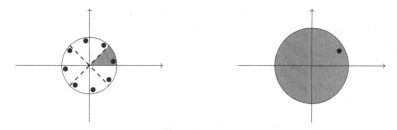

Fig. 1.2.   The behaviour of the map $a_m z^m$ for $m = 8$. The map is 8-to-one. The eight points on the left are mapped to the same point on the right.

Proof. We assume again that $z_0 = 0$ and $f(z_0) = 0$ and we consider $\tilde{f}(z) = a_m z^m$, with $a_m = |a_m|e^{i\beta}$. A complex number $z = \rho e^{i\phi}$ is mapped into $\tilde{f}(z) = \rho^m |a_m|e^{i(m\phi+\beta)}$. Notice that the argument ϕ is multiplied by a factor m and augmented by β. Consider a ball of radius ϵ centred in $z = 0$ and divide it into m regions of angular extension $2\pi/m$. Each of them is mapped to a full ball of radius $\epsilon^m |a_m|$ (see Figure 1.2). Thus the image of the ball of radius ϵ covers m-times the ball of radius $\epsilon^m |a_m|$. The function $\tilde{f}$ is m-to-one and maps open sets into open sets. f has the same local behaviour as $\tilde{f}$. □

The local behaviour discussed above can be used to prove some general properties of holomorphic maps. First of all, we have the *open mapping theorem*: a holomorphic function that is not identically constant maps open sets into open sets.

Proof. Consider an open set Ω in the domain of holomorphicity of f and a point $z_0 \in \Omega$. We just showed that, both for $f'(z_0) = 0$ and $f'(z_0) \neq 0$, there exists an open neighbourhood U of z_0 that is mapped into an open set $f(U) \subset f(\Omega)$. Thus $f(\Omega)$ contains an open neighbourhood of any of its points $f(z_0)$ and therefore is an open set. This result does not hold for constant functions. In this case, the function maps the entire plane $\mathbb{C}$ into a point. □

The second result is the following. If f is holomorphic and bijective in a domain Ω then $f'(z) \neq 0$ for all $z \in \Omega$ and f is a conformal map. Its inverse exists and it is holomorphic in $f(\Omega)$.

Proof. Being bijective, the function f cannot be constant. We saw that, if $f'(z_0) = 0$ at a point z_0, f is not bijective in the neighbourhood of z_0. Since this contradicts the hypothesis, $f'(z) \neq 0$ for all $z \in \Omega$. Then, in a neighbourhood of an arbitrary point z, f is conformal and has a holomorphic inverse. Since f is bijective, this local property is promoted to a global one, namely valid in the whole domain Ω. $\square$

The last result is the *maximum modulus theorem* (for a different proof see Petrini *et al.* (2017), Section 2.8.2): if f is holomorphic in a domain Ω and is not constant, $|f|$ has no local maximum in Ω (and no local minimum in Ω unless $|f| = 0$). Similarly, $\operatorname{Re} f$ and $\operatorname{Im} f$ have no local extrema in Ω.

Proof. The proof is very simple. Recall first that the modulus $|f(z_0)|$ gives the distance between $f(z_0)$ and the origin. By the open mapping theorem, given any point $z_0 \in \Omega$, there exists an open neighbourhood U of z_0 such that $f(U)$ is an open set. As such it contains at least a neighbourhood of $f(z_0)$. Since a neighbourhood of $f(z_0)$ contains points that are closer to the origin than $f(z_0)$, $|f(z)| < |f(z_0)|$, and points that are more distant, $|f(z)| > |f(z_0)|$, it follows that $|f(z_0)|$ is not a local maximum nor minimum of $|f(z)|$. The only exception is $|f(z_0)| = 0$, which is necessarily a minimum since $|f(z)| \geq 0$ by definition of absolute value. Similarly, $\operatorname{Re} f(z)$ and $\operatorname{Im} f(z)$ can be greater and smaller than $\operatorname{Re} f(z_0)$ and $\operatorname{Im} f(z_0)$, respectively. Then $\operatorname{Re} f(z)$ and $\operatorname{Im} f(z)$ cannot have local extrema in z_0. $\square$

Example 1.2. Consider the function $f(z) = z^2$. Since $f'(z) = 2z$, in the neighbourhood of any point $z \neq 0$ f is bijective and conformal, with holomorphic inverse. In the neighbourhood of the origin $z = 0$, where it has an expansion (1.2) with $m = 2$, f is not bijective and doubles the angles. We can see all this by considering the image of a selected set of curves, namely the set of half-lines L originating from the origin and the set of circles C centred in the origin. These sets are mapped into themselves by f. The half-lines L and the circles C intersect orthogonally at any point $z \neq 0$ and so do their images, since the map f is conformal for $z \neq 0$. On the other hand, two half-lines intersecting in the origin at an angle θ are mapped into half-lines intersecting in the origin at an angle 2θ. In particular, $f(z)$ maps the first quadrant into the upper half-plane (see Figure 1.3). Notice that f is locally bijective but it is not globally bijective and, indeed, the inverse function $g(w) = \sqrt{w}$ is multivalued on $\mathbb{C}$. In any simply connected domain

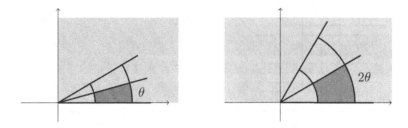

Fig. 1.3. The function $f(z) = z^2$.

$f(\Omega)$ not containing the origin, g is single valued and holomorphic and it provides a holomorphic determination of the square root function.

Example 1.3. Consider the function $f(z) = 1/z$. The function is not holomorphic in $z = 0$. Since $f'(z) = -1/z^2$, in the neighbourhood of all other points f is bijective and conformal, with holomorphic inverse. f is its own inverse. It maps the set of half-lines L originating from the origin and the set of circles C centred in the origin into themselves preserving their orthogonality. A half-line of angle θ, $L_\theta = \{z = \rho e^{i\theta} \mid \rho \in (0, \infty)\}$ is mapped into a half-line of angle $-\theta$, $L_{-\theta} = \{z = e^{-i\theta}/\rho \mid \rho \in (0, \infty)\}$. A circle of radius r, $C_r = \{z = r e^{i\phi} \mid \phi \in [0, 2\pi)\}$ is mapped into a circle of radius $1/r$, $C_{1/r} = \{z = e^{-i\phi}/r \mid \phi \in [0, 2\pi)\}$.

Example 1.4. Consider the function $f(z) = e^z$. Since $f'(z) = e^z$, f is everywhere locally bijective and conformal, with holomorphic inverse. Being conformal, it sends orthogonal curves into orthogonal curves. In particular, it sends the grid of lines parallel to the real and imaginary axis into a grid of lines and circles (see Figure 1.4). A line parallel to the real axis, $y = a$, is mapped into a half-line $L_a = \{z = e^x e^{ia} \mid x \in \mathbb{R}\}$ originating from the origin and intersecting the real axis at angle a. A line parallel to the imaginary axis, $x = b$, is mapped into a circle of radius e^b, $C_{e^b} = \{z = e^b e^{iy} \mid y \in \mathbb{R}\}$. The limiting case of the line with $b = -\infty$ is mapped to the origin and the angles are not preserved at this point; indeed the function e^z has an essential singularity at $z = \infty$. The exponential function is not globally bijective. It maps all the points differing by integer multiples of $2\pi i$ into the same point, $e^{z+2n\pi i} = e^z$, $n \in \mathbb{N}$. Any horizontal strip of height 2π (for

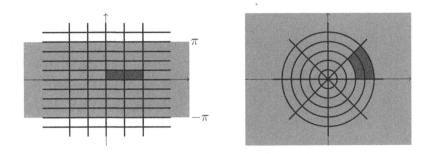

Fig. 1.4.   The function $f(z) = e^z$.

example the region $\{\operatorname{Im} z \in [-\pi, \pi)\}$ in Figure 1.4) is mapped into the whole complex plane. The inverse function $g(w) = \ln w$ is multivalued. In any simply connected domain $f(\Omega)$ not containing the origin, g is single valued and holomorphic and gives a holomorphic determination of the logarithm.

Example 1.5. The *level sets* of the real and imaginary part of a holomorphic function $w = f(z)$, namely

$$\operatorname{Re} f(z) = const, \qquad \operatorname{Im} f(z) = const, \tag{1.10}$$

intersect orthogonally in any point where $f'(z) \neq 0$. Indeed, in such points, the inverse function $f^{-1}(w)$ is holomorphic and conformal, and the level sets are the counter-images of the orthogonal lines $\operatorname{Re} w = const$ and $\operatorname{Im} w = const$.

Example 1.6. The *linear fractional transformations*[2]

$$F(z) = \frac{az + b}{cz + d} \tag{1.11}$$

with $a, b, c, d \in \mathbb{C}$ and $ad - bc \neq 0$ are conformal maps for all points $z \neq -d/c$, since in such points $F'(z) = (ad - bc)/(cz + d)^2$ is non-zero. Particular examples are translations $w = z + \alpha$, dilations $w = \beta z$ and the inversion $w = 1/z$. Any linear fractional transformation is a combination of these three operations. For instance, if $c = 0$, (1.11) is a dilation followed by a

---

[2]They are also called *Möbius transformations*.

translation. If $c \neq 0$, by writing

$$F(z) = \frac{bc - ad}{c^2(z + d/c)} + \frac{a}{c} \,, \tag{1.12}$$

we see that $F(z)$ is a combination of a translation, a dilation, an inversion and a further translation.

It is often convenient to extend $F(z)$ to the Riemann sphere $\hat{\mathbb{C}} = \mathbb{C} \cup \{\infty\}$, by defining $F(-d/c) = \infty$ and $F(\infty) = a/c$ (if $c = 0$ clearly $F(\infty) = \infty$). One can easily see that the linear fractional transformations are the unique holomorphic bijections from the Riemann sphere to itself.

The linear fractional transformations have a number of interesting properties. First, they map the set of all lines and circles of the plane into itself. This is obvious for translations and dilations, and it can be easily checked also for inversions (see Exercise 1.2). Notice that the set of all lines and circles of the plane can be identified with the set of all circles on the Riemann sphere, since a line can be seen as a circle on the Riemann sphere passing through the point at infinity. Secondly, given any two ordered triples of distinct points (z_1, z_2, z_3) and (w_1, w_2, w_3), there exists one and only one transformation such that $w_i = F(z_i)$, $i = 1, 2, 3$. This can be done by steps. We can map the points z_1, z_2 and z_3 to the points 0, 1 and ∞ by the *cross ratio*

$$B(z, z_1, z_2, z_3) = \frac{(z - z_1)(z_2 - z_3)}{(z - z_3)(z_2 - z_1)}. \tag{1.13}$$

Then by a combination of B and its inverse we can map any triple of points into any other triple (Exercise 1.3).

It is interesting to notice that the set of all linear fractional transformations form a group. Indeed, it is easily checked that the inverse of a transformation and the composition of two of them are still linear fractional transformations. We can identify this group with a group of matrices as follows. We associate F with the matrix

$$\hat{F} = \begin{pmatrix} a & b \\ c & d \end{pmatrix}. \tag{1.14}$$

The inverse and the composition of transformations are associated with the inverse and the product of the corresponding matrices (Exercise 1.4). The coefficients a, b, c and d in F can be rescaled by a non-vanishing common complex factor λ without changing the transformation and, therefore, they can be normalised in such a way that $ad - bc = 1$. After this normalisation, we still have the freedom of changing sign to all the a, b, c, d without affecting F. Thus the group of linear fractional transformations is isomorphic to $SL(2, \mathbb{C})/\mathbb{Z}_2$.[3] Another useful representation of this group comes from the identification $\hat{\mathbb{C}} = \mathbb{P}^1_{\mathbb{C}}$ (see Petrini *et al.*, 2017, Section 1.4).[4] The transformation (1.11) can be identified with the linear invertible transformation (linear automorphisms) of $\mathbb{P}^1_{\mathbb{C}}$ to itself that acts on the homogeneous coordinates as $(z_1, z_2) \longrightarrow (az_1 + bz_2, cz_1 + dz_2)$. By identifying $\mathbb{C}$ with the subset of points $(z, 1)$ in $\mathbb{P}^1_{\mathbb{C}}$, we have indeed

$$(z, 1) \longrightarrow (az + b, cz + d) = \left(\frac{az + b}{cz + d}, 1\right) = (F(z), 1), \qquad (1.15)$$

since $(z_1, z_2) \sim (\lambda z_1, \lambda z_2)$ in $\mathbb{P}^1_{\mathbb{C}}$. Thus the group of linear fractional transformations can be identified with the group of linear automorphisms of $\mathbb{P}^1_{\mathbb{C}}$.

1.2. The Riemann Mapping Theorem

Sometimes a problem concerning holomorphic functions can be solved by mapping a domain Ω into a simpler one with a holomorphic transformation. The *Riemann mapping theorem* states that any simply connected domain Ω that is not the entire plane can be bi-holomorphically mapped into the open unit disc.[5]

A *bi-holomorphic map* $f(z)$ from Ω to $f(\Omega)$ is a one-to-one, holomorphic map whose inverse is holomorphic. Since f is bijective, $f'(z) \neq 0$ for

---

[3]$SL(2, \mathbb{C})$ is the group of two-by-two complex matrices with determinant 1. $\mathbb{Z}_2$ is the subgroup $\{\mathbb{I}, -\mathbb{I}\}$.

[4]$\mathbb{P}^1_{\mathbb{C}}$ is the set of equivalences classes of pairs of complex numbers (z_1, z_2) under the identification $(z_1, z_2) \sim (\lambda z_1, \lambda z_2)$ with $\lambda \in \mathbb{C} - \{0\}$.

[5]For a proof, see Bak and Newman (2010) and Rudin (1987).

$z \in \Omega$, as proved in Section 1.1. Therefore f is also conformal. For this reason a bi-holomorphic map is also simply called a *conformal map* and two domains that can be bi-holomorphically mapped into each other are called *conformally equivalent.* A consequence of the Riemann mapping theorem is that all simply connected regions of the plane (with the exception of $\mathbb{C}$ itself) are conformally equivalent.

Notice that the theorem asserts the existence of a conformal mapping but it does not help finding it. We do not prove the theorem here but we give some examples.

Example 1.7. The upper half-plane $H = \{z = x + iy \in \mathbb{C} \,|\, y > 0\}$ is conformally equivalent to the open unit disk $D = \{w \in \mathbb{C} \,|\, |w| < 1\}$. The conformal map that connects the two domains is given by

$$w = f(z) = \frac{z - i}{z + i} \tag{1.16}$$

and is an example of linear fractional transformation (see Example 1.6). Let us show that $f(z)$ is bi-holomorphic in H and maps the upper half-plane into the unit disk. The function $f(z)$ is holomorphic in H since its only singularity is a pole at $z = -i$, which sits in the lower half-plane. Since $f'(z) = 2i/(z + i)^2 \neq 0$, f is also conformal in H. It is easy to see that f maps the real axis $z = x \in \mathbb{R}$ into the unit circle. Indeed $w = (x - i)/(x + i)$ is a complex number of modulus one since it is the ratio of a number and its complex conjugate. The upper half-plane is a domain bounded by the real axis and it should be mapped into a domain bounded by the unit circle. There are two such domains: the interior and the exterior of the unit circle. To determine which one is the image of H it is enough to look at the image of a point. Since $z = i$ is mapped into $w = 0$, we see that the upper half-plane H is mapped into the unit disk D. Alternatively we can look at the orientation of the boundary. As shown in Figure 1.5, going from left to right on the real axis we have the upper half-plane on the left. Following the images of $z = 0$, $z = 1$ and $z = \infty$ we go along the unit circle counterclockwise and we have the unit disk on the left. Finally, it is easy to check that the map is bijective.

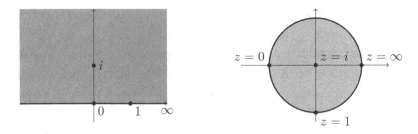

Fig. 1.5. The mapping of the real axis into the unit circle.

1.3. Applications of Conformal Mapping

The method of conformal mapping and the Riemann mapping theorem are useful in a large set of problems of mathematical physics involving the Laplace equation in two dimensions. In this chapter, we discuss some examples in electrostatics and fluid mechanics.

1.3.1. *The Dirichlet problem for the Laplace equation*

Recall that a *harmonic* function $u(x, y)$ of two real variables is a solution of the two-dimensional Laplace equation

$$\Delta u = \frac{\partial^2 u}{\partial x^2} + \frac{\partial^2 u}{\partial y^2} = 0. \tag{1.17}$$

We want to find a solution u that is harmonic in a domain Ω, continuous on the closure of Ω, $\bar{\Omega}$, and takes a prescribed value on the boundary of Ω,

$$u|_{\partial \Omega} = g, \tag{1.18}$$

where g is a continuous real function defined on $\partial \Omega$. This is called *Dirichlet problem* for the Laplace equation.

The solution of the Dirichlet problem in a simply connected, bounded domain Ω is unique. Suppose that we have two solutions u_1 and u_2 of the Laplace equation in Ω satisfying $u_1 = u_2 = g$ on $\partial \Omega$. The difference $u_1 - u_2$ is a harmonic function in Ω and it is therefore the real part of a holomorphic function.[6] The maximum modulus theorem tells us that $u_1 - u_2$ cannot

---

[6] A function $u(x, y)$ is harmonic in a simply connected domain Ω if and only if it is the real part of a holomorphic function (see Petrini *et al.*, 2017, Section 2.9).

have local extrema. The maximum and minimum of $u_1 - u_2$ are then on the boundary $\partial\Omega$ and they are zero, since $u_1 = u_2$ on $\partial\Omega$. But a continuous function whose maximum and minimum in $\bar{\Omega}$ are zero is identically zero. Then $u_1 = u_2$ everywhere in Ω.

If Ω is the unit disk, we will see in Example 5.7 that the solution of the Dirichlet problem is given by the *Poisson integral formula*

$$u(re^{i\theta}) = \frac{1}{2\pi} \int_0^{2\pi} P_r(\theta - t)g(t)\mathrm{d}t, \tag{1.19}$$

where $re^{i\theta}$ parameterises the unit disk, with $0 \le r < 1$ and $\theta \in [0, 2\pi)$, and e^{it} with $t \in [0, 2\pi)$ parameterises the unit circle. The function $P_r(\phi)$ is the *Poisson kernel*

$$P_r(\phi) = \sum_{n=-\infty}^{\infty} r^{|n|}e^{in\phi} = \frac{1 - r^2}{1 - 2r\cos\phi + r^2} = \mathrm{Re}\left(\frac{1 + re^{i\phi}}{1 - re^{i\phi}}\right). \tag{1.20}$$

Writing $w = re^{i\theta}$ and

$$u(w) = \mathrm{Re}\left(\frac{1}{2\pi} \int_0^{2\pi} \frac{1 + we^{-it}}{1 - we^{-it}} g(t)\mathrm{d}t\right) = \mathrm{Re}f(w), \tag{1.21}$$

we see that u is the real part of a holomorphic function[7] and hence it is harmonic.

Given the Poisson formula, we can find the solution of the Dirichlet problem for any simply connected domain $\Omega \ne \mathbb{C}$ by using the theory of conformal mapping. It is enough to find the conformal transformation $w(z)$ that sends Ω into the unit disk. Then $u(z) = \mathrm{Re}f(w(z))$, with f given in (1.21), is harmonic and is the solution of the Dirichlet problem on Ω.

Example 1.8 (Dirichlet problem in the upper half-plane). As we saw in Example 1.7, a point $z = x + iy$ of the upper half-plane is mapped into a point of the unit disk by the conformal map $w = (z - i)/(z + i)$. In particular, a point of the real axis with abscissa v is mapped to the point on the unit circle $e^{it} = (v - i)/(v + i)$. We can use this map to solve

---

[7]The integral in (1.21) depends holomorphically on w.

the Dirichlet problem in the upper half-plane. A solution[8] that reduces to $g(v)$ on the real axis is given by

$$u(x, y) = \mathrm{Re}\, f(w(z)) = \frac{1}{\pi} \int_{-\infty}^{+\infty} \frac{y}{(x - v)^2 + y^2}\, g(v) dv\,, \tag{1.22}$$

where we replaced $w = (z - i)/(z + i)$ and $e^{it} = (v - i)/(v + i)$ in (1.21).

1.3.2. *Fluid mechanics*

The motion of an infinitesimal element of a two-dimensional fluid is specified by its velocity $\mathbf{v}(x, y, t)$. We consider a *perfect* fluid, which is characterised by being *steady* (time-independent),

$$\mathbf{v} = (v_1(x, y), v_2(x, y)), \tag{1.23}$$

incompressible (the fluid volume does not change as it moves)

$$\mathrm{div}\, \mathbf{v} = \frac{\partial v_1}{\partial x} + \frac{\partial v_2}{\partial y} = 0, \tag{1.24}$$

and *irrotational* (the fluid has no vorticity)

$$\mathrm{rot}\, \mathbf{v} = \frac{\partial v_2}{\partial x} - \frac{\partial v_1}{\partial y} = 0. \tag{1.25}$$

The conditions (1.24) and (1.25) can be identified with the Cauchy–Riemann conditions for the function

$$f(z) = v_1(x, y) - i v_2(x, y). \tag{1.26}$$

$f(z)$ is therefore holomorphic. In any simply connected domain Ω, $f(z)$ has a holomorphic primitive $\chi(z)$,[9] $f(z) = \chi'(z)$. The function $\chi(z)$ is called the *complex potential function* of the fluid. The real and imaginary parts of $\chi(z)$

$$\chi(z) = \phi(x, y) + i\psi(x, y) \tag{1.27}$$

---

[8]The general solution is $u(x, y) + Cy$, where $u(x, y)$ is given in (1.22) and C is an arbitrary constant. Since the upper half-plane is not compact, the uniqueness theorem for the Dirichlet problem mentioned above does not hold.

[9]See Petrini *et al.* (2017).

are called *potential* and *stream function* of the fluid. Since

$$v_1(x, y) - iv_2(x, y) = f(z) = \chi'(z) = \frac{\partial \phi}{\partial x} - i\frac{\partial \phi}{\partial y}, \qquad (1.28)$$

we see that $\mathbf{v} = \nabla \phi$, so ϕ is indeed a potential for the velocity field.

The level sets of the functions ϕ and ψ,

$$\phi(x, y) = const, \qquad \psi(x, y) = const \qquad (1.29)$$

are called *equipotential curves* and *streamlines* of the fluid, respectively. As level sets of the real and imaginary part of a holomorphic function, they are mutually orthogonal at all points where $\mathbf{v} \neq 0$ (see Example 1.5). Since the velocity field $\mathbf{v} = \nabla \phi$ is orthogonal to the equipotential curves,[10] it is everywhere tangent to the streamlines.

A typical problem in fluid mechanics involves finding the motion of a fluid inside a domain Ω bounded by a solid vessel. The natural boundary condition for such a problem is that the field $\mathbf{v}$ is tangent to the boundary $\partial\Omega$. In other words, $\partial\Omega$ should be a streamline of the fluid.

Example 1.9 (Flow with constant velocity). The simplest example of perfect fluid is given by a constant velocity field. The function $f(z) = a \in \mathbb{R}$ describes a fluid moving in the horizontal direction. The complex potential function is $\chi(z) = c + az$ where c is an arbitrary constant.

Example 1.10 (Flow around a disk). We want to determine the flow of a perfect fluid in the upper half-plane past a solid disk of unit radius (see Figure 1.6) by mapping it to the constant flow of Example 1.9. Consider the function

$$w = g(z) = z + \frac{1}{z}. \qquad (1.30)$$

---

[10]Given a curve on the plane defined by $w(x, y) = const$, where $w(x, y)$ is a real function of two variables x and y, at each point on the curve we define the vector $\mathbf{dr} = (dx, dy)$ parallel to it. The differential of $w(x, y)$ is defined as the scalar product $dw = (\nabla w, \mathbf{dr})$, where $\nabla w = (\partial_x w, \partial_y w)$ is the gradient of w. Since on the curve $w(x, y) = const$ we have $dw = 0$, it follows that $(\nabla w, \mathbf{dr}) = 0$, and therefore the gradient is orthogonal to the curve.

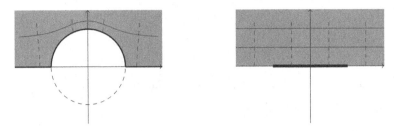

Fig. 1.6.   The function $z + 1/z$ maps the flow around a disk into the linear flow in the upper half-plane. Streamlines are indicated by solid lines and equipotential curves by dashed lines.

g maps the unit circle $z = e^{it}$, $t \in [0, 2\pi)$, into the segment $[-2, 2]$ of the real axis: $g(e^{it}) = e^{it} + e^{-it} = 2\cos(t)$. The points e^{it} and e^{-it} are mapped to the same point of the segment $[-2, 2]$. g maps conformally both the interior and the exterior of the unit circle into the complex plane with the segment $[-2, 2]$ removed. This can also be seen by inverting $w = z + 1/z$. The counter-images of a point w

$$z = \frac{1}{2}(w \pm \sqrt{w^2 - 4}) \tag{1.31}$$

have product one, so either both values are on the unit circle or one is inside and the other outside. Moreover, g maps the real axis into the lines $(-\infty, -2]$ and $[2, \infty)$ since $g(x) = x + 1/x$ for $x \in \mathbb{R}$. Combining all this information, we see that $g(z)$ maps the part of the upper half-plane which is outside the unit circle into the upper half-plane, as in Figure 1.6. We can thus use $g^{-1}(w)$ to conformally map the constant velocity fluid of Example 1.9 into a flow in the upper plane around a disk. The equipotential curves and the streamlines are depicted in Figure 1.6. The transformation $g(z)$ is known as *Joukowski transform* and it is used, for example, in aerodynamics: it maps circles not centred in the origin in a wide range of airfoil shapes (see Figure 1.7).

1.3.3.   *Electrostatics*

Problems in electrostatics in the plane can be treated in an analogous way. The fluid velocity is replaced by the electric field. A static electric field

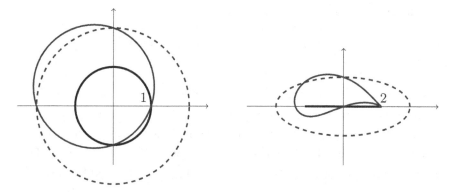

Fig. 1.7.   The Joukowski function $z + 1/z$ maps the unit circle centred in the origin into a segment, a circle of radius $r > 1$ centred in the origin into an ellipse, and a circle passing through $z = 1$ and containing $z = -1$ into an airfoil shape.

$\mathbf{E}(x, y)$ satisfies

$$\operatorname{div} \mathbf{E} = \frac{\partial E_1}{\partial x} + \frac{\partial E_2}{\partial y} = 0, \tag{1.32}$$

in any region without sources, and

$$\operatorname{rot} \mathbf{E} = \frac{\partial E_2}{\partial x} - \frac{\partial E_1}{\partial y} = 0. \tag{1.33}$$

The conditions (1.32) and (1.33) can be identified with the Cauchy–Riemann conditions for the function

$$f(z) = -E_1(x, y) + iE_2(x, y). \tag{1.34}$$

$f(z)$ is therefore holomorphic. In any simply connected domain Ω, $f(z)$ has a holomorphic primitive $\chi(z)$, $f(z) = \chi'(z)$. The real part $\phi(x, y)$ of $\chi(z)$

$$\chi(z) = \phi(x, y) + i\psi(x, y) \tag{1.35}$$

is the *electrostatic potential* of the field and we have

$$-E_1(x, y) + iE_2(x, y) = f(z) = \chi'(z) = \frac{\partial \phi}{\partial x} - i\frac{\partial \phi}{\partial y}, \tag{1.36}$$

so that[11] $\mathbf{E} = -\nabla \phi$.

---

[11]We choose the signs in (1.34) differently than in the previous subsection in order to match the standard definition of ϕ.

The level sets of the functions ϕ and ψ

$$\phi(x, y) = const, \qquad \psi(x, y) = const \tag{1.37}$$

are called *equipotential curves* and *flux lines*, respectively. As in the previous subsection, they are mutually orthogonal and the field $\mathbf{E}$ is everywhere tangent to the flux lines.

A typical electrostatic problem involves finding the electric field inside a domain Ω bounded by a capacitor that is kept at constant potential. The natural boundary condition for such a problem is that the field $\mathbf{E}$ is orthogonal to the boundary $\partial\Omega$. In other words, $\partial\Omega$ should be an equipotential curve.

Example 1.11 (Capacitor of infinite length). Consider a capacitor consisting of two horizontal, infinite plates located at $y = a$ and $y = b$ and kept at constant potentials V_a and V_b, respectively. The function

$$\chi(z) = V_a + \frac{(V_b - V_a)(-iz - a)}{(b - a)}$$

is holomorphic and respects the boundary conditions. The corresponding electric field $\mathbf{E} = (0, -(V_b - V_a)/(b - a))$ is obtained from $f(z) = \chi'(z)$ through (1.34). It is constant and vertical, as expected from physical intuition (see the left part of Figure 1.8).

Example 1.12 (Capacitor of finite length). Consider the problem of finding the electric field between two semi-infinite plates kept at a constant

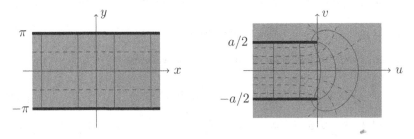

Fig. 1.8. Mapping of the region between two infinite plates into the region outside two semi-infinite plates. Flux lines are indicated with solid lines and equipotential curves with dashed lines.

potential. We use the holomorphic function

$$w = g(z) = \frac{a}{2\pi}(1 + z + e^z), \qquad (1.38)$$

with a real, to map the problem to that of Example 1.11. Set $z = x + iy$ and $w = u + iv$. g maps the real axis into the real axis and the horizontal lines $y = \pm\pi$ into

$$w = \frac{a}{2\pi}(1 \pm i\pi + x + e^{x\pm i\pi}) = \frac{a}{2\pi}(1 \pm i\pi + x - e^x), \qquad x \in \mathbb{R} \quad (1.39)$$

which are semi-infinite lines with $v = \pm a/2$: when x goes from $-\infty$ to $+\infty$, u grows from $-\infty$ to 0 and then its goes back to $-\infty$. The lines $y = \pm\pi$ are folded on themselves in the mapping. The horizontal open strip $-\pi < y < \pi$ in the z-plane is mapped into the w-plane with two semi-infinite cuts at $v = \pm a/2$ as in Figure 1.8. We can thus use $g(z)$ to conformally map the constant electric field of Example 1.11 into the electric field between two semi-infinite capacitors. The equipotential curves and the flux lines are depicted in Figure 1.8.

1.4. Exercises

Exercise 1.1. Consider $w = f(z) = e^z - 1$. Using the Lagrange formula, find the first four Taylor coefficients of the expansion of $f^{-1}(w)$ around $w = 0$ and check the result by explicitly inverting $f(z)$.

Exercise 1.2. Show that a linear fractional transformation maps the set of lines and circles of the plane into itself.

Exercise 1.3. Show that
(a) the cross ratio (1.13) of four points $B(z_1, z_2, z_3, z_4)$ is invariant under a linear fractional transformation: $z_i \to F(z_i)$;
(b) the linear fractional transformation that maps the points (z_1, z_2, z_3) into (w_1, w_2, w_3) is obtained by solving $B(w, w_1, w_2, w_3) = B(z, z_1, z_2, z_3)$.

Exercise 1.4. Show that the composition of two linear fractional transformations of the form (1.11) corresponds to the product of the associated matrices (1.14). Show that, for $ad - bc = 1$, the inverse of F in (1.11), $F^{-1}(w) = (-dw + b)/(cw - a)$, corresponds to the matrix $\hat{F}^{-1}$.

Exercise 1.5. Find a linear fractional transformation that maps $(1, 0, i)$ to $(i, 1, 2)$.

Exercise 1.6. Find the level sets of the real and imaginary parts of the holomorphic functions

$$\text{(a)} \quad \ln z, \qquad \text{(b)} \quad z^2, \qquad \text{(c)} \quad \ln(z + \sqrt{z^2 - 1}).$$

Exercise 1.7. Find the image of the set $\{\text{Re } z > \text{Re } z_0\, , \text{Im } z > 0\}$ under the map $f(z) = \frac{z - z_0}{z - \bar{z}_0}$ with $z_0 \in \mathbb{C}$.

Exercise 1.8. Find the image of the strip $\{-\frac{\pi}{2} < \text{Re } z < \frac{\pi}{2}\}$ under the map $f(z) = 6 \sin z$.

Exercise 1.9. Consider the Joukowski map $f(z) = z + \frac{1}{z}$ and the airfoil obtained as the image of the circle centred in $(-1 + i)/2$ and radius $\sqrt{5}/2$ (see Figure 1.7). Find the flow past such airfoil of a fluid that is moving at 45 degrees in the plane.

Exercise 1.10. Find a function $u(x, y)$ that is harmonic on the first quadrant and satisfies $u(x, 0) = 1$, for all $x \geq 0$, and $u(0, y) = 0$, for all $y \geq 0$.

2

Laplace Transform

Many functions in analysis are defined by integral representations of the form

$$f(x) = \int_{-\infty}^{+\infty} K(x,y)g(y)\mathrm{d}y, \tag{2.1}$$

where g is an assigned function and $K(x,y)$ is usually called the *kernel* of the *integral transform* $f(x)$. Integral transforms are very useful in solving certain types of differential and integral equations. A well-known example is the Fourier transform (see Petrini *et al.*, 2017). In this chapter, we analyse another example of integral transform, the Laplace transform, which is relevant for several practical applications, like for instance signal theory and the study of circuits in electronics.

2.1. The Laplace Transform

Consider a function f that is locally integrable on $[0, \infty)$, namely that is integrable on any compact subset of $[0, \infty)$. Its *Laplace transform* is defined as the integral[1]

$$\mathcal{L}[f](s) = \int_{0}^{+\infty} \mathrm{d}x\, f(x)e^{-sx}, \tag{2.2}$$

---

[1] Equation (2.2) should be properly called the *unilateral* or *one-sided* Laplace transform. By extending the integration in (2.2) to the whole $\mathbb{R}$ we can also define the *bilateral* or *two-sided* Laplace transform. In this book Laplace transform will always refer to (2.2).

where s is a complex number. We will often simply denote the Laplace transform of a function $f(x)$ with a capital letter $F(s)$. The integral (2.2) exists whenever $f(x)$ is locally integrable on $[0, \infty)$ and of *exponential order at infinity*, namely there exist two real numbers α and M, and a point x_0 such that

$$|f(x)| < Me^{\alpha x} \qquad \forall\, x > x_0. \tag{2.3}$$

The infimum α_0 of the values α for which inequality (2.3) holds is called the *convergence abscissa* of $f(x)$. The Laplace transform $F(s)$ exists in the domain $\mathrm{Re}(s) > \alpha_0$ and is holomorphic there.

Proof. The Laplace transform exists for those values of s such that the function $g(x) = f(x)e^{-sx} \in L^1[0, \infty)$.[2] $g(x)$ is locally integrable on $[0, \infty)$ since $f(x)$ is. Moreover $|g(x)| = |f(x)|e^{-\mathrm{Re}(s)x} < Me^{-(\mathrm{Re}(s)-\alpha)x}$ for any $\alpha > \alpha_0$ and $x > x_0$ because of (2.3). Therefore $g(x)$ is integrable on $[0, \infty)$ if $\mathrm{Re}(s) > \alpha_0$. The derivative of $F(s)$ is computed as

$$F'(s) = \frac{d}{ds} \int_0^{+\infty} dx\, f(x)e^{-sx} = -\int_0^{+\infty} dx\, x f(x)e^{-sx}. \tag{2.4}$$

This expression is well defined because the integrand is again integrable on $[0, \infty)$ if $\mathrm{Re}(s) > \alpha_0$.[3] $\qquad\square$

The Laplace transform can be seen as a holomorphic generalisation of the Fourier transform.[4] The relation between the two transforms is obtained by writing $s = \alpha + ip$ in (2.2) and extending the integral to the whole real axis

$$F(s) = \int_0^{+\infty} dx\, f(x)e^{-sx} = \int_{-\infty}^{+\infty} dx(\theta(x)f(x)e^{-\alpha x})e^{-ipx}, \tag{2.5}$$

---

[2]Recall that $L^1[0, \infty)$ is the vector space of the functions f such that $\int_0^{+\infty} |f(x)|dx < +\infty$. See Petrini *et al.* (2017) or Appendix A for a brief review.

[3]We could exchange derivation and integration by the dominated convergence theorem since the derivative of the integrand, in a sufficiently small neighbourhood of s, is uniformly dominated by a function belonging to $L^1[0, \infty)$: $|xf(x)e^{-sx}| < x|f(x)|e^{-\alpha x} = xe^{-(\alpha-\tilde\alpha)x}|f(x)|e^{-\tilde\alpha x} < Mxe^{-(\alpha-\tilde\alpha)x}$ where $\alpha_0 < \tilde\alpha < \alpha < \mathrm{Re}(s)$ and we used (2.3).

[4]See Appendix A for conventions and definitions.

where $\theta(x)$ is the Heaviside function: $\theta(x) = 0$ for $x < 0$ and $\theta(x) = 1$ for $x > 0$. We see that the Laplace transform $F(\alpha + ip)$, considered as a function of p at fixed α, is the Fourier transform of the function $\sqrt{2\pi}\theta(x)f(x)e^{-\alpha x}$. This immediately gives us an *inversion formula* for the Laplace transform. If $F(\alpha + ip)$ is integrable as a function of p in its domain of definition $\alpha > \alpha_0$, then

$$f(x)\,\theta(x) = \mathcal{L}^{-1}[F] = \frac{1}{2\pi i}\int_{\alpha-i\infty}^{\alpha+i\infty}F(s)e^{sx}\mathrm{d}s, \tag{2.6}$$

where the integral can be taken on any vertical line $\mathrm{Re}(z) = \alpha > \alpha_0$.

Proof. We use the Fourier inversion formula (A.12). Applying it to the function $\sqrt{2\pi}\theta(x)f(x)e^{-\alpha x}$ in (2.5) we obtain

$$f(x)\theta(x)e^{-\alpha x} = \frac{1}{2\pi}\int_{-\infty}^{+\infty}F(\alpha+ip)e^{ipx}\mathrm{d}p. \tag{2.7}$$

Multiplying both sides by $e^{\alpha x}$ and using $s = \alpha + ip$, one obtains (2.6). Since $F(s)$ is holomorphic for $\mathrm{Re}(s) > \alpha_0$, the integrals over all vertical lines[5] give the same result and the right-hand side of (2.6) does not depend on α, provided $\alpha > \alpha_0$. □

The inversion formula can be generalised to cases where $F(s)$ is not integrable on the vertical lines $\mathrm{Re}(s) = \alpha$, provided $f(x)$ is sufficiently regular. One can show that, if $f(x)$ is piece-wise C^1,

$$\lim_{R\to\infty}\frac{1}{2\pi i}\int_{\alpha-iR}^{\alpha+iR}F(s)e^{sx}\mathrm{d}s = \begin{cases} f(x) & f \text{ continuous in } x, \\ \dfrac{f_+(x) + f_-(x)}{2} & f \text{ has a jump in } x. \end{cases}$$

The conditions for the validity of the previous formula are very similar to the conditions of Dirichlet's theorem for the Fourier transform (see Petrini *et al.*, 2017, Section 8.2.2).

---

[5]By Cauchy's theorem, the integral over the rectangle of vertices $\alpha_1 - iR$, $\alpha_1 + iR$, $\alpha_2 + iR$, $\alpha_2 - iR$ with $\alpha_0 < \alpha_1 < \alpha_2$ is zero. Since $F(\alpha \pm iR)$ goes to zero for $R \to \infty$, the integrals over the horizontal segments $(\alpha_1 + iR, \alpha_2 + iR)$ and $(\alpha_2 - iR, \alpha_1 - iR)$ vanish in this limit. We conclude that the integrals over the vertical lines $\mathrm{Re}(s) = \alpha_1$ and $\mathrm{Re}(s) = \alpha_2$ are equal.

Notice that, if we start with a function $f(x)$ that is defined on $\mathbb{R}$ and not only on $[0, \infty)$, the inverse Laplace transform gives back a function that vanishes for $x < 0$.

Example 2.1. The Laplace transform of $f(x) = e^{ax}$, with $a \in \mathbb{R}$, is

$$F(s) = \int_0^{+\infty} \mathrm{d}x e^{(a-s)x} = \frac{1}{s-a}. \tag{2.8}$$

Since $e^{ax} < e^{\alpha x}$ for all $\alpha > a$, the convergence abscissa of $f(x)$ is a. Indeed $F(s)$ is holomorphic in the half-plane $\mathrm{Re}(s) > a$. From (2.8) we see that $F(s)$ can be analytically continued to $\mathbb{C} - \{a\}$ and has a simple pole at $s = a$. It is also interesting to see how the inversion formula works. Equation (2.6) becomes

$$\theta(x)e^{ax} = \frac{1}{2\pi i} \int_{\alpha-i\infty}^{\alpha+i\infty} \frac{e^{sx}}{s-a} \mathrm{d}s, \tag{2.9}$$

with $\alpha > a$. We can explicitly check this formula by evaluating the integral on the right-hand side of (2.9) using the residue theorem. We close the contour of integration with a large semi-circle with centre in $s = \alpha > a$. By Jordan's lemma (see Appendix A), for $x < 0$ we need to choose a semi-circle that lies to the right of the vertical line $\mathrm{Re}(s) = \alpha$. Since $F(s)$ is holomorphic for $\mathrm{Re}(s) > a$, there is no singularity inside the integration contour and the result is zero. This is a general fact for all Laplace transforms and it explains operatively why the inverse Laplace transform is always zero for $x < 0$. For $x > 0$ we need to choose a large semi-circle that lies to the left of the vertical line $\mathrm{Re}(s) = \alpha$. The contour now contains the simple pole $s = a$ and the residue theorem gives e^{ax}, in agreement with the left-hand side of (2.9). Notice that this computation relies on the fact that we can analytically continue $F(s)$.

Example 2.2. The Laplace transform of x^n is

$$\mathcal{L}[x^n](s) = (-1)^n \frac{d^n}{ds^n} \int_0^{+\infty} e^{-sx} \mathrm{d}x = (-1)^n \frac{d^n}{ds^n} \frac{1}{s} = \frac{n!}{s^{n+1}}. \tag{2.10}$$

Since for large values of x, $x^n \leq e^{\epsilon x}$ for any $\epsilon > 0$, the convergence abscissa of x^n is $\alpha_0 = 0$. We see indeed that $L[x^n](s)$ is holomorphic for $\mathrm{Re}(s) > 0$

and it has a pole at $s = 0$. $F(s)$ can be analytically continued to $\mathbb{C} - \{0\}$. We can use (2.10) to extend the Laplace transform to power series. Given a function defined via its Taylor expansion

$$f(x) = \sum_{n=0}^{\infty} c_n x^n, \qquad (2.11)$$

we can write

$$\mathcal{L}[f](s) = \sum_{n=0}^{\infty} \frac{n!\, c_n}{s^{n+1}}. \qquad (2.12)$$

Notice however that it is not granted that this expression converges.

2.1.1. *Properties of the Laplace transform*

We list here the main properties of the Laplace transform. Others are given in Exercise 2.2. Given two functions f, g that admit a Laplace transform, we have

(i) *Linearity*

$$\mathcal{L}[\alpha f + \beta g](s) = \alpha \mathcal{L}[f](s) + \beta \mathcal{L}[g](s), \qquad \alpha, \beta \in \mathbb{C}; \qquad (2.13)$$

(ii) *Transform of derivatives*

$$\mathcal{L}[f'](s) = s\mathcal{L}[f](s) - f(0); \qquad (2.14)$$

$$\mathcal{L}[f^{(n)}](s) = s^n \mathcal{L}[f](s) - \sum_{k=0}^{n-1} s^{n-k-1} f^{(k)}(0); \qquad (2.15)$$

(iii) *Multiplication*

$$\mathcal{L}[x^n f](s) = (-1)^n \frac{d^n}{ds^n} \mathcal{L}[f](s); \qquad (2.16)$$

(iv) *Convolution*

$$\mathcal{L}[f * g](s) = \mathcal{L}[f](s) \cdot \mathcal{L}[g](s), \qquad (2.17)$$

where f and g vanish for $x < 0$. This last equation is valid for $\mathrm{Re}(s) > \max\{\alpha_f, \alpha_g\}$, where α_f and α_g are the convergence abscissae of f and g.

Recall that the convolution of two functions f and g is defined as $(f * g)(x) = \int_{\mathbb{R}} f(x - y)g(y)dy$. For functions vanishing for $x < 0$ it reduces to

$$(f * g)(x) = \int_0^x dy f(x - y)g(y). \tag{2.18}$$

Proof. Property (i) is obvious. The first line of property (ii) can be obtained integrating by parts

$$\mathcal{L}[f'](s) = f(x)e^{-sx}\big|_0^{+\infty} + s \int_0^{+\infty} dx\, f(x)e^{-sx} = s\mathcal{L}[f](s) - f(0), \tag{2.19}$$

where the boundary term at infinity vanishes because $f(x)e^{-sx}$ is integrable on $[0, +\infty)$. The second line of (ii) easily follows by iteration. Property (iii) follows from the fact that $(-1)^n \frac{d^n}{ds^n} e^{-sx} = x^n e^{-sx}$ and we can exchange integral and derivative. Finally, using (2.18), we can write

$$\mathcal{L}[f * g](s) = \int_0^{+\infty} dx\, e^{-sx} \int_0^x dy\, f(x - y)g(y) = \int_D dx\, dy\, e^{-sx} f(x - y)g(y),$$

where D is the domain $\{x \geq 0, 0 \leq y \leq x\}$ in $\mathbb{R}^2$. Under the change of variables $\eta = x - y, \theta = y$, of unit Jacobian, D is mapped into $D' = \{\eta \geq 0, \theta \geq 0\}$ so that

$$\mathcal{L}[f * g](s) = \int_{D'} d\eta\, d\theta\, e^{-s(\eta+\theta)} f(\eta)g(\theta) = \mathcal{L}[f](s) \cdot \mathcal{L}[g](s), \tag{2.20}$$

thus proving property (iv). $\qquad\qquad\qquad\qquad\qquad\qquad\qquad\qquad\qquad\square$

2.1.2. *Laplace transform of distributions*

The Laplace transform can be extended to a class of distributions.[6] Since we are interested in the unilateral transform, we consider only distributions with support contained in $[0, \infty)$.[7] Moreover, as for functions, distributions need to have a nice behaviour at infinity in order for the Laplace transform

---

[6]See Appendix A for a summary of definitions and notations. See also Petrini *et al.* (2017) for more details about distributions.

[7]This means that T gives zero when acting on test functions with support in $(-\infty, 0)$.

to exist. A distribution $T \in \mathcal{D}'(\mathbb{R})$, with support in $[0, \infty)$, is of *exponential order at infinity* if there exists a real number α such that $e^{-\alpha x}T \in \mathcal{S}'(\mathbb{R})$. The infimum α_0 of the values of α such that $e^{-\alpha x}T$ is a tempered distribution is called the *convergence abscissa* of T.

The Laplace transform of a distribution of exponential order at infinity and support contained in $[0, \infty)$ is defined as

$$\mathcal{L}[T](s) = \langle T, \phi_s \rangle, \tag{2.21}$$

where $\phi_s(x) = e^{-sx}$. The Laplace transform (2.21) is a function of s and is holomorphic in the domain $\text{Re}(s) > \alpha_0$.

Proof. With some abuse of language we write $\langle T, e^{-sx} \rangle$ for the action of T on $\phi_s(x) = e^{-sx}$. Notice that e^{-sx} is not of compact support, and, therefore, the action of $T \in \mathcal{D}'(\mathbb{R})$ on it is not well defined. However, we can define $\mathcal{L}[T](s) = \langle e^{-\alpha x}T, \theta_R(x)e^{-(s-\alpha)x} \rangle$ where $\theta_R(x)$ is any C^∞ function identically equal to 1 for $x > 0$ and identically vanishing for $x < R < 0$, and $\alpha_0 < \alpha < \text{Re}(s)$. The function $\theta_R(x)e^{-(s-\alpha)x}$ is C^∞, identically zero for large negative x and rapidly decreasing for large positive x and, therefore, is a test function in $\mathcal{S}(\mathbb{R})$. Since $e^{-\alpha x}T$ is a tempered distribution, $\mathcal{L}[T]$ is well defined and is obviously independent of the choice of θ_R and α. With this definition (2.21) is well defined. It follows then, from the continuity of T and of the pairing $\langle \cdot, \cdot \rangle$, that $\mathcal{L}[T](s)$ is continuous and differentiable in s. $\qquad \square$

Among the distributions admitting a Laplace transform there is the class of tempered distributions with support in $[0, \infty)$. Indeed, since a tempered distribution T is at most of algebraic growth at infinity, it satisfies $e^{-\alpha x}T \in \mathcal{S}'(\mathbb{R})$ for all $\alpha > 0$. Its Laplace transform is then defined at least in the domain $\text{Re}(s) > 0$. If T also vanishes at infinity like $e^{-\alpha x}$ with $\alpha > 0$, meaning that $e^{\alpha x}T \in \mathcal{S}'(\mathbb{R})$, then its Laplace transform is defined in the domain $\text{Re}(s) > -\alpha$. Distributions with compact support satisfy the condition above with $\alpha = \infty$ and their Laplace transform is therefore an entire function.[8]

---

[8] An entire function is a function of a complex variable that is holomorphic on the whole complex plane.

Example 2.3. Both distributions[9] $\theta(x)$ and $\delta(x - x_0)$ are of exponential order. We have[10]

$$\mathcal{L}[\theta](s) = \int_0^{+\infty} e^{-sx}\mathrm{d}x = \frac{1}{s},$$

and

$$\mathcal{L}[\delta_{x_0}](s) = \int_0^{+\infty} \delta(x - x_0)e^{-sx}\mathrm{d}x = e^{-sx_0}, \tag{2.22}$$

$$\mathcal{L}[\delta_{x_0}^{(n)}](s) = \int_0^{+\infty} \delta^{(n)}(x - x_0)e^{-sx}\mathrm{d}x = s^n e^{-sx_0}, \tag{2.23}$$

at least for $x_0 > 0$. Here $\delta^{(n)}$ denotes the nth derivative of δ. Clearly, there is an ambiguity for $x_0 = 0$ since the delta function is localised at the extremum of integration. We can define the Laplace transform of $\delta(x)$ and its derivatives as the $x_0 \to 0$ limit of the previous expressions.[11]

The properties of the Laplace transform discussed in Section 2.1.1 extend to the distributional Laplace transform. However, in property (ii) we have to replace the ordinary derivative with the distributional one, as neatly shown by the following example.

Example 2.4. Let us consider a function with a jump at the point $x = x_0 > 0$. It can be written as

$$f(x) = f_-(x)\theta(x_0 - x) + f_+(x)\theta(x - x_0), \tag{2.24}$$

where $f_\pm$ are smooth differentiable functions in x_0. The function $f'(x)$ exists except for $x = x_0$ and it is locally integrable. Its Laplace transform,

$$\int_0^{+\infty} f'(x)e^{-sx}\mathrm{d}x = sL(s) - f(0) - e^{-sx_0}[f_+(x_0) - f_-(x_0)], \tag{2.25}$$

---

[9] Here $\theta(x)$ is considered as the regular distribution acting on test functions as

$$\langle \theta, \phi \rangle = \int_{-\infty}^{+\infty} \mathrm{d}x \, \theta(x)\phi(x) = \int_0^{+\infty} \mathrm{d}x \, \phi(x).$$

[10] It is clear that, if the Laplace transform exists for a certain function f, then it also exists in the sense of distributions, and the two coincide.

[11] The reader must be aware that this definition is not in general consistent with other common prescriptions, like for instance the fact that $\int_0^{+\infty} \delta(x)\mathrm{d}x = 1/2$, which emphasises the parity of the Dirac delta.

contains an extra term with respect to property (ii) which comes from integrating by parts. On the contrary, if we take the Laplace transform of the distributional derivative

$$f'_-(x)\theta(x_0 - x) + f'_+(x)\theta(x - x_0) + \delta(x - x_0)[f_+(x_0) - f_-(x_0)],$$

and we integrate by parts, we find

$$s \int_0^{+\infty} f(x)e^{-sx}dx - f(0) = s\mathcal{L}[T](s) - f(0), \qquad (2.26)$$

in agreement with property (ii). Additional subtleties arise when the jump is in the origin, similarly to Example 2.3.

2.2.   Integral Transforms and Differential Equations

We saw in Section 2.1.1 that the Laplace transform maps derivation into multiplication by the conjugate variable. This property has interesting applications to the solution of ordinary and partial linear differential equations with constant coefficients. From the examples below we will see that the Laplace transform is particularly useful to solve differential equations with initial conditions. The Fourier transform has the same feature of mapping derivation into multiplication and can be similarly used. We will see many examples in the rest of the book.

Example 2.5. Consider the first-order ordinary differential equation

$$\frac{dx(t)}{dt} + ax(t) = f(t), \qquad (2.27)$$

with initial condition $x(0) = 0$. We want to compute its Laplace transform. From the linearity of the Laplace transform and the expression (2.14) for the derivative, we obtain the algebraic equation

$$\mathcal{L}\left[\frac{dx(t)}{dt} + ax(t) - f(t)\right] = sX(s) - x(0) + aX(s) - F(s) = 0, \qquad (2.28)$$

where $X(s)$ and $F(s)$ are the Laplace transforms of x and f, respectively. The solution of (2.28) is $X(s) = F(s)/(s + a)$, since $x(0) = 0$. Then $x(t)$

can be found by inverting the Laplace transform

$$x(t) = \mathcal{L}^{-1}\left[\frac{1}{s+a}\cdot F(s)\right] = \mathcal{L}^{-1}\left[\frac{1}{s+a}\right] * f = \int_0^t e^{-a(t-\tau)}f(\tau)d\tau.$$

In the last step we used the convolution theorem (2.17) and the result of Example 2.1 to evaluate the inverse Laplace transform of $1/(s+a)$. Notice that the solution for $x(t)$ is what one would obtain using the well-known method of variation of arbitrary constants (see Section 4.2).

Example 2.6 (Driven harmonic oscillator). We want to study the motion of a mass m attached to a spring of constant k and subjected to a friction force $-\alpha\frac{dx(t)}{dt}$, with $\alpha > 0$, and an exterior force $f(t)$. We assume that at the instant $t = 0$ the mass has position x_0[12] and speed v_0. This amounts to solving Newton's equation

$$m\frac{d^2x(t)}{dt^2} + \alpha\frac{dx(t)}{dt} + kx(t) = f(t), \tag{2.29}$$

with initial conditions

$$x(0) = x_0, \quad \dot{x}(0) = v_0, \tag{2.30}$$

where the $\dot{x}$ denotes derivation with respect to time. Setting $\alpha = 0$ and $f(t) = 0$ one recovers the standard harmonic oscillator. To solve equation (2.29) we take its Laplace transform. Using (2.15) for the transform of the nth derivative and the initial conditions (2.30), we find

$$(s^2 + \eta s + \omega_0^2)X(s) - (s+\eta)x_0 - v_0 = F(s)/m, \tag{2.31}$$

where $\omega_0^2 = k/m$ is the angular frequency (or pulsation) of the system, $\eta = \alpha/m$, and $X(s)$ and $F(s)$ are the Laplace transforms of x and f. The polynomial $Z(s) = s^2 + \eta s + \omega_0^2$ is sometimes called the *transfer function* of the system. The solution of (2.31),

$$X(s) = \frac{(\eta + s)x_0 + v_0}{Z(s)} + \frac{F(s)}{mZ(s)}, \tag{2.32}$$

---

[12]The motion is only along the direction x and we choose the origin of our coordinate system to coincide with the equilibrium position of the mass.

is holomorphic on $\mathbb{C}$ but for the simple poles $s_\pm = (-\eta \pm \sqrt{\eta^2 - 4\omega_0^2})/2$ at the zeros of $Z(s)$. Note that when $\eta = 2\omega_0$, the two poles merge in a double pole at $s_0 = -\omega_0$. The solution $x(t)$ is given by the inverse Laplace transform of (2.32)

$$x(t) = x_h(t) + x_p(t)$$
$$= \frac{1}{2\pi i} \int_\gamma ds\ e^{st} \frac{(\eta + s)x_0 + v_0}{Z(s)} + \frac{1}{2\pi i m} \int_\gamma ds\ e^{st} \frac{F(s)}{Z(s)},$$

where γ is a vertical line crossing the real axis to the right of Res$_+$. The first integral can be easily evaluated using the residue theorem. Consider first the case $\eta \neq 2\omega_0$ where the integrand function has two simple poles. The resulting solution is

$$x_h(t) = \frac{\theta(t)}{s_+ - s_-}[e^{s_+t}(\eta x_0 + v_0 + s_+ x_0) - e^{s_-t}(\eta x_0 + v_0 + s_- x_0)]$$

$$= \theta(t)e^{-\frac{\eta t}{2}}\left[\frac{x_0\eta + 2v_0}{\sqrt{\eta^2 - 4\omega_0^2}}\sinh\left(\frac{t}{2}\sqrt{\eta^2 - 4\omega_0^2}\right)\right.$$

$$\left. + x_0\cosh\left(\frac{t}{2}\sqrt{\eta^2 - 4\omega_0^2}\right)\right]. \tag{2.33}$$

For $\eta = 2\omega_0$, the function has a double pole and the solution becomes

$$x_h(t) = \theta(t)e^{s_0 t}[x_0 + t(\eta x_0 + v_0 + s_0 x_0)]$$

$$= \theta(t)e^{-\omega_0 t}[t(\omega_0 x_0 + v_0) + x_0]. \tag{2.34}$$

The second integral can be evaluated as in Example 2.5

$$x_p(t) = \frac{1}{m}\int_0^t d\tau\ \chi(t - \tau)f(\tau), \tag{2.35}$$

where the function χ is the inverse Laplace transform of $1/Z(s)$

$$\chi(t) = \frac{1}{2\pi i}\int_\gamma ds\ e^{st}\frac{1}{s^2 + \eta s + \omega_0^2}. \tag{2.36}$$

The explicit form of χ can again be computed using the residue theorem. We find

$$\chi(t) = \theta(t)\frac{e^{s_+t} - e^{s_-t}}{s_+ - s_-} = \theta(t)\frac{2e^{-\eta t/2}}{\sqrt{\eta^2 - 4\omega_0^2}}\sinh\left(\frac{t}{2}\sqrt{\eta^2 - 4\omega_0^2}\right) \tag{2.37}$$

when $Z(s)$ has two simple zeros in $s_\pm$, and

$$\chi(t) = \theta(t)te^{s_0 t} = \theta(t)te^{-\omega_0 t} \tag{2.38}$$

when $Z(s)$ has a double zero in s_0. The two terms x_h and x_p have different physical interpretations. The first term x_h is completely determined by the choice of initial conditions and corresponds to the general solution of the homogeneous equation associated to (2.29). It describes a transient behaviour, which only depends on the initial conditions and vanishes exponentially with time due to the friction force.[13] The second term x_p corresponds to the particular solution of the inhomogeneous equation (2.29). It describes a steady state depending only on the external driving force, with no memory of the initial conditions. The function χ appearing in (2.35) is called the *linear response function* of the system, and determines the response of the system to an external force. The form of (2.35) is consistent with the physical principle of causality, which requires that the response of the system at a time t can only depend on the values of the external force at times $t' < t$, and indeed $\chi(t - t') = 0$ for $t' > t$. The linear response function χ is an example of *Green function*, which will be discussed extensively in Chapter 6. In applications to statistical mechanics, χ is also called *susceptibility*.

Example 2.7 (RLC circuit). The driven harmonic oscillator equation (2.29) is the most general second-order inhomogeneous ordinary differential equation with constant coefficients and can be used to describe many other physical systems, for example an RLC circuit. Consider a capacitor (C), a resistor (R) and an inductor (L) connected in series. The evolution in time of the charge on the plates of the capacitor is described by the differential

---

[13]The transient behaviour actually depends on the relation between η and ω_0. For $\eta > 2\omega_0$, x_h is a sum of terms of the form $e^{s_\pm t}$ with $s_\pm = -\eta/2 \pm \sqrt{\eta^2/4 - \omega_0^2}$ and decays to the equilibrium position with no oscillations (overdamping behaviour). For $\eta = 2\omega_0$ the decay is as in (2.34) and it is the fastest possible (critical damping). For $\eta < 2\omega_0$, x_h is a sum of terms of the form $e^{-\eta t/2}e^{\pm it\sqrt{\omega_0^2 - \eta^2/4}}$ and oscillates while decaying (underdamping).

equation

$$L\frac{d^2}{dt^2}q(t) + R\frac{d}{dt}q(t) + \frac{1}{C}q(t) = V(t), \tag{2.39}$$

where $V(t)$ is the external voltage applied to the circuit. We denote the initial conditions as

$$q(0) = q_0, \quad \dot{q}(0) = i(0) = i_0, \tag{2.40}$$

where i is the current flowing through the circuit. The resolution of (2.39) is the same as in Example 2.6 with the replacements

$$x \to q, \quad m \to L, \quad \eta \to R/L, \quad \omega_0^2 \to 1/LC, \quad v_0 \to i_0, \quad f \to V, \tag{2.41}$$

and can be used to study the circuit. We consider two different configurations.

(a) Charging the capacitor. Suppose the circuit is initially open and that the capacitor is not charged ($q_0 = i_0 = 0$). We close the circuit on a generator of alternated current with voltage $V(t) = Ve^{i\omega t}$ for $t \geq 0$. The solution q_h of the homogeneous equation is identically zero since the initial conditions are zero. The solution of the inhomogeneous equation is instead given by (2.35), which for the given external source reads

$$q_p(t) = \frac{V}{L}\int_0^t d\tau \chi(t-\tau)e^{i\omega\tau}. \tag{2.42}$$

We need to distinguish between the case of simple and double zeros of the transfer function, which, with the replacements (2.41), is now given by $Z(s) = s^2 + Rs/L + 1/LC$. For $R^2 \neq 4L/C$ we have simple zeros $s_\pm = -\frac{R}{2L} \pm \frac{1}{2L}\sqrt{R^2 - \frac{4L}{C}}$ and using (2.37) we find

$$q_p(t) = \frac{V/Le^{i\omega t}}{(s_+ - i\omega)(s_- - i\omega)} + \frac{V/L}{(s_+ - s_-)}\left[\frac{e^{s_+ t}}{s_+ - i\omega} - \frac{e^{s_- t}}{s_- - i\omega}\right]. \tag{2.43}$$

For $R^2 = 4L/C$ we have a double zero $s_0 = -R/2L$ and, plugging (2.38) into (2.42), we find

$$q_p(t) = \frac{V}{L}\frac{1}{(s_0 - i\omega)^2}\left[e^{i\omega t} + e^{s_0 t}(s_0 t - i\omega t - 1)\right]. \tag{2.44}$$

As for the harmonic oscillator in Example 2.6, the behaviour (2.43) depends on the sign of $\Delta = R^2 - \frac{4L}{C}$ (see footnote 13). For both (2.44) and (2.43) the exponentials $e^{s_k t}$ become negligible for large t, since $\mathrm{Re}\, s_k$ is always negative, and $q(t)$ becomes proportional to the voltage of the generator.

(b) Discharging the capacitor. Suppose the circuit is powered for $t < 0$ by a generator that provides a constant voltage V so that the charge on the capacitor is $q_0 = VC$. At $t = 0$ we switch off the generator so that $V(t) = 0$ for $t > 0$ and $i_0 = 0$. The equation for $t > 0$ is homogeneous and $q_p = 0$. The only contribution to $q(t)$ comes from the transient term q_h. In the case of simple zeros of $Z(s)$, $R^2 \neq 4L/C$, we use (2.33) and we find

$$q_h(t) = \frac{VC}{s_+ - s_-} \left[e^{s_+ t} \left(s_+ + \frac{R}{L} \right) - e^{s_- t} \left(s_- + \frac{R}{L} \right) \right]. \qquad (2.45)$$

In the case of a double zero of $Z(s)$, $R^2 = 4L/C$, we find from (2.34)

$$q_h(t) = VC e^{-Rt/2L} \left[1 + \frac{Rt}{2L} \right].$$

In both cases, since $\mathrm{Re}\, s_k < 0$, the solution vanishes for $t \to \infty$ as expected from a process of discharge.

2.3. Exercises

Exercise 2.1. Check the inversion formula for Example 2.2.

Exercise 2.2. Prove the following properties of the Laplace transform

(a) translation: $\mathcal{L}[e^{ax} f](s) = \mathcal{L}[f](s - a)$;

(b) dilatation: $\mathcal{L}[f(ax)](s) = \frac{1}{a}\mathcal{L}[f]\left(\frac{s}{a}\right), a > 0$;

(c) transform of the integral: $\mathcal{L}[\int_0^t f(\tau)d\tau](s) = \frac{1}{s}\mathcal{L}[f](s)$. [Hint: consider $1 * f$];

(d) integration: $\mathcal{L}[\frac{f(x)}{x}](s) = \int_s^{+\infty} \mathcal{L}[f](u)du$.

Exercise 2.3. Find the Laplace transform of the functions

$$\begin{array}{ll} \text{(a) } x^p, \quad p > -1, & \text{(c) } \cos wx, \quad w \in \mathbb{R}, \\ \text{(b) } e^{ax}, \quad a \in \mathbb{C}, & \text{(d) } \sin wx, \quad w \in \mathbb{R}. \end{array}$$

Exercise 2.4. Find the Laplace transform of a periodic, locally integrable function $f(x+L) = f(x)$. Use this result to compute the Laplace transform of the sawtooth function $f(x) = x$ for $x \in [0, 1)$ and $f(x + 1) = f(x)$.

* Exercise 2.5. Prove that, if $F(s)$ is the Laplace transform of $f(t)$, $F(\sqrt{s})$ is the Laplace transform of $g(t) = (4\pi t^3)^{-1/2} \int_0^{+\infty} y e^{-y^2/(4t)} f(y) dy$. Use this result to find the inverse Laplace transform of the functions

$$\text{(a) } e^{-a\sqrt{s}}, \quad \text{(b) } \frac{e^{-a\sqrt{s}}}{\sqrt{s}}, \quad a \geq 0.$$

Exercise 2.6. Use the Laplace transform to solve the equation $x'(t) + 3x(t) = e^{\alpha t}$ with $x(0) = 1$, where $x'(t)$ is the derivative of x with respect to t and $\alpha \in \mathbb{R}$. Discuss also the special case $\alpha = -3$.

Exercise 2.7. Use the Laplace transform to solve the system of linear differential equations

$$y'(t) - x'(t) + y(t) + 2x(t) = e^t,$$
$$y'(t) + x'(t) + x(t) = e^{2t},$$
$$x(0) = y(0) = 1.$$

Exercise 2.8. Use the Laplace transform to solve the integral equation

$$\sin t + \int_0^t f(x) e^{t-x} dx = t.$$

Exercise 2.9. Solve the partial differential equation $\partial_x^2 u = \partial_t^2 u$ on the strip $S = \{0 \leq x \leq \pi, t \geq 0\}$ with the boundary conditions

$$u(0,t) = u(\pi, t) = 0,$$
$$u(x,0) = \sin x, \quad \partial_t u(x,0) = 0.$$

[Hint: solve an ordinary differential equation for the Laplace transform in the variable t, $U(x,s) = \int_0^{+\infty} u(x,t) e^{-st} dt$.]

Exercise 2.10. Find the Laplace transform of the *Laguerre polynomials* $L_n(t) = \frac{e^t}{n!} \frac{d^n}{dt^n}(t^n e^{-t})$.

3

Asymptotic Expansions

It is not always possible to solve exactly a differential equation or an integral. However, in many cases, it is enough to have an estimate of the behaviour of the solution of a differential equation for large (or small) value of the independent variable. Similarly, it is often sufficient to know the behaviour of an integral when a parameter in the integrand is taken to be large (or small). The theory of asymptotic expansion allows to determine these behaviours. What typically one can do is to expand the equation or the integral in an appropriate power series for large values of the relevant variable or parameter. Most often, the expansions so obtained are not convergent. However even divergent series, when they satisfy some basic requirements, can be useful and can be a good approximation of a function in a certain regime.

In this chapter we mostly discuss integrals, while examples of asymptotic expansions of solutions of differential equations will be discussed in Chapter 7. There are various methods to estimate integrals that depend on a parameter. In this book we will discuss three approaches: the Laplace, the stationary phase and the steepest descent methods. They are mostly different version of the same idea, and it is not uncommon in physics to group all of them under the name of saddle-point method. The saddle-point method is widely used in statistical mechanics to evaluate partition functions and in quantum mechanics and quantum field theory to evaluate path integrals.

In the latter context, when the small parameter is $\hbar$, it is also referred as semi-classical approximation.

All the methods we mentioned apply both to real and complex variables. As we will see in many examples, some of the most interesting applications come when one considers complex variables.

3.1. Asymptotic Series

A very common question in physics and mathematics is how to approximate a function f in a given neighbourhood. A typical example is provided by the Taylor expansion of both real and complex functions. A function f that is C^N on an interval $I \subset \mathbb{R}$ can be approximated around any point $x_0 \in I$ by

$$f(x) = f(x_0) + f'(x_0)(x - x_0) + \cdots + \frac{f^{(N)}(x_0)}{N!}(x - x_0)^N + R_N(x), \quad (3.1)$$

where the *remainder* $R_N(x)$ is $o(x - x_0)^N$, namely it goes to zero faster[1] than $(x - x_0)^N$. Even if the function f is $C^\infty(I)$ it is not guaranteed that its Taylor series converges. This only happens when the function is analytic in x_0.[2]

What matters for us is that a power series, even if not convergent, can provide a good approximation of the function f. It is best to look at concrete examples.

Example 3.1. Consider the *exponential integral function*

$$E_1(x) = -Ei(-x) = \int_x^{+\infty} dt \frac{e^{-t}}{t}, \quad x > 0. \quad (3.2)$$

---

[1] Given two functions $h(t)$ and $g(t)$ from $\mathbb{R}$ to $\mathbb{R}$, we say that $h = o(g)$ at the point t_0 if $\lim_{t \to t_0} h(t)/g(t) = 0$, $h = O(g)$ at the point t_0 if $\lim_{t \to t_0} h(t)/g(t)$ is finite, and $h \sim g$ at the point t_0 if $\lim_{t \to t_0} h(t)/g(t) = 1$. The same definitions apply to functions from $\mathbb{R}^n$ to $\mathbb{R}$ and to functions from $\mathbb{C}$ to $\mathbb{C}$.

[2] A function f that is C^∞ on an interval $I \subset \mathbb{R}$ is analytic at a point $x_0 \in I$ if its Taylor series in x_0 converges in a neighbourhood U of x_0 and the sum coincides with f.

We can build an expansion of $E_1(x)$ for large x by successive integrations by parts

$$E_1(x) = \frac{e^{-x}}{x} - \int_x^\infty dt \frac{e^{-t}}{t^2}$$

$$= e^{-x}\left[\frac{1}{x} - \frac{1}{x^2} + \frac{2}{x^3} + \cdots + (-1)^{N-1}\frac{(N-1)!}{x^N}\right] + R_N(x), \quad (3.3)$$

where we defined the remainder

$$R_N(x) = (-1)^N N! \int_x^{+\infty} dt \frac{e^{-t}}{t^{N+1}}. \tag{3.4}$$

For any $x > 0$ the series

$$\sum_{n=1}^\infty a_n(x) = \sum_{n=1}^\infty (-1)^{n-1}\frac{(n-1)!}{x^n}e^{-x} \tag{3.5}$$

is divergent, as one can check from the ratio a_{n+1}/a_n (see (A.2))

$$\frac{a_{n+1}(x)}{a_n(x)} = -\frac{n}{x} \xrightarrow[n\to\infty]{} \infty. \tag{3.6}$$

However the partial sum $S_N(x) = e^{-x}\sum_{n=1}^N (-1)^{n-1}\frac{(n-1)!}{x^n}$ gives a good approximation of the function $E_1(x)$ for large x. Indeed, integrating by parts (3.4) one finds

$$|R_N(x)| = \left|\frac{N!}{x^{N+1}}e^{-x} - (N+1)! \int_x^{+\infty} dt \frac{e^{-t}}{t^{N+2}}\right| \leq \frac{N!}{x^{N+1}}e^{-x}, \tag{3.7}$$

so that, for any fixed N,

$$\left|\frac{R_N(x)}{(N-1)!e^{-x}x^{-N}}\right| \leq \frac{N}{x} \xrightarrow[x\to\infty]{} 0. \tag{3.8}$$

This implies that the remainder $R_N(x)$ is small compared to the last term in the partial sum $S_N(x)$, and, provided $N \leq x$, $S_N(x)$ is a good approximation of the function $E_1(x)$.

The example above illustrates the behaviour of *asymptotic series*. Let $\phi_n(x)$ be a sequence of functions such that $\phi_{n+1}(x) = o(\phi_n(x))$ for $x \to x_0$.

If, for any $N \in \mathbb{N}$, we can express a given function $f(x)$ as

$$f(x) = \sum_{k=0}^{N} a_k \phi_k(x) + o(\phi_N(x)), \tag{3.9}$$

then the infinite sum $\sum_{k=0}^{\infty} a_k \phi_k(x)$ is an asymptotic series for f at the point x_0. Equivalently this means that

$$\lim_{x \to x_0} \phi_N(x)^{-1} R_N(x) = 0, \tag{3.10}$$

where $f(x) - \sum_{k=0}^{N} a_k \phi_k(x) = R_N(x)$ is the remainder of the series. We then write $f(x) \sim \sum_{k=0}^{\infty} a_k \phi_k(x)$. Typical examples of asymptotic series are given by power series where $\phi_k(x) = (x - x_0)^k$ if $x_0 \neq \infty$ or $\phi_k(x) = 1/x^k$ for $x_0 = \pm\infty$.

Asymptotic series are not, in general, convergent. For an asymptotic series the remainder $R_N(x)$ is small compared with the last term kept in the sum (3.9)

$$R_N(x) = o(\phi_N(x)), \tag{3.11}$$

for $x \to x_0$ at fixed N. This is different from the notion of convergence of a series at the point x, which requires that $R_N(x) \to 0$ when $N \to \infty$ at fixed x.

Notice that, while for a given set of functions ϕ_k the asymptotic expansion is unique, a function f can have many different expansions with respect to different sets of ϕ_k.

Example 3.2. Consider $f(x) = \tan x$. For $x \to 0$ we have

$$\tan x \sim x + \frac{x^3}{3} + \frac{2}{15} x^5 + \cdots$$

$$\sim \sin x + \frac{(\sin x)^3}{2} + \frac{3}{8}(\sin x)^5 + \cdots . \tag{3.12}$$

Conversely, two different functions f and g can have the same asymptotic expansion. Indeed, if $l(x)$ is subleading compared to $f(x)$, then the functions $f(x)$ and $g(x) = f(x) + l(x)$ have the same asymptotic series.

Example 3.3. Given any function with asymptotic expansion $f(x) \sim \sum_{n=0}^{\infty} a_n x^n$ for $x \to 0$, we have

$$f(x) \sim f(x) + e^{-1/x^2} \sim \sum_{n=0}^{\infty} a_n x^n. \tag{3.13}$$

This is because, when x goes to 0, $e^{-1/x^2} = o(x^n)$ for all $n \in \mathbb{N}$.

Given the asymptotic series $\sum_{n=0}^{\infty} a_k \phi_k$ one may define its *sum* as the equivalence class of all functions that have the same asymptotic expansion $f \sim \sum_{n=0}^{\infty} a_k \phi_k$. One can also show that any asymptotic series has a sum (see Erdélyi, 2012).

Asymptotic series with respect to a sequence ϕ_k can be summed, and, under reasonable assumptions,[3] multiplied and integrated term by term (see Erdélyi, 2012; Bender and Orszag, 1978). On the contrary, the derivation term by term of an asymptotic series does not necessarily give another asymptotic series, since terms that are subleading can become relevant after differentiation.

Example 3.4. Consider a function $f(x)$ with an asymptotic expansion $f(x) \sim \sum_{n=0}^{\infty} a_n x^n$ near $x = 0$. The function $g(x) = f(x) + e^{-1/x^2} \cos e^{1/x^2}$ has the same expansion, $g(x) \sim f(x)$ since $\lim_{x \to 0} g(x)/f(x) = 1$. However

$$g'(x) = f'(x) + \frac{2}{x^3} e^{-1/x^2} \cos e^{1/x^2} + \frac{2}{x^3} \sin e^{1/x^2}, \tag{3.14}$$

where the last term is now of order $O(1/x^3)$. Therefore $g'(x) \not\sim f'(x)$.

Asymptotic series can also be used to approximate holomorphic functions. In particular they are relevant for functions with branch points or essential singularities, when the Laurent series cannot be defined or has an infinite number of terms (and hence does not provide a useful approximation). However, on the complex plane, asymptotic series for large or small z make sense only in certain regions of $\mathbb{C}$, typically *sectors* $S = \{z \in \mathbb{C} \,|\, \alpha \leq \arg z \leq \beta\}$ or just lines $L = \{z \in \mathbb{C} \,|\, \arg z = \alpha, |z| > a\}$. To see how this works it is useful to look at some examples.

---

[3]For asymptotic power series multiplication is straightforward, while for integration some mild additional assumptions are still required.

Example 3.5. We want to understand the behaviour near $z = \infty$ of the function $f(z) = \sinh z^2$. $f(z)$ has an essential singularity in $z = \infty$. Its Laurent expansion in $w = 1/z$ around zero is given by

$$\frac{e^{\frac{1}{w^2}} - e^{-\frac{1}{w^2}}}{2} = \sum_{n=0}^{\infty} \frac{1}{(2n+1)!} \frac{1}{w^{4n+2}} = \frac{1}{w^2} + \frac{1}{6}\frac{1}{w^6} + \cdots. \quad (3.15)$$

This is not an asymptotic expansion, since the terms grow with n for small w. The asymptotic behaviour of $f(z)$ depends on the sign of $\operatorname{Re} z^2$. We thus divide $\mathbb{C}$ into the four regions

$$S_1 = \left\{ z \in \mathbb{C} \, \middle| \, -\frac{\pi}{4} < \arg z < \frac{\pi}{4} \right\}, \quad S_2 = \left\{ z \in \mathbb{C} \, \middle| \, \frac{\pi}{4} < \arg z < \frac{3\pi}{4} \right\},$$

$$S_3 = \left\{ z \in \mathbb{C} \, \middle| \, \frac{3\pi}{4} < \arg z < \frac{5\pi}{4} \right\}, \quad S_4 = \left\{ z \in \mathbb{C} \, \middle| \, \frac{5\pi}{4} < \arg z < \frac{7\pi}{4} \right\}.$$

In S_1 and S_3, $\operatorname{Re} z^2 > 0$ and e^{-z^2} is exponentially small compared to e^{z^2}, while in S_2 and S_4, $\operatorname{Re} z^2 < 0$ and e^{-z^2} dominates. The asymptotic behaviour of $f(z)$ is then

$$f(z) = \begin{cases} e^{z^2}/2, & z \in S_1, S_3, \\ e^{-z^2}/2, & z \in S_2, S_4. \end{cases} \quad (3.16)$$

We see that the function has different asymptotic behaviours in different sectors. This is called the *Stokes phenomenon* and it is typical of functions with essential singularities. Indeed we know that the limit of a function with an essentially singularity depends on the direction along which we approach the point. The half-lines $\arg z = \pm\pi/4, \pm 3\pi/4$ where the asymptotic behaviour of f changes are called *anti-Stokes lines*. The half-lines $\arg z = 0, \pm\pi/2, \pi$ where the subdominant term is as small as possible compared to the dominant one are called *Stokes lines*.[4]

Due to the Stokes phenomenon, asymptotic expansions cannot be, in general, analytically continued, even if the corresponding holomorphic functions can. This is well illustrated by Example 3.6 in the next section.

---

[4]Many mathematics books use a different nomenclature: what is a Stokes line for us corresponds to an anti-Stokes line for them and vice versa.

3.2. Laplace and Fourier Integrals

A typical problem of asymptotic analysis is to estimate the value of an integral for large (or small) value of a parameter. We consider first integrals of the form

$$\int_a^b e^{zt} g(t)\mathrm{d}t, \tag{3.17}$$

where $[a, b]$ is an interval in $\mathbb{R}$ and z is a complex parameter that we want to take large along some direction in the complex plane. Integrals of this kind contain, in particular, the Fourier and Laplace transforms. An asymptotic expansion of integrals of the form (3.17) can be obtained by repeatedly integrating by parts, or by expanding $g(t)$ in series and integrating term by term.

Let us start with the Laplace transform of a function $g(t)$,

$$f(z) = \int_0^{+\infty} e^{-zt} g(t)\mathrm{d}t. \tag{3.18}$$

As discussed in Section 2.1, this integral exists for Re $z > \alpha_0$, where α_0 is the infimum of the numbers α such that $e^{-\alpha t} g(t) \in L^1[0, \infty)$. We are interested in the behaviour of $f(z)$ when $|z| \to \infty$. In order to stay in the region where (3.18) is well defined, we restrict to the sector $S_\epsilon = \{-\pi/2 + \epsilon <$ arg $z < \pi/2 - \epsilon\}$ for $\epsilon > 0$. For any ϵ, this sector intersects the right half-plane where the integral (3.18) exists. For large $|z|$ with Re $z > 0$, the integrand is exponentially suppressed by the term $|e^{-zt}| = e^{-\text{Re}\, zt}$ and it is substantially different from zero only for the values of t such that Re $zt \ll 1$. Thus in the limit $|z| \to \infty$, the dominant contribution to (3.18) comes from the integration region near $t = 0$, and it is sufficient to know the behaviour of $g(t)$ near $t = 0$ to give an estimate of the integral. If $g(t)$ is C^∞ in the neighbourhood of $t = 0$, we can expand it in Taylor series in t and integrate term by term[5]

$$\int_0^{+\infty} e^{-zt} g(t)\mathrm{d}t = \sum_{k=0}^\infty \frac{g^{(k)}(0)}{k!} \int_0^{+\infty} e^{-zt} t^k \mathrm{d}t = \sum_{k=0}^\infty \frac{g^{(k)}(0)}{z^{k+1}}. \tag{3.19}$$

---

[5]Integrals of the type $\int_0^{+\infty} e^{-zt} t^k \mathrm{d}t$ are easily evaluated with the trick $\int_0^{+\infty} e^{-zt} t^k \mathrm{d}t = (-1)^k d^k/dz^k \int_0^{+\infty} e^{-zt}\mathrm{d}t = k!/z^{k+1}$. One can also use the identity (A.15).

The same expression can be obtained by repeated integrations by parts (see Exercise 3.1). The expansion in (3.19) is the asymptotic expansion of $f(z)$ for large $|z|$.

There is a general theorem, known as *Watson's lemma*, that, under certain conditions on $g(t)$, gives the asymptotic expansion of the integral (3.18). Consider a continuous function g such that $e^{-\alpha_0 t}g(t) \in L^1[0, \infty)$ for some α_0 and it has an asymptotic expansion in the neighbourhood of $t = 0$ given by $g(t) \sim \sum_{m=0}^{\infty} c_m t^{a_m}$, where $a_m = \alpha + \beta m$ with $\operatorname{Re}\alpha > -1$ and $\operatorname{Re}\beta > 0$. Then

$$f(z) \sim \sum_{m=0}^{\infty} \frac{c_m \Gamma(a_m + 1)}{z^{a_m+1}} \tag{3.20}$$

is an asymptotic expansion of (3.18) valid for $z \in S_\epsilon$. Equation (3.20) is obtained by substituting the expansion of $g(t)$ in (3.18) and integrating term by term using (A.15).

Example 3.6. Consider the Laplace integral[6]

$$I(z) = \int_0^{+\infty} \frac{e^{-zt}}{1 + t^4} dt. \tag{3.21}$$

The convergence abscissa of $1/(1 + t^4)$ is $\alpha_0 = 0$, so that $I(z)$ exists for $\operatorname{Re} z > 0$. Its asymptotic expansion in any sector with $\arg z \in (-\pi/2, \pi/2)$ is easily obtained by expanding the denominator of the integrand as a geometric series, and integrating term by term

$$I(z) \sim \sum_{n=0}^{\infty} (-1)^n \int_0^{+\infty} e^{-zt} t^{4n} dt = \sum_{n=0}^{\infty} (-1)^n \frac{(4n)!}{z^{4n+1}}. \tag{3.22}$$

It is interesting to see what happens to the asymptotic expansion (3.22) when we analytically continue the function $I(z)$ to other regions of the complex plane. This can be done, for example, by promoting t to a complex variable and rotating the integration contour of an angle $-\pi/2$ in the complex plane. The function

$$\tilde{I}(z) = \int_0^{-i\infty} \frac{e^{-zt}}{1 + t^4} dt = -i \int_0^{+\infty} \frac{e^{izy}}{1 + y^4} dy, \tag{3.23}$$

---

[6]This and few other examples in this book are taken from Pradisi (2012).

is now defined for $\arg z \in (0, \pi)$. The functions $I(z)$ and $\tilde{I}(z)$ are closely related but not equal. Since, in rotating the contour from the real axis to the negative imaginary one, we encounter a pole in $t = e^{-i\pi/4}$, the two integrals cannot be really deformed one into the other. However, we can relate the two integrals by considering the integral $I_R(z)$ of $f(t) = e^{-zt}/(1+t^4)$ on the closed, positively oriented curve consisting of the segment $[0, R]$ on the real axis, the quarter of circle of radius R in the fourth quadrant and the segment $[-iR, 0]$ on the imaginary axis. This integral can be evaluated by the residue theorem and gives $I_R(z) = 2\pi i \mathrm{Res}[f, e^{-i\pi/4}] = \pi/2 \exp(5\pi i/4 - ze^{-\pi i/4})$. In the limit of large R the integral on the quarter of circle goes to zero by Jordan's lemma. The integrals over the two segments reduce to $I(z)$ and $\tilde{I}(z)$, and we obtain

$$I(z) = \tilde{I}(z) - I_R(z) = \tilde{I}(z) - \frac{\pi}{2} e^{\frac{5\pi i}{4} - ze^{-i\pi/4}}. \tag{3.24}$$

We can use (3.24) as the definition of $I(z)$ in the second quadrant and this gives the analytical continuation of $I(z)$ from the sector $\arg z \in (-\pi/2, \pi/2)$ to the sector $\arg z \in (-\pi/2, \pi)$. Equation (3.24) also allows to derive an asymptotic expansion of $I(z)$ on the whole sector $\arg z \in (-\pi/2, \pi)$. The integral (3.23) defining $\tilde{I}(z)$ has an asymptotic expansion in the sector $\arg z \in (0, \pi)$ which is obtained as before[7]

$$\tilde{I}(z) \sim \sum_{n=0}^{\infty} \int_0^{-i\infty} (-1)^n e^{-zt} t^{4n} dt = \sum_{n=0}^{\infty} (-1)^n \frac{(4n)!}{z^{4n+1}}. \tag{3.25}$$

Since the exponential term $\exp(-ze^{-i\pi/4})$ is suppressed in the sector $\arg z \in (\pi/2, 3\pi/4)$, equations (3.24) and (3.25) imply that (3.22) is a good asymptotic expansion for $I(z)$ also in the sector $\arg z \in (\pi/2, 3\pi/4)$. On the other hand, in the sector $(3\pi/4, \pi)$ the term $\exp(-ze^{-i\pi/4})$ becomes exponentially large for $|z| \to \infty$, since $\mathrm{Re}(-ze^{-i\pi/4})$ is positive. Thus, in the sector $(3\pi/4, \pi)$, $I(z)$ is asymptotic, up to a constant, to the exponential $\exp(-ze^{-i\pi/4})$. This is another example of Stokes phenomenon and the line $\arg z = 3\pi/4$ is an anti-Stokes line similarly to Example 3.5. We see that,

---

[7]Use again footnote 5 and notice that for $\arg z \in (0, \pi)$ the integral $\int_0^{-i\infty} e^{-zt} dt = e^{-zt}/(-z)|_0^{-i\infty} = 1/z$ is convergent.

although $I(z)$ can be analytically continued to a larger domain of definition, its asymptotic expansion cannot.

The previous results can be immediately generalised to integrals of the form

$$f(z) = \int_a^b e^{-zt} g(t) dt, \tag{3.26}$$

with z large and $a < b$. The presence of the exponential implies that the integral (3.26) is dominated by the region of integration around $t = a$. If g is C^∞ in the neighbourhood of $t = a$, expanding it in Taylor series around $t = a$, we find

$$f(z) \sim e^{-za} \sum_{k=0}^\infty \frac{g^{(k)}(a)}{k!} \int_a^b e^{-z(t-a)}(t-a)^k dt$$

$$\sim e^{-za} \sum_{k=0}^\infty \frac{g^{(k)}(a)}{k!} \int_0^{+\infty} e^{-z\tau} \tau^k d\tau = e^{-za} \sum_{k=0}^\infty \frac{g^{(k)}(a)}{z^{k+1}}. \tag{3.27}$$

Notice that we have extended the integral in $\tau = t - a$ from $[0, b-a]$ to $[0, \infty)$ since the error committed in this way is $O(e^{-bz})$ and, therefore, exponentially small compared to the terms in (3.27) (see Exercise 3.2). For more general functions that are continuous but not C^∞, one can use Watson's lemma. If, for $t \to a$, $g(t) \sim \sum_{m=0}^\infty c_m(t-a)^{a_m}$ with $a_m = \alpha + \beta m$ and $\operatorname{Re} \alpha > -1$ and $\operatorname{Re} \beta > 0$, then

$$f(z) \sim e^{-za} \sum_{m=0}^\infty \frac{c_m \Gamma(a_m + 1)}{z^{a_m+1}}. \tag{3.28}$$

We can also consider integrals of Fourier type,

$$f(x) = \int_a^b e^{ixt} g(t) dt, \tag{3.29}$$

where x is a real variable and $g(t) \in L^1[a, b]$. We want to write the asymptotic expansion for large x. The *Riemann–Lebesgue lemma* tells us that integrals of this form go to zero for large x (see Petrini *et al.*, 2017). This is due to the large cancellations introduced in the integral by the rapid variation of the exponential for large x. Since e^{ixt} oscillates, we cannot claim anymore that the integrand is dominated by particular values of t, as we did for Laplace integrals. However, we can always use the trick of integrating by parts repeatedly, as in Example 3.1 and Exercise 3.1. If g is sufficiently regular, we obtain

$$f(x) = \frac{e^{itx}g(t)}{ix}\Big|_a^b - \frac{1}{ix}\int_a^b e^{ixt}g'(t)dt$$

$$= \sum_{m=0}^N \frac{(-1)^m}{(ix)^{m+1}}(e^{ibx}g^{(m)}(b) - e^{iax}g^{(m)}(a))$$

$$+ \frac{(-1)^{N+1}}{(ix)^{N+1}}\int_a^b e^{ixt}g^{(N+1)}(t)dt.$$

This is an asymptotic expansion since the reminder is $o(1/x^{N+1})$ by the Riemann–Lebesgue lemma.

3.3. Laplace's Method

We now want to analyse integrals of the form

$$f(x) = \int_a^b e^{x\phi(t)}g(t)dt, \tag{3.30}$$

with $\phi(t)$ and $g(t)$ real functions, for large values of the real positive parameter x. As in the previous section, the integral is dominated, for large x, by the region near the point t where $\phi(t)$ is maximum. This can be an endpoint of the interval $[a, b]$ or an interior point.

If $\phi'(t) \neq 0$ in $[a, b]$, we can perform a change of variable $\tau = -\phi(t)$ and reduce the integral to the form (3.26), which can be estimated using Watson's lemma.

Of particular interest for applications to physics is the case where $\phi(t)$ has a critical point c, $\phi'(c) = 0$, on the interval $[a, b]$. Suppose, for simplicity, that $\phi(t)$ is a C^2 function with its maximum in $t = c \in (a, b)$: $\phi'(c) = 0$ and $\phi''(c) < 0$. We can split the integral into the sum of the integrals over the segments $[a, c - \epsilon]$, $[c - \epsilon, c + \epsilon]$ and $[c + \epsilon, b]$. As in the previous section, the integrals over $[a, c - \epsilon]$ and $[c + \epsilon, b]$ are exponentially smaller than the integral over $[c - \epsilon, c + \epsilon]$, so we can concentrate on the latter. For very small ϵ, we can expand ϕ and g in Taylor series in the neighbourhood of $t = c$ and keep only the first terms

$$\phi(t) = \phi(c) + \frac{\phi''(c)}{2}(t-c)^2 + \cdots ,$$

$$g(t) = g(c) + \cdots . \tag{3.31}$$

Notice that there is no linear term in ϕ because c is a maximum. We also assume that $g(c) \neq 0$. The leading behaviour of the integral is thus[8]

$$f(x) \sim e^{x\phi(c)}g(c) \int_{\mathbb{R}} e^{x\phi''(c)(t-c)^2/2}dt = \sqrt{\frac{2\pi}{x|\phi''(c)|}}e^{x\phi(c)}g(c). \tag{3.32}$$

Notice that to perform the Gaussian integration in the last line we have extended the integral from $[c - \epsilon, c + \epsilon]$ to $\mathbb{R}$. This is possible since the difference between the two integrals is exponentially small with respect to the dominant term $e^{x\phi(c)}$. This procedure to compute the leading behaviour of (3.30) for large x is called *Laplace's method*.

We can also go to higher order in the expansion of $f(x)$ by keeping more terms in the Taylor expansion of ϕ and g

$$\phi(t) = \phi(c) + \frac{\phi''(c)}{2}(t-c)^2 + \frac{\phi'''(c)}{6}(t-c)^3 + \cdots ,$$

$$g(t) = g(c) + g'(c)(t-c) + \cdots .$$

Making the change of variable $y = (t - c)\sqrt{|\phi''(c)|x}$, we obtain

$$f(x) \sim e^{x\phi(c)} \int_{\mathbb{R}} e^{x\phi''(c)(t-c)^2/2 + x\phi'''(c)(t-c)^3/6 + \cdots}(g(c) + g'(c)(t-c) + \cdots)dt$$

$$= \frac{e^{x\phi(c)}}{\sqrt{x|\phi''(c)|}} \int_{\mathbb{R}} e^{-y^2/2 + \frac{\phi'''(c)y^3}{6\sqrt{x|\phi''(c)|^3}} + \cdots} \left[g(c) + g'(c)\frac{y}{\sqrt{|\phi''(c)|x}} + \cdots \right] dy$$

$$= \frac{e^{x\phi(c)}}{\sqrt{x|\phi''(c)|}} \int_{\mathbb{R}} e^{-y^2/2} \left[g(c) + \sum_{n=1}^{\infty} \frac{c_n(y)}{x^{n/2}} \right] dy, \tag{3.33}$$

---

[8]Recall that $\phi''(c) < 0$.

where $c_n(y) = (-1)^n c_n(-y)$ are polynomials obtained by multiplying the expansion of g and the exponential in power series of $1/\sqrt{x}$. By integrating term by term and observing that only integrals with even powers of y are non-zero by parity, we find an asymptotic expansion for $f(x)$ of the form

$$f(x) \sim \sqrt{\frac{2\pi}{x|\phi''(c)|}} e^{x\phi(c)} g(c) \left(1 + \frac{a_1}{x} + \frac{a_2}{x^2} + \cdots\right). \tag{3.34}$$

In practical applications it is often convenient to perform an explicit change of variables $\phi(t) - \phi(c) = -u^2$ in order to simplify the expansion of the integrand, as shown in Example 3.7.

Laplace's method can also be applied when the maximum of ϕ is in $c = a$ or $c = b$, the only difference being that the integrals in (3.32) and (3.33) are now on $[0, \infty)$ or $(-\infty, 0]$.

It is worth noticing that the expansions above can be also obtained with the methods of the previous section. Indeed, even if $\phi(t)$ is not monotonic, we can always break the interval $[a, b]$ into the union of sub-intervals where ϕ is monotonic. By performing an appropriate change of variables, each of the integrals over the sub-intervals can be written in the form (3.26) and estimated using Watson's lemma (see Exercise 3.3).

Example 3.7. We want to study the asymptotic behaviour of the integral

$$f(x) = \int_{-\infty}^{+\infty} e^{-x \cosh t} dt, \tag{3.35}$$

for large real x. The integral is of the type (3.30) with $\phi(t) = -\cosh t$ and $g(t) = 1$. The function $\phi(t)$ has a single maximum at $t = 0$. The first term in the asymptotic expansion is obtained using (3.32) with $\phi(0) = -1$ and $\phi''(0) = -1$,

$$f(x) \sim \sqrt{\frac{2\pi}{x}} e^{-x}. \tag{3.36}$$

We can also obtain the full asymptotic series by expanding the integrand around $t = 0$. It is convenient to change variable and set $\cosh(t) - \cosh(0) = y^2$ in order to have Gaussian integrals. The integral becomes (see (A.17))

$$\int_{-\infty}^{+\infty} e^{-x(1+y^2)} \frac{2dy}{\sqrt{2+y^2}} = \sqrt{2}e^{-x} \int_{-\infty}^{+\infty} e^{-xy^2} \sum_{n=0}^{\infty} \frac{(-1)^n \Gamma(n+1/2)}{\Gamma(1/2)\Gamma(n+1)} \frac{y^{2n}}{2^n} dy.$$

Using (A.16), we finally find

$$f(x) \sim \sqrt{\frac{2}{\pi x}} e^{-x} \sum_{n=0}^{\infty} (-1)^n \frac{\Gamma(n+\frac{1}{2})^2}{\Gamma(n+1)2^n x^n}. \tag{3.37}$$

Example 3.8. We study the asymptotic behaviour of the *Euler Gamma function*

$$\Gamma(x) = \int_{0}^{+\infty} e^{-y} y^{x-1} dy. \tag{3.38}$$

We can write

$$\Gamma(x+1) = \int_{0}^{+\infty} e^{-y} y^x dy = \int_{0}^{+\infty} e^{-y+x\ln y} dy = x^{x+1} \int_{0}^{+\infty} e^{x(-t+\ln t)} dt,$$

where we defined $y = xt$. To evaluate the previous integral for large x we compare it with (3.30). We have $\phi(t) = -t+\ln t$ and $g(t) = 1$. The function $\phi(t)$ has its maximum at $t = 1$ where $\phi(1) = -1$ and $\phi''(1) = -1$. Applying (3.32), we find

$$\Gamma(x+1) \sim \sqrt{\frac{2\pi}{x}} x^{x+1} e^{-x}. \tag{3.39}$$

For $x = n \in \mathbb{N}$, since $\Gamma(n+1) = n!$, we find the well-known *Stirling formula*, $n! \sim \sqrt{2\pi n}\, n^n\, e^{-n}$, which is indeed valid for large n. The next order of the expansion is discussed in Exercise 3.4.

3.4.   Stationary Phase Method

We now want to study the large x behaviour of integrals of the form

$$f(x) = \int_{a}^{b} e^{ix\phi(t)} g(t)dt, \tag{3.40}$$

where $\phi(t)$ and $g(t)$ are real functions and $x > 0$.

The discussion is similar to that in the previous section. If $\phi(t)$ is monotonic and $\phi'(t) \neq 0$ in $[a, b]$, we can perform a change of variable $\tau = \phi(t)$

and reduce the problem to the study of the asymptotic behaviour of an integral of Fourier type discussed in Section 3.2. If $\phi(t)$ is not monotonic, the integral is dominated by the critical points of the function. The idea is that, when $\phi(t)$ varies as a function of t, the integrand oscillates, and it oscillates the more rapidly the bigger x is. These rapid oscillations tend to give a vanishing result for the integral, similarly to the Riemann–Lebesgue lemma. If instead, $\phi(t)$ is constant in t, the integrand does not oscillate and the integral is not vanishing. Therefore, we expect the integral (3.40) to take its main contribution from the region where $\phi(t)$ is varying the least, namely the region near the zeros of $\phi'(t)$.

More explicitly, suppose that there is a point $c \in (a, b)$ where $\phi'(c) = 0$ and $\phi''(c) \neq 0$. This time, c can either be a maximum or a minimum. As in the previous section, we can split the integral into the sum of three integrals over the segments $[a, c - \epsilon]$, $[c - \epsilon, c + \epsilon]$ and $[c + \epsilon, b]$. We can evaluate the asymptotic behaviour of the integrals over $[a, c - \epsilon]$ and $[c + \epsilon, b]$ as in Section 3.2, discovering that they are $O(1/x)$. As we will see, the integral over $[c - \epsilon, c + \epsilon]$ is of order $O(1/\sqrt{x})$, so we can again focus on it. As before, in the neighbourhood of $t = c$, we expand ϕ and g in Taylor series as in (3.31) and keep only the first terms. The leading behaviour of the integral is now

$$f(x) \sim e^{ix\phi(c)} \int_{c-\epsilon}^{c+\epsilon} e^{ix\phi''(c)(t-c)^2/2} g(c)\,\mathrm{d}t \sim e^{ix\phi(c)} g(c) \int_{\mathbb{R}} e^{ix\phi''(c)(t-c)^2/2}\,\mathrm{d}t$$

$$= g(c) e^{ix\phi(c)} \sqrt{\frac{2}{x|\phi''(c)|}} \int_{\mathbb{R}} e^{i\,\mathrm{sign}(\phi''(c))s^2}\,\mathrm{d}s,$$

where we set $s = \sqrt{x|\phi''(c)|}(t - c)^2/2$ and we assumed that $g(c) \neq 0$. The error made by extending the integral from $[c - \epsilon, c + \epsilon]$ to $\mathbb{R}$ is again of order $O(1/x)$ and is consistent with the previous approximations. The final integral is known as a Fresnel's integral, and can be evaluated by rotating the contour in the complex plane, $s = v e^{\pm i\pi/4}$

$$\int_{\mathbb{R}} e^{\pm is^2}\,\mathrm{d}s = e^{\pm i\pi/4} \int_{\mathbb{R}} e^{-v^2}\,\mathrm{d}v = e^{\pm i\pi/4}\sqrt{\pi}. \tag{3.41}$$

Then, we obtain for the asymptotic behaviour of $f(x)$

$$f(x) \sim g(c)\sqrt{\frac{2\pi}{x|\phi''(c)|}}e^{ix\phi(c)+i\frac{\pi}{4}\operatorname{sign}(\phi''(c))}. \tag{3.42}$$

The result is of order $O(1/\sqrt{x})$, justifying the approximations made. This is called the *stationary phase method*. Notice that, since we have neglected terms of order $1/x$, the stationary phase method can only provide the leading approximation of the integral (3.40).

Example 3.9. Consider the integral

$$I(x) = \int_0^1 \cos(x(t^3 - t))\mathrm{d}t = \operatorname{Re}\int_0^1 e^{ix(t^3-t)}\mathrm{d}t. \tag{3.43}$$

The last integral in the expression above is of the form (3.40) with $\phi(t) = t^3 - t$ and $g(t) = 1$. In the interval $[0,1]$ the function $\phi(t)$ has a minimum in $t = 1/\sqrt{3}$ with $\phi(1/\sqrt{3}) = -2/(3\sqrt{3})$ and $\phi''(1/\sqrt{3}) = 2\sqrt{3}$. Using (3.42) we find

$$\int_0^1 e^{ix(t^3-t)}\mathrm{d}t \sim \sqrt{\frac{\pi}{\sqrt{3}x}}e^{-i\frac{2x}{3\sqrt{3}}+i\frac{\pi}{4}}, \tag{3.44}$$

so that

$$I(x) \sim \sqrt{\frac{\pi}{\sqrt{3}x}}\cos\left(\frac{\pi}{4} - \frac{2x}{3\sqrt{3}}\right). \tag{3.45}$$

To understand why the stationary phase method works, it is instructive to plot the integrand in the interval $[0,1]$. We see from Figure 3.1

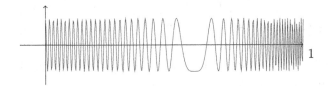

Fig. 3.1.   Plot of $\cos(x(t^3 - t))$ in the interval $[0,1]$ for $x = 400$.

that the function rapidly oscillates everywhere except near the stationary point $t = 1/\sqrt{3}$. Positive and negative values of the cosine cancel in most of the interval and only the contribution of a small region near $t = 1/\sqrt{3}$ matters.

3.5. Saddle-Point Method

Finally we want to analyse integrals of the form

$$f(x) = \int_\gamma e^{xh(z)} g(z) \mathrm{d}z, \tag{3.46}$$

where x is a large real variable, $h(z)$ and $g(z)$ are holomorphic functions and γ is a curve in the complex plane contained in the domain of holomorphicity of h and g.

The integrand contains the exponential term

$$e^{xh(z)} = e^{x \operatorname{Re} h(z)} e^{ix \operatorname{Im} h(z)} = e^{xu} e^{ixv}, \tag{3.47}$$

where u and v are the real and imaginary parts of the holomorphic function $h(z)$. In general, the term e^{ixv} oscillates and leads to large cancellations in the integral. The idea is then to deform the contour γ, inside the region of holomorphicity of h and g, into a curve along which the phase v is constant. If the curve γ is also such that e^{xu} has a maximum, we can use Laplace's method for estimating the integral.

The right curves are the ones passing through the critical points of h, where $h'(z) = 0$. Indeed, by the Cauchy–Riemann conditions, $h'(z) = \partial_x u - i\partial_y u = i(\partial_x v - i\partial_y v)$, and a critical point z_0 of $h(z)$ is also a critical point for u and v. However, as discussed in Section 1.1, the real and imaginary parts of a holomorphic function have no local minima nor maxima. Therefore z_0 cannot be a maximum for u: in a neighbourhood of z_0, u will increase in some directions and decrease in some others. z_0 is called a *saddle point*. The *order* of the saddle point is $m - 1$, if m is the order of the first non-zero derivative $h^{(m)}(z_0) \neq 0$.

The behaviour of u in a neighbourhood of a saddle point can be further characterised. Consider first the points in the neighbourhood where

$h'(z) \neq 0$. It follows from Example 1.5 that the level sets $u = const$ and $v = const$ are orthogonal. This means that the gradient ∇u is parallel to the curves of constant phase, $v = const$, and therefore these are also the curves where u varies the most. Notice that there is only one curve of constant phase passing by a point with $h'(z) \neq 0$. This is not the case for a saddle point z_0. If z_0 is of order $m - 1$, in its neighbourhood the function $h(z)$ is m-to-one (see Section 1.1). Then there are m different curves of constant phase intersecting at the saddle point. These can be seen as $2m$ paths originating from z_0. On such paths u varies the most, increasing along m of them and decreasing along the others. The paths where u decreases the most are called *paths of steepest descent*.

The strategy to evaluate (3.46) is the following. We first find the saddle point, $h'(z_0) = 0$, nearest to the integration contour. We deform γ, if possible, until it passes through z_0 along a curve of steepest descent. Since u has a maximum along this curve and the phase of $e^{ix \, \mathrm{Im} \, h(z)}$ is constant, we can use the Laplace's method to find the asymptotic behaviour of the integral. This method is known as the *method of steepest descent* or the *saddle-point method*. In cases where we cannot deform the contour γ without encountering singularities or leaving the domain of holomorphicity of the integrand, we can always try to deform γ into a contour with constant phase, even if there is no saddle point on it, and expand the integrand around the maximum of u on the curve. The various possibilities are explained in the following examples.

Example 3.10 (The Airy function). Consider the *Airy function* defined by

$$Ai(y) = \frac{1}{\pi} \int_0^{+\infty} \cos\left(yt + \frac{t^3}{3}\right) dt = \frac{1}{2\pi} \int_{-\infty}^{+\infty} e^{i\left(yt + \frac{t^3}{3}\right)} dt, \quad (3.48)$$

for real y. One can show that the above integral exists in improper sense. We can also extend the Airy function to the complex plane as

$$Ai(y) = \frac{1}{2\pi} \int_{\gamma_1} e^{i(yw + \frac{w^3}{3})} dw, \quad (3.49)$$

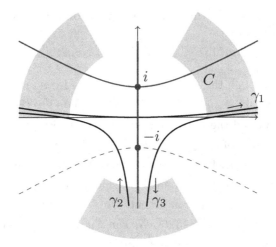

Fig. 3.2. The curves through $z = \pm i$ are the curves of constant phase for $h(z) = i(z + z^3/3)$. The curves of steepest descent are the hyperbola C for $z = i$ and the imaginary axis for $z = -i$. They are denoted with solid lines. The curves γ_i enter in the definition of the Airy and Bairy functions. The shaded area indicates the asymptotic regions where the integrals converge.

where the contour γ_1 starts at left infinity with argument $2\pi/3 < \arg w < \pi$ and ends at right infinity with argument $0 < \arg w < \pi/3$, so that $\mathrm{Re}(iw^3) < 0$ for large $|w|$ and the integral converges (see Figure 3.2). It is easy to see that the Airy function satisfies the differential equation (see Exercise 3.14)

$$Ai''(y) - yAi(y) = 0. \tag{3.50}$$

The Airy function is an entire function on $\mathbb{C}$. Let us consider its asymptotic behaviour for large, real and positive y. It is useful to perform the change of variables $y = x^{2/3}$, $w = x^{1/3}z$ so that

$$Ai(x^{2/3}) = \frac{x^{1/3}}{2\pi} \int_{\gamma_1} e^{ix(z + \frac{z^3}{3})} dz. \tag{3.51}$$

The critical points of $h(z) = i(z + z^3/3)$ are at $z = \pm i$. We can then use the saddle-point method. We need to find the curves with constant phase.

Setting $z = \zeta + i\eta$, we see that the curves with constant phase are

$$\mathrm{Im}\, h(z) = \frac{\zeta^3}{3} - \zeta\eta^2 + \zeta = const. \tag{3.52}$$

Those passing through the saddle points $\pm i$, have $\mathrm{Im}\, h(z) = \mathrm{Im}\, h(\pm i) = 0$ and are given by $\zeta(\zeta^2 - 3\eta^2 + 3) = 0$. They correspond to the hyperbola C, $\zeta^2 - 3\eta^2 + 3 = 0$, and the imaginary axis $\zeta = 0$ (see Figure 3.2). It is easy to see that the curve of steepest descent through $z = i$ is the hyperbola C in the upper half-plane. From Figure 3.2 we see that we can continuously deform γ_1 into C without encountering singularities and staying in the region where the integral is convergent. On C we have $\mathrm{Im}\, h(z) = 0$ and $h(z) = \mathrm{Re}\, h(z) = \eta(\eta^2 - 3\zeta^2 - 3)/3$. We can parameterise the hyperbola as $\eta = \cosh t$, $\zeta = \sqrt{3} \sinh t$. We thus obtain

$$Ai(x^{2/3}) = \frac{x^{1/3}}{2\pi} \int_{-\infty}^{+\infty} e^{\frac{x \cosh t(\cosh^2 t - 9 \sinh^2 t - 3)}{3}} (\sqrt{3} \cosh t + i \sinh t) dt.$$

We can estimate this last integral with Laplace's method. As expected, the exponent has a maximum at the saddle point, which is $t = 0$, so that formula (3.32) gives

$$Ai(x^{2/3}) \sim \frac{1}{2\sqrt{\pi}} \frac{1}{x^{1/6}} e^{-\frac{2}{3}x}, \tag{3.53}$$

or, in terms of the original variable $y = x^{2/3}$,

$$Ai(y) \sim \frac{1}{2\sqrt{\pi}} \frac{1}{y^{1/4}} e^{-\frac{2}{3}y^{3/2}}. \tag{3.54}$$

We can even go to higher orders and obtain the asymptotic expansion of $Ai(y)$ (see Exercise 3.15)

$$Ai(y) \sim \frac{1}{2\pi y^{1/4}} e^{-\frac{2}{3}y^{3/2}} \sum_{m=0}^{\infty} \frac{\Gamma(3m + 1/2)}{(2m)!(-9y^{3/2})^m}. \tag{3.55}$$

The asymptotic behaviour of the Airy function depends on the direction in which y goes to infinity.[9] The expansion (3.55) has been derived for positive

---

[9]This also explains the apparent contradiction that $Ai(y)$, which is an entire function, has an asymptotic expansion in fractional powers of y, typical of multivalued functions.

y but is also valid in the sector $|\arg y| < \pi$. On the real negative axis we have instead, for $y \ll 0$ (see Exercise 3.16)[10]

$$Ai(y) \sim \frac{1}{\sqrt{\pi}|y|^{1/4}} \cos\left(\frac{2}{3}|y|^{3/2} - \frac{\pi}{4}\right). \tag{3.56}$$

This is another example of Stokes phenomenon.

Other linearly independent solutions of the Airy equation (3.50) can be obtained by replacing the contour γ_1 in (3.49) with other contours lying in the region where the integral is convergent, for example γ_2 or γ_3 in Figure 3.2. Two linearly independent solutions can be obtained by choosing any pair of contours among the γ_i. Obviously $\int_{\gamma_1} + \int_{\gamma_2} + \int_{\gamma_3} = 0$. A common choice for the second linearly independent solution is given by the *Bairy function* and uses a linear combination of the contours γ_1 and γ_2[11]

$$Bi(y) = -\frac{i}{2\pi} \int_{\gamma_1} e^{i\left(yw + \frac{w^3}{3}\right)} dw - \frac{i}{\pi} \int_{\gamma_2} e^{i\left(yw + \frac{w^3}{3}\right)} dw$$

$$= \frac{1}{\pi} \int_0^{+\infty} \left(\exp\left(yt - \frac{t^3}{3}\right) + \sin\left(yt + \frac{t^3}{3}\right)\right) dt. \tag{3.57}$$

The Bairy function behaves as

$$Bi(y) \sim \frac{1}{\sqrt{\pi}} \frac{1}{y^{1/4}} e^{+\frac{2}{3}y^{3/2}}, \tag{3.58}$$

for large positive y, and as

$$Bi(y) \sim -\frac{1}{\sqrt{\pi}|y|^{1/4}} \sin\left(\frac{2}{3}|y|^{3/2} - \frac{\pi}{4}\right), \tag{3.59}$$

---

[10]Strictly speaking, (3.56) is not an asymptotic expansion. Both the Airy function and the right-hand side of (3.56) have infinitely many zeros for large negative y and their positions are not coincident. Therefore the limit for $y \to -\infty$ of the ratio of the Airy function and right-hand side of (3.56) cannot be one, although the graphs are very close. We should better write $Ai(y) = \frac{1}{2\sqrt{\pi}} \frac{1}{y^{1/4}} e^{-\frac{2}{3}y^{3/2}} w_1(y) \pm \frac{i}{2\sqrt{\pi}} \frac{1}{y^{1/4}} e^{\frac{2}{3}y^{3/2}} w_2(y)$, where $w_i(y) \sim \sum_{n=0}^{\infty} c_n^i y^{-3n/2}$ with $c_0^i = 1$. These expressions are valid for $\frac{\pi}{3} < \arg y < \frac{5\pi}{3}$ (+ sign) and $-\frac{5\pi}{3} < \arg y < -\frac{\pi}{3}$ (− sign), as discussed in Bender and Orszag (1978) (see also Exercise 7.15). When $\arg y = \pm\pi$ the two exponentials are of same order and combine into (3.56).

[11]The integral in the second line is obtained deforming the contour γ_1 to the real axis and the contour γ_2 as the sum of the negative imaginary semi-axis and the negative real semi-axis.

for large negative y (see Exercise 3.16).[12] Notice that the Bairy function blows up for large positive y. One can see that, in this limit, its behaviour is determined by the other saddle point $z = -i$.

Example 3.11. Consider the integral

$$f(x) = \int_0^{+\infty + i\epsilon} e^{ix\left(t + \frac{t^3}{3}\right)} dt, \tag{3.60}$$

for large real x, where the contour starts at $t = 0$ and goes to infinity in the upper half-plane staying infinitesimally close to the real axis. The integrand function is the same as in Example 3.10 and we use the results obtained there. Recall that the saddle points are in $t = \pm i$. We can deform the contour to pass through the saddle point $t = i$. For example, we can take the contour made by the segment $[0, i]$ and the part of the hyperbola C lying in the first quadrant (see Figure 3.2). The integral over the half hyperbola is computed as in the previous example and goes like $e^{-2x/3} x^{-1/2}$. The integral over the segment $[0, i]$ becomes

$$\int_0^i e^{ix\left(t + \frac{t^3}{3}\right)} dt = i \int_0^1 e^{x\left(-u + \frac{u^3}{3}\right)} du, \tag{3.61}$$

with the parameterisation $t = iu$. The function at exponent, $u^3/3 - u$, is monotonic in $[0, 1]$ with maximum in $u = 0$. We expect then the region of integration around $u = 0$ to dominate. We can explicitly derive the first few terms of the asymptotic expansion by expanding the integrand in power series in u and integrating term by term. It is convenient to define a new

---

[12] Relation (3.58) is valid for $|\arg y| < \frac{\pi}{3}$. Relation (3.59) is again imprecise since the Bairy function and the right-hand side of (3.59) have different zeros. We should better write $Bi(y) = \pm \frac{i}{2\sqrt{\pi}} \frac{1}{y^{1/4}} e^{-\frac{2}{3}y^{3/2}} w_1(z) + \frac{1}{2\sqrt{\pi}} \frac{1}{y^{1/4}} e^{\frac{2}{3}y^{3/2}} w_2(z)$, where $w_i(y) \sim \sum_{n=0}^{\infty} c_n^i y^{-3n/2}$ with $c_0^i = 1$. These expressions are valid for $\frac{\pi}{3} < \arg y < \frac{5\pi}{3}$ (+ sign) and $-\frac{5\pi}{3} < \arg y < -\frac{\pi}{3}$ (− sign), as discussed in Bender and Orszag (1978) (see also Exercise 7.15). When $\arg y = \pm \pi$ the two exponentials are of same order and combine into (3.59).

variable $\tau(u) = u^3/3 - u$, thus obtaining

$$-i \int_{-2/3}^{0} \frac{e^{x\tau}}{u(\tau)^2 - 1} d\tau = i \int_{-2/3}^{0} e^{x\tau} \left(1 + \tau^2 + \cdots\right) d\tau$$

$$= \frac{i}{x} \left(1 + \frac{2}{x^2} + \cdots\right) + O(e^{-2x/3}), \quad (3.62)$$

where we used $u(\tau) = -\tau + O(\tau^3)$ and we discarded the exponentially small terms coming from the extremum of integration $-2/3$. Since the integral over the hyperbola is also exponentially small, (3.62) is the asymptotic expansion of $f(x)$.

Example 3.12. Consider now the integral

$$f(x) = \int_0^1 \frac{dt}{t^{1/3}} e^{ix(t+t^2)}, \quad (3.63)$$

for large real x. We extend the integrand to the complex plane allowing a cut for the cubic root $t^{1/3}$ along the real negative axis. The integral is of the form (3.46) with $h(t) = i(t+t^2)$. The saddle point, $h'(t) = 0$, is $t = -1/2$ and lies on the cut. Let us find the contours with constant phase. Setting $t = u + iv$ we find that the constant phase contours $\operatorname{Im} h(z) = u + u^2 - v^2 = const$ are a family of hyperbolae. In this example, we cannot deform in a simple way the contour so that it passes through the saddle point. We can, however, deform the contour to become the combination of two curves of constant phase, $\gamma_1 - \gamma_2$ (see Figure 3.3). On γ_1 and γ_2 the real part of the function

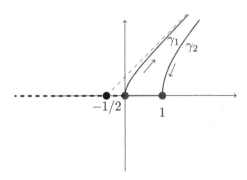

Fig. 3.3. The line $v = u + 1/2$ and γ_i are curves of constant phase for $h(t) = i(t + t^2)$.

$h(t)$ has no critical points but it nevertheless decreases away from the points $t = 0$ and $t = 1$, which are then the points of maximum on the curves. The main contribution to the integral comes from these points. Consider first γ_1. On γ_1, $\text{Im}\, h(t) = 0$. It is convenient to change variable in the integral and set $\rho = h(0) - h(t) = -i(t + t^2)$. Notice that ρ is real along the entire curve γ_1. The integral over γ_1 then becomes an integral over the real axis for the variable ρ and we obtain

$$\int_{\gamma_1} \frac{dt}{t^{1/3}} e^{ix(t+t^2)} = \int_0^{+\infty} \frac{id\rho}{\sqrt{1+4i\rho}} \frac{e^{-\rho x}}{\left(-\frac{1}{2} + \frac{1}{2}\sqrt{1+4i\rho}\right)^{1/3}}, \qquad (3.64)$$

where, in inverting the relation $\rho = -i(t + t^2)$, t has been chosen to lie in the upper half-plane. The main contribution comes from $\rho \sim 0$ and to extract it we expand the denominator of the integrand in power series in ρ. At leading order, we obtain

$$i \int_0^{+\infty} d\rho\, e^{-i\frac{\pi}{6}} \frac{e^{-\rho x}}{\rho^{1/3}} \left(1 + O(\rho)\right) \sim e^{i\frac{\pi}{3}} \frac{\Gamma(2/3)}{x^{2/3}} \left(1 + O\left(\frac{1}{x}\right)\right). \qquad (3.65)$$

The integral on γ_2 is treated similarly setting $\rho = h(1) - h(t)$. One finds

$$\int_{\gamma_2} \frac{dt}{t^{1/3}} e^{ix(t+t^2)} = \frac{i}{3} e^{2ix} \left(\frac{1}{x} + \cdots\right). \qquad (3.66)$$

The phase factor e^{2ix} comes from the constant phase $\text{Im}\, h(t) = \text{Im}\, h(1) = 2$ along the curve. Combining the two contributions, we find the asymptotic expansion

$$f(x) \sim e^{i\frac{\pi}{3}} \frac{\Gamma(2/3)}{x^{2/3}} - \frac{i}{3} e^{2ix} \frac{1}{x} + \cdots. \qquad (3.67)$$

3.6. Exercises

Exercise 3.1. Derive (3.19) by repeatedly integrating by parts.

Exercise 3.2. Show that the contributions to the integral (3.27) coming from extending the integration in τ from $[0, b - a]$ to $[0, +\infty)$ are subleading with respect to the result of the integration around the origin.

Exercise 3.3. Derive (3.32) by splitting the integral over $[a, b]$ into a sum of integrals over sub-intervals where $\phi(t)$ is monotonic. With a change of variables, write all of these integrals in the form (3.26) and use Watson's lemma.

Exercise 3.4. Find the next-to-leading order in the Stirling formula of Example 3.8.

Exercise 3.5. Consider the *Euler Beta function*

$$B(p, q) = \int_0^1 t^{p-1} (1 - t)^{q-1} dt.$$

(a) Show that it is related to the Euler Gamma function by $B(p, q) = \Gamma(p)\Gamma(q)/\Gamma(p + q)$.

(b) Compute the leading and the next-to-leading terms in the asymptotic expansion of $B(x+1, x+1)$ for $x \to +\infty$. Obtain the same result using Stirling's formula of Example 3.8 and the relation derived in (a).

Exercise 3.6. Calculate the first two terms of the asymptotic expansion for $x \to +\infty$ of

$$I(x) = \int_0^{\frac{\pi}{2}} \sin^3 t \, e^{x \cos t} \, dt.$$

Compare the result with the exact one, showing that the asymptotic series does not capture the exponentially suppressed contributions.

Exercise 3.7. Using Laplace's method, compute the leading and the next-to-leading terms in the asymptotic expansion for $x \to +\infty$ of

$$I(x) = \int_{-1}^1 \frac{e^{x(t^3 - 3t)}}{1 + t^2} \, dt.$$

Exercise 3.8. The Bessel function $J_0(x)$ (see Example 7.9) has the following integral representation

$$J_0(x) = \frac{2}{\pi} \int_0^{\frac{\pi}{2}} d\theta \cos(x \cos \theta).$$

Find the leading term of the asymptotic expansion of $J_0(x)$ for $x \to +\infty$.

Exercise 3.9. Using the stationary phase method, calculate the leading term of

$$I(x) = \int_1^{+\infty} dt \, \frac{e^{ix(t^2 - 2t)}}{t^2 + 1}$$

for $x \to +\infty$. Repeat the same calculation using the saddle-point method.

Exercise 3.10. Find the asymptotic expansion of $\int_0^1 e^{ixt^3} dt$ for large positive x.

Exercise 3.11. Derive the asymptotic expansion for $x \to +\infty$ of

$$I(x) = \int_{-i\infty}^{+i\infty} e^{-x(\pi z - z^2)} \left(1 - \sin z\right) dz.$$

Exercise 3.12. Compute the leading term of the asymptotic expansion for $x \to +\infty$ of the integral

$$I(x) = \frac{1}{2\pi i} \int_\gamma \frac{e^{i\frac{\pi}{2}(t - \frac{1}{t})}}{t^{n+1}} dt,$$

where γ is a simple, closed curve encircling anti-clockwise the origin of the complex plane.

* **Exercise 3.13.** Consider the integral

$$I(x) = \frac{1}{2\pi i} \int_{a-i\infty}^{a+i\infty} ds \, e^{x(s - \sqrt{s})} g(s),$$

where $a > 1/4$ is real. Discuss the asymptotic expansion of $I(x)$ for $x \to +\infty$, in the two cases $g(s) = \frac{1}{s}$ and $g(s) = \frac{1}{s^2 + \omega^2}$, with $\omega \in \mathbb{R}$.

Exercise 3.14. Show that the Airy function (3.49) satisfies the differential equation (3.50).

* **Exercise 3.15.** Find the asymptotic expansion (3.55) of the Airy function.

* **Exercise 3.16.** Saddle points and Stokes phenomenon. By setting $y = x^{2/3} e^{i\phi}$, $w = x^{1/3} z$ in (3.49) and (3.57), we can study the asymptotic behaviour of the Airy and Bairy functions for large $|y|$ in various directions of the plane. Call $z_{L,R}$ the two saddle points of $h(z) = i(ze^{i\phi} + z^3/3)$. The Stokes (anti-Stokes) lines are the direction ϕ in the complex plane where $\operatorname{Im} h(z_L) = \operatorname{Im} h(z_R)$ ($\operatorname{Re} h(z_L) = \operatorname{Re} h(z_R)$).

(a) Study the saddle points of h and the curves of steepest descent when ϕ varies.

(b) Derive the asymptotic behaviours (3.55) and (3.56) of the Airy function for y complex.

(c) Derive the asymptotic behaviours (3.58) and (3.59) of the Bairy function.

(d) Find the Stokes and anti-Stokes lines and explain their meaning.

* Exercise 3.17. Consider the Airy equation $u''(y) - yu(y) = 0$.

(a) Show that the general solution is given by $u(y) = cAi(y) + dAi(\omega y)$ where c and d are arbitrary complex numbers and $\omega = e^{2\pi i/3}$;

(b) Use (3.55) to find the leading asymptotic behaviour of $u(y)$ for generic c and d in the sectors $(-\pi, -\pi/3)$, $(-\pi/3, \pi/3)$ and $(\pi/3, \pi)$;

(c) Find c and d such that $u(y) = Bi(y)$ and derive the asymptotic expansions of $Bi(y)$ in the sectors $(-\pi, -\pi/3)$, $(-\pi/3, \pi/3)$ and $(\pi/3, \pi)$.

PART II
Differential Equations

Introduction

This part is devoted to the study of properties of ordinary (ODE) and partial differential equations (PDE) of interest in physics. A general ordinary differential equation has the form

$$G(x, y(x), y'(x), \ldots, y^{(n)}(x)) = 0,$$

where the unknown y is a function of one independent variable only, $y^{(k)}(x) = \frac{d^k y(x)}{dx^k}$ and G is an arbitrary function. The order of the highest derivative is called *order* of the equation. If the unknown is a function of n independent variables x_i, $u = u(x_1, \ldots, x_n)$, the equation is called a partial differential equation,

$$G(x_1, \ldots, x_n, \partial_{x_1} u, \ldots, \partial_{x_n} u, \partial^2_{x_1 x_1} u, \ldots, \partial^2_{x_1 x_n} u, \ldots) = 0,$$

where G is again a generic function. A classical example of ordinary differential equation in physics is Newton's equation describing the motion of a particle of mass m under the force f

$$m \frac{d^2 x(t)}{dt^2} = f(t, x(t)).$$

An important example of partial differential equation is Schrödinger equation

$$i\hbar \frac{d\psi}{dt} = -\frac{\hbar^2}{2m} \sum_{i=1}^{3} \frac{\partial^2 \psi}{\partial x_i^2} + V\psi,$$

71

which describes the time evolution of the wave function of a particle in the potential $V(\mathbf{x}, t)$ (see Chapter 9). In this case the unknown $\psi(\mathbf{x}, t)$ depends on the three independent space directions $\mathbf{x}$ and the time t.

Differential equations are divided into two large classes, linear and nonlinear. A differential equation that depends linearly on the unknown functions and their derivatives is called *linear*. If this is not the case, the equations is *nonlinear*. Consider, for instance, Riccati equation

$$y'(x) = q_0(x) + q_1(x)y(x) + q_2(x)y^2(x),$$

where q_0, q_1 and q_2 are arbitrary functions of the independent variable x. For generic q_i the equation is nonlinear, while for $q_2 = 0$ it reduces to a linear one. Many equations describing physical problems are linear. This is the case for the Schrödinger, Laplace, heat or wave equations we will discuss in this part. We will also see that linear and nonlinear equations exhibit very different properties.

Solving a differential equation implies an integration and therefore the solutions will depend on a given number of arbitrary parameters, or *integration constants*. To determine a particular solution, one has to impose extra conditions on the problem. The most common conditions are of initial or of boundary type. Consider, for simplicity, an ordinary differential equation of order n written in *normal form*, namely

$$y^{(n)}(x) = F(x, y(x), \ldots, y^{(n-1)}(x)),$$

where the function y is defined on the interval $[a, b] \subseteq \mathbb{R}$. In the *Cauchy problem*, or *initial value problem*, one solves the equation with the condition that the unknown y and its $n-1$ derivatives take fixed values at a given point $x_0 \in [a, b]$. For a *boundary problem*, one solves the ordinary differential equation imposing that y and some of its derivates take, for instance, fixed values at the boundary of the interval. In physical problems, the choice of initial or boundary conditions is dictated by the physical properties of the system we want to describe. A general question one can ask is whether there exists a solution to a Cauchy or boundary problem and whether this is unique. For an ordinary differential equation, under mild conditions on its coefficients, one can prove the existence and uniqueness of the solutions to the Cauchy problem in the neighbourhood of the initial point x_0. For boundary problems and partial differential equations things are more complicated. We will discuss examples of Cauchy problems in Chapter 4 and boundary problems in Chapter 5.

We will give a brief discussion of the main properties of ordinary and partial linear equations in Chapter 4, mostly focusing on second-order equations as many differential equations relevant for physics are of this kind. However, this book does not aim at giving an exhaustive treatment of differential equations and their solution methods. We are rather interested in showing how notions like Hilbert spaces, distributions, integral transforms or asymptotic series naturally emerge from the study of differential equations.

As we will see in Chapter 5, boundary problems for ordinary differential equations are often formalised in terms of eigenvalue problems for self-adjoint operators[1] L on a Hilbert space $L^2(\Omega)$

$$L\mathbf{u} = \lambda\mathbf{u}.$$

For self-adjoint operators the numbers λ are real and, in general, they consist of two sets, a discrete one $\{\lambda_n\}$, with $n \in \mathbb{N}$, and a continuous one $\{\lambda \in I \subseteq \mathbb{R}\}$. The spectral theorem for self-adjoint operators states that the corresponding eigenvectors $\mathbf{u}_n$ and $\mathbf{u}_\lambda$ form a (generalised) orthonormal basis in $L^2(\Omega)$. Many examples of orthogonal polynomials appear in the solution of boundary problems as basis of eigenvectors $\mathbf{u}_n$ for the corresponding operators L. Orthogonal polynomials, special functions and asymptotic series also emerge in the study of the solutions of ordinary differential equations in the neighbourhood of a point (see Chapter 7).[2]

---

[1] We refer to Petrini *et al.* (2017) for an accurate discussion of linear operators in Hilbert spaces and the spectral theorem (see also Appendix A for a brief summary of the basic notions).

[2] For general properties of special functions we refer to Andrews *et al.* (1999) and Bateman *et al.* (1953).

An entire chapter is devoted to Green functions. As explained in Chapter 6, they provide a general method to determine particular solutions of inhomogeneous linear, ordinary or partial, differential equations. As we will see, Green functions are an important tool in many fields of physics, and are also connected with the spectral properties of linear operators.

4

The Cauchy Problem for Differential Equations

Cauchy problems are very natural in physics. The typical example is the solution of Newton's equation in classical mechanics, which is a second-order equation for the position of a particle. We know indeed that the motion of a particle is uniquely specified by its initial position and velocity. The Cauchy problem is also natural for some partial differential equations, like the heat and wave equations.

An important result about Cauchy problems for ordinary differential equations is the existence and uniqueness theorem, which states that, under mild assumptions, a Cauchy problem always admits a unique solution in a neighbourhood of the point x_0 where the initial conditions are given.

4.1. Cauchy Problem for Ordinary Differential Equations

Consider an ordinary differential equation of order n in normal form

$$y^{(n)}(x) = F(x, y(x), \ldots, y^{(n-1)}(x)). \qquad (4.1)$$

The Cauchy problem for (4.1) consists in finding an n-times differentiable solution $y(x)$ in some interval $(x_0 - \delta, x_0 + \delta)$ around a point x_0 that satisfies

$$y(x_0) = y_0, \quad y'(x_0) = y_1, \quad \ldots, \quad y^{(n-1)}(x_0) = y_{n-1}. \qquad (4.2)$$

By defining $w_k = y^{(k)}$, with $k = 0, \ldots, n-1$ we can reduce (4.1) to a system of n first-order differential equations

$$
\begin{cases}
w_0'(x) = w_1(x) \\
\quad \vdots \\
w_{n-2}'(x) = w_{n-1}(x) \\
w_{n-1}'(x) = F(x, w_0(x), w_1(x), \ldots, w_{n-1}(x)).
\end{cases}
\tag{4.3}
$$

Then, without loss of generality, we can focus on the Cauchy problem for a system of first-order equations

$$
\begin{cases}
\dfrac{d\mathbf{w}}{dx} = \mathbf{f}(x, \mathbf{w}(x)), \\[2mm]
\mathbf{w}(x_0) = \mathbf{w}_0,
\end{cases}
\tag{4.4}
$$

where $\mathbf{w}(x)$ and $\mathbf{f}$ are functions with values in $\mathbb{R}^n$. The Cauchy problem (4.1)–(4.2) is a particular case of the system (4.4) with $\mathbf{w} = (w_0, \ldots, w_{n-1})$, $\mathbf{f} = (w_1, w_2, \ldots, F)$ and $\mathbf{w}_0 = (y_0, \ldots, y_{n-1})$.

To study the Cauchy problem (4.4) it is convenient to rewrite it as an integral equation

$$
\mathbf{w}(x) = \mathbf{w}_0 + \int_{x_0}^{x} \mathbf{f}(x', \mathbf{w}(x')) dx'.
\tag{4.5}
$$

We can define the map

$$
T(\mathbf{w}(x)) = \mathbf{w}_0 + \int_{x_0}^{x} \mathbf{f}(x', \mathbf{w}(x')) dx',
\tag{4.6}
$$

from the space of functions where $\mathbf{w}(x)$ lives, to itself. A solution of equation (4.5) is equivalent to the existence of a *fixed point* $T(\mathbf{w}) = \mathbf{w}$ for the map T.

There is a simple and general theorem that deals with fixed points of maps of this type. This is the *Banach–Caccioppoli contraction theorem* or *fixed-point theorem*. Let X be a normed space (see Appendix A and Petrini *et al.* (2017), Section 5.2). A *contraction* is a map $T : X \to X$ satisfying

$$||T(\mathbf{x}) - T(\mathbf{y})|| \leq \alpha ||\mathbf{x} - \mathbf{y}||, \tag{4.7}$$

for all $\mathbf{x}, \mathbf{y} \in X$ and $0 \leq \alpha < 1$. Notice that T is not required to be linear.[1] The theorem states that, if X is complete and T is a contraction in X, the equation $T(\mathbf{x}) = \mathbf{x}$ has a unique solution that can be obtained as

$$\mathbf{x} = \lim_{n \to \infty} T^n(\mathbf{x}_0), \tag{4.8}$$

where $\mathbf{x}_0$ is any vector in X. Here T^n denotes the n-fold iterated action of T.

Proof. Starting with a given $\mathbf{x}_0$, let us define by iteration $\mathbf{x}_{n+1} = T(\mathbf{x}_n)$. Since, by definition of $\mathbf{x}_n$,

$$||\mathbf{x}_{n+1} - \mathbf{x}_n|| = ||T(\mathbf{x}_n) - T(\mathbf{x}_{n-1})|| \leq \alpha ||\mathbf{x}_n - \mathbf{x}_{n-1}|| \leq \cdots \leq \alpha^n ||\mathbf{x}_1 - \mathbf{x}_0||,$$

we find, for $m > n$,

$$||\mathbf{x}_m - \mathbf{x}_n|| \leq ||\mathbf{x}_m - \mathbf{x}_{m-1}|| + \cdots + ||\mathbf{x}_{n+1} - \mathbf{x}_n||$$

$$\leq (\alpha^{m-1} + \cdots + \alpha^n)||\mathbf{x}_1 - \mathbf{x}_0|| \leq \frac{\alpha^n}{1-\alpha} ||\mathbf{x}_1 - \mathbf{x}_0||, \tag{4.9}$$

where we used $0 \leq \alpha < 1$. When m and n go to infinity, the right-hand side of the previous equation goes to zero, since $\alpha < 1$. The sequence $\{\mathbf{x}_n\}$ is then a Cauchy sequence in X. Since X is complete, $\{\mathbf{x}_n\}$ converges to $\mathbf{x} \in X$. We can write

$$||T(\mathbf{x}) - \mathbf{x}|| \leq ||T(\mathbf{x}) - \mathbf{x}_n|| + ||\mathbf{x}_n - \mathbf{x}|| \leq ||T(\mathbf{x}) - T(\mathbf{x}_{n-1})|| + ||\mathbf{x}_n - \mathbf{x}||$$

$$\leq \alpha ||\mathbf{x} - \mathbf{x}_{n-1}|| + ||\mathbf{x}_n - \mathbf{x}||,$$

and, taking the limit $n \to \infty$, we find that $T(\mathbf{x}) = \mathbf{x}$. We need to show that $\mathbf{x}$ is the unique solution. We prove it by contradiction. Suppose T has two distinct fixed points, $T(\mathbf{x}) = \mathbf{x}$ and $T(\mathbf{y}) = \mathbf{y}$. Then we find $||\mathbf{x} - \mathbf{y}|| = ||T(\mathbf{x}) - T(\mathbf{y})|| \leq \alpha ||\mathbf{x} - \mathbf{y}||$, which is impossible since $\alpha < 1$. $\qquad\square$

---

[1]Indeed, in the theory of differential equations we are interested in nonlinear maps. Note, however, that a simple example of contraction is given by a bounded linear operator of norm strictly less than one.

We now use this result to show that the Cauchy problem

$$
\begin{cases}
\dfrac{d\mathbf{w}}{dx} = \mathbf{f}(x, \mathbf{w}(x)), \\[2mm]
\mathbf{w}(x_0) = \mathbf{w}_0,
\end{cases}
\tag{4.10}
$$

where $\mathbf{f}$ is defined in $R = \{a \leq x \leq b, c_i \leq w_i \leq d_i\} \subset \mathbb{R} \times \mathbb{R}^n$, has a unique solution in a neighbourhood of x_0, provided $\mathbf{f}$ is continuous in R and satisfies the *Lipschitz condition* with respect to the variable $\mathbf{w}$

$$
|\mathbf{f}(x, \mathbf{w}_{(1)}) - \mathbf{f}(x, \mathbf{w}_{(2)})| \leq K|\mathbf{w}_{(1)} - \mathbf{w}_{(2)}|, \qquad (x, \mathbf{w}_{(i)}) \in R \quad (4.11)
$$

for some constant $K \geq 0$. Here $|\mathbf{f}|$ denotes the Euclidean norm of a vector in $\mathbb{R}^n$. Moreover, the solution is also a continuous function of the initial condition $\mathbf{w}_0$.[2]

Proof. Consider the space $\mathbf{C}[a, b]$ of continuous vector-valued functions on the interval $[a, b]$ endowed with the sup-norm $\|\mathbf{u}\|_{\text{sup}} = \sup_{[a,b]}|\mathbf{u}(x)|$. One can show that $\mathbf{C}[a, b]$ is complete (see Petrini *et al.*, 2017, Section 6.1). Since $\mathbf{f}$ is continuous, $T(\mathbf{w}) = \mathbf{w}_0 + \int_{x_0}^x \mathbf{f}(x', \mathbf{w}(x'))dx'$ can be seen as a map from $\mathbf{C}[a, b]$ to itself. Consider the subspace of $\mathbf{C}[a, b]$ defined by

$$
S_\epsilon = \{\mathbf{u} \in \mathbf{C}[a, b] \,|\, |\mathbf{u}(x) - \mathbf{w}_0| \leq \epsilon M \text{ for } x \in [x_0 - \epsilon, x_0 + \epsilon]\}, \tag{4.12}
$$

where M is the supremum of $|\mathbf{f}|$ in R, and $\epsilon > 0$. It is easy to see that S_ϵ is a closed subspace of $\mathbf{C}[a, b]$. Indeed, if a sequence $\{\mathbf{u}_n\} \in S_\epsilon$ converges to an element $\mathbf{u}$ of $\mathbf{C}[a, b]$, the condition $|\mathbf{u}_n(x) - \mathbf{w}_0| \leq \epsilon M$ for all $x \in [x_0 - \epsilon, x_0 + \epsilon]$ implies $|\mathbf{u}(x) - \mathbf{w}_0| \leq \epsilon M$ for all $x \in [x_0 - \epsilon, x_0 + \epsilon]$. Therefore S_ϵ contains all its accumulation points and it is closed. A closed subset of a complete space is also complete. T maps S_ϵ into itself since $|T(\mathbf{u}) - \mathbf{w}_0| = |\int_{x_0}^x \mathbf{f}(x', \mathbf{u}(x'))dx'| \leq M\epsilon$ for $x \in [x_0 - \epsilon, x_0 + \epsilon]$. Moreover, for sufficiently small ϵ, T is a contraction in S. Indeed, using the Lipschitz condition, we find

$$
\|T(\mathbf{u}_{(1)}) - T(\mathbf{u}_{(2)})\|_{\text{sup}} = \sup_{[x_0-\epsilon, x_0+\epsilon]} \left| \int_{x_0}^x (\mathbf{f}(x', \mathbf{u}_1(x')) - \mathbf{f}(x', \mathbf{u}_2(x')))dx' \right|
$$

$$
\leq K\epsilon \sup_{[x_0-\epsilon, x_0+\epsilon]} |\mathbf{u}_{(1)}(x) - \mathbf{u}_{(2)}(x)| = K\epsilon \|\mathbf{u}_{(1)} - \mathbf{u}_{(2)}\|_{\text{sup}},
$$

---

[2] If $\mathbf{f}$ depends in a continuous way on a parameter α, the solution is also continuous with respect to it.

and (4.7) is satisfied with $\alpha = K\epsilon$. For sufficiently small ϵ, $\alpha < 1$. We can now use the fixed-point theorem for T to conclude that the equation $T(\mathbf{w}) = \mathbf{w}$ has a unique solution in S_ϵ. This is equivalent to saying that (4.10) has a unique solution in the interval $[x_0 - \epsilon, x_0 + \epsilon]$. □

The requirement that the function is Lipschitz guarantees that the solution of the Cauchy problem is unique. If $\mathbf{f}$ is only continuous in R, the existence of a solution is still guaranteed in the neighbourhood of x_0 (Peano's theorem) but uniqueness is not.

Example 4.1. Consider the Cauchy problem

$$y'(x) = \sqrt{2|y(x)|}, \qquad y(0) = 0. \tag{4.13}$$

The function $f(x, y) = \sqrt{2|y|}$ does not satisfy the Lipschitz condition in $y = 0$ since $|f(x, y) - f(x, 0)|/|y| = \sqrt{2/|y|}$ is not bounded near $y = 0$. For any interval $[-a, b]$ with $a, b > 0$, the function

$$y(x) = \begin{cases} -\dfrac{(x + a)^2}{2}, & x \leq -a, \\[2mm] 0, & -a \leq x \leq b, \\[2mm] \dfrac{(x - b)^2}{2}, & x \geq b, \end{cases} \tag{4.14}$$

is a C^1 solution of the Cauchy problem. We see that there are infinitely many solutions, depending on a and b, in any interval around $x = 0$, no matter how small. This example is due to Peano.

On the other hand, if $\mathbf{f} \in C^1(R)$ the Lipschitz condition is certainly satisfied. Therefore, whenever $\mathbf{f} \in C^1$ in the neighbourhood of (x_0, y_0), the Cauchy problem has a unique continuous and differentiable solution in the same neighbourhood. However, the solution does not necessarily extend to the entire interval where $\mathbf{f}$ is defined and continuously differentiable, even if $\mathbf{f} \in C^\infty(R)$ in the neighbourhood of (x_0, y_0).

Example 4.2. Consider the Cauchy problem

$$y'(x) = y^2(x), \qquad y(0) = c. \tag{4.15}$$

The function $f(x, y) = y^2$ is $C^\infty(\mathbb{R}^2)$ and is also analytic. However, the unique solution of the Cauchy problem is $y(x) = c/(1 - cx)$, which is C^∞

in a neighbourhood of $x = 0$ but it is singular at $x = 1/c$. Notice that the singularity depends on the initial condition. This example illustrates a common behaviour of nonlinear equations, whose solutions can have singularities even when the coefficients of the equation are regular. The singularities are called *spontaneous singularities* and their position depend on the initial conditions. This is a major difference with respect to linear equations, for which the solutions can develop singularities only where the coefficients of the equation are singular.

Under certain further conditions the solution of the Cauchy problem in a given neighbourhood can be extended to a finite interval. For instance, one can show that, if $\mathbf{f}$ is continuous and satisfies the Lipschitz condition for all $x \in [a, b]$ and all $\mathbf{w} \in \mathbb{R}^n$, the Cauchy problem (4.10) has a unique solution on the whole interval $[a, b]$. This, in particular, applies to linear differential equations.

Example 4.3. A *linear ordinary differential equation* of order n has the form

$$\sum_{k=0}^{n} a_k(x) y^{(k)}(x) = g(x), \tag{4.16}$$

where $g(x)$ is the *source term*. The Cauchy problem requires finding a solution that satisfies (4.2). Using the trick (4.3), we can reduce the equation to the form (4.4) with

$$\mathbf{f}(x, \mathbf{w}(x)) = \left\{ w_1(x), w_2(x), \ldots, \frac{1}{a_n(x)} \left(g(x) - \sum_{k=0}^{n-1} a_k(x) w_k(x) \right) \right\},$$

where $\mathbf{w}(x) = \{y(x), y^{(1)}(x), \ldots, y^{(n-1)}(x)\}$ and $\mathbf{w}_0 = \{y_0, y_1, \ldots, y_{n-1}\}$. One can easily see that, if $a_k(x)$ and $g(x)$ are continuous on a finite interval $[a, b]$ and $a_n(x) \neq 0$ for all $x \in [a, b]$, $\mathbf{f}$ is also continuous and satisfies the Lipschitz condition for all $x \in [a, b]$ and all $\mathbf{w} \in \mathbb{R}^n$.[3] It follows that there exists a unique continuous and n-times differentiable solution on the entire interval $[a, b]$. In other words, the solution can develop singularities

---

[3] A short computation shows that (4.11) is satisfied by taking, for instance, $K = \sqrt{1 + n^2 c^2}$ with $c = \max_{0 \leq k \leq n-1} (\max_{[a,b]} |a_k(x)/a_n(x)|)$.

only where the coefficients are singular. Moreover, if the coefficients are analytic also the solution is. One way to see it is to extend the equation to the complex plane and use the general results that will be discussed in Chapter 7.

4.2. Second-Order Linear Ordinary Differential Equations

Many interesting equations that one encounters in physics are second-order and linear. We recall here some of their properties and methods of solutions that will be needed in the rest of the book.

Consider the second-order linear ordinary differential equation

$$a_2(x)y''(x) + a_1(x)y'(x) + a_0(x)y(x) = 0. \tag{4.17}$$

Since the source term is zero, the equation is called *homogeneous*. We assume that the coefficients $a_i(x)$ are continuous and $a_2(x) \neq 0$ in an interval $[a, b]$. Given two solutions $y_1(x)$ and $y_2(x)$, also $\alpha y_1(x) + \beta y_2(x)$ is a solution for all numbers α, β. Therefore, the space of solutions of equation (4.17) is a vector space. This vector space has dimension two.

Proof. We can construct an explicit basis in the vector space of solutions using the existence and uniqueness theorem. Consider a point $x_0 \in [a, b]$ and let $y_1(x)$ be the unique solution to the Cauchy problem with initial conditions

$$y_1(x_0) = 1, \qquad y_1'(x_0) = 0, \tag{4.18}$$

and $y_2(x)$ the unique solution to the Cauchy problem

$$y_2(x_0) = 0, \qquad y_2'(x_0) = 1. \tag{4.19}$$

As discussed in the previous section, y_1 and y_2 are continuous and twice-differentiable in $[a, b]$. We claim that the generic solution $y(x)$ of (4.17) can be written as a linear combination $y(x) = y(x_0)y_1(x) + y'(x_0)y_2(x)$. Indeed, the two functions $y(x)$ and $y(x_0)y_1(x) + y'(x_0)y_2(x)$ are both solutions of (4.17) and they satisfy, by construction, the same initial conditions in x_0. Therefore, by the existence and uniqueness theorem, they must coincide on the whole interval $[a, b]$. $\qquad \square$

Two linearly independent solutions of equation (4.17) are called a *funda-mental set of solutions*. We can detect whether two solutions $y_1(x)$ and $y_2(x)$ are linearly independent by computing their *Wronskian*

$$W(x) = \det \begin{pmatrix} y_1(x) & y_2(x) \\ y_1'(x) & y_2'(x) \end{pmatrix} = y_1(x)y_2'(x) - y_2(x)y_1'(x). \quad (4.20)$$

Indeed, if $y_1(x)$ and $y_2(x)$ are linearly dependent, $y_2(x) = \alpha y_1(x)$ for some constant α and the Wronskian is zero. The Wronskian satisfies some simple and non-trivial identities. By taking a derivative of (4.20), and using the fact that y_1 and y_2 solve (4.17), we obtain

$$W' = y_1 y_2'' - y_2 y_1'' = -\frac{1}{a_2}(y_1 a_1 y_2' - y_2 a_1 y_1'(x)) = -\frac{a_1}{a_2}W. \quad (4.21)$$

This is a first-order differential equation whose solution is

$$W(x) = C \exp\left(-\int_{x_0}^{x} \frac{a_1(y)}{a_2(y)} dy \right), \quad (4.22)$$

where C is an arbitrary constant. By changing the integration constant C, we can shift the lower integration limit to any arbitrary number. We see that either $W(x)$ is identically zero, for $C = 0$, or it never vanishes in the interval $[a, b]$. It follows that, if the two-dimensional vectors (y_1, y_1') and (y_2, y_2') are linearly independent at a point, they are linearly independent at all points of the interval $[a, b]$.

Consider now the Cauchy problem

$$a_2(x)y''(x) + a_1(x)y'(x) + a_0(x)y(x) = 0, \quad \begin{cases} y(x_0) = y_0, \\ y'(x_0) = y_0'. \end{cases} \quad (4.23)$$

The general solution of (4.17) can be written in terms of a fundamental set of solutions, $y_1(x)$ and $y_2(x)$, as

$$y(x) = c_1 y_1(x) + c_2 y_2(x), \quad (4.24)$$

with arbitrary constants c_1 and c_2. The initial conditions impose

$$c_1 y_1(x_0) + c_2 y_2(x_0) = y_0,$$
$$c_1 y_1'(x_0) + c_2 y_2'(x_0) = y_0', \quad (4.25)$$

and we just need to solve this inhomogeneous system of linear algebraic equations for the constants c_1 and c_2. The system has always a solution.

Indeed, the determinant of the coefficients of c_1 and c_2 is the Wronskian, which is different from zero since $y_1(x)$ and $y_2(x)$ are linearly independent.[4]

We review now some simple equations that can be solved in terms of elementary functions.

Example 4.4 (Equations with constant coefficients). The linearly independent solutions of a linear equation with constant coefficients

$$a_2 y''(x) + a_1 y'(x) + a_0 y(x) = 0, \qquad a_i \in \mathbb{C} \tag{4.26}$$

is easily found using the ansatz $y(x) = e^{\alpha x}$. Substituting it into the equation we find

$$(a_2 \alpha^2 + a_1 \alpha + a_0) e^{\alpha x} = 0, \tag{4.27}$$

which is satisfied if α is a solution of the *characteristic equation*

$$a_2 \alpha^2 + a_1 \alpha + a_0 = 0. \tag{4.28}$$

If the characteristic equation has two distinct complex solutions α_1 and α_2, two linearly independent solutions of (4.26) are $y_1(x) = e^{\alpha_1 x}$ and $y_2(x) = e^{\alpha_2 x}$. If the equation has two coincident solutions $\alpha_1 = \alpha_2$, one can easily check that two linearly independent solutions are $y_1(x) = e^{\alpha_1 x}$ and $y_2(x) = x e^{\alpha_1 x}$ (see Exercise 4.4).

Example 4.5 (Euler equation). The fundamental set of solutions of the linear equation

$$a_2 x^2 y''(x) + a_1 x y'(x) + a_0 y(x) = 0, \qquad a_i \in \mathbb{C} \tag{4.29}$$

is found substituting the ansatz $y(x) = x^\alpha$ into the equation. We find

$$(a_2 \alpha(\alpha - 1) + a_1 \alpha + a_0) x^\alpha = 0, \tag{4.30}$$

which is satisfied if α is a solution of the *characteristic equation*

$$a_2 \alpha(\alpha - 1) + a_1 \alpha + a_0 = 0. \tag{4.31}$$

If the characteristic equation has two distinct complex solutions α_1 and α_2, (4.29) has two linearly independent solutions $y_1(x) = x^{\alpha_1}$ and $y_2(x) =$

---

[4]Notice that the same argument would fail for different type of conditions, like for example conditions at different points x_0 and x_1, $y(x_0) = y_0$, $y(x_1) = y_1$, as we will discuss extensively in Chapter 5.

x^{α_2}. If the equation has two coincident solutions $\alpha_1 = \alpha_2$, the two linearly independent solutions are $y_1(x) = x^{\alpha_1}$ and $y_2(x) = \ln x\, x^{\alpha_1}$ (see Exercise 4.4).

When the coefficients in (4.17) are generic functions, there is no general method to find explicit solutions. However, it is interesting to note that, once one of the solutions, $y_1(x)$, is known, we can find the second one using the Wronskian. The method works as follows. Using (4.22), we find

$$\frac{d}{dx}\left(\frac{y_2}{y_1}\right) = \frac{y_1 y_2' - y_2 y_1'}{y_1^2} = \frac{W}{y_1^2} = \frac{C}{y_1^2} e^{-\int_{x_0}^x \frac{a_1(w)}{a_2(w)} dw}, \tag{4.32}$$

so that, by an integration, we find[5]

$$y_2(x) = C y_1(x) \int_{x_0}^x \frac{e^{-\int_{x_0}^t \frac{a_1(w)}{a_2(w)} dw}}{y_1^2(t)} dt. \tag{4.33}$$

Consider now the second-order linear *inhomogeneous* ordinary differential equation

$$a_2(x) y''(x) + a_1(x) y'(x) + a_0 y(x) = g(x). \tag{4.34}$$

If $y_1(x)$ and $y_2(x)$ are solutions of (4.34), $y_1(x) - y_2(x)$ is a solution of the homogeneous equation (4.17). Therefore, to determine the general solution it is sufficient to find a particular solution of the inhomogeneous equation and add to it the general solution of the corresponding homogeneous equation.

There are various methods for finding a particular solution of (4.34). One is the method of *variation of arbitrary constants*. It consists in looking for a solution of (4.34) of the form

$$y(x) = c_1(x) y_1(x) + c_2(x) y_2(x), \tag{4.35}$$

where $y_1(x)$ and $y_1(x)$ are two linearly independent solutions of the corresponding homogeneous equation (4.17), and the coefficients $c_1(x)$ and $c_2(x)$ now depend on the point x. By substituting the ansatz (4.35) in (4.34), one obtains a simple first-order system of equations for $c_1(x)$ and $c_2(x)$ that

---

[5]We are assuming that $y_2(x_0) = 0$, which can be always obtained by adding to $y_2(x)$ a term proportional to $y_1(x)$.

can always be solved (see Exercise 4.5). A second method for finding a particular solution of (4.34) involves the use of Green functions and it is extensively discussed in Chapter 6.

As a final remark, let us mention that it is sometimes useful to put a second-order linear differential equation in *canonical form*,

$$\tilde{y}''(x) + c(x)\tilde{y}(x) = h(x), \tag{4.36}$$

namely with the coefficient of the second derivative equal to one and without the first-order derivative term. Any equation of the form (4.34) can be always reduced to the form (4.36), with the change of variable

$$\tilde{y}(x) = e^{\int_{x_0}^{x} \frac{a_1(t)}{2a_2(t)} dt} y(x). \tag{4.37}$$

The results of this section can be extended to equations of order n

$$\sum_{k=0}^{n} a_k(x) y^{(k)}(x) = g(x), \tag{4.38}$$

where $a_k(x)$ and $g(x)$ are arbitrary functions of the independent variable x. The homogeneous equation associated with (4.38) is obtained setting $g(x) = 0$ in (4.38) and has n linearly independent solutions. The Wronskian is defined as

$$W(x) = \det \begin{pmatrix} y_1(x) & y_2(x) & \cdots & y_n(x) \\ y_1'(x) & y_2'(x) & \cdots & y_n'(x) \\ \vdots & \vdots & \cdots & \vdots \\ y_1^{(n-1)}(x) & y_2^{(n-1)}(x) & \cdots & y_n^{(n-1)}(x) \end{pmatrix}, \tag{4.39}$$

and it is zero if the solutions $y_i(x)$ are linearly dependent. The solution of the inhomogeneous equation (4.38) is obtained by adding to a particular solution the general solution of the corresponding homogeneous equation.

4.3. Second-Order Linear Partial Differential Equations

The study of partial differential equations is more involved than for ordinary ones, and even the theory of linear equations has many different facets. Since most of the interesting partial differential equations in physics are second-order and linear we focus on those.

Consider the most general linear, second-order, partial differential equation in $\mathbb{R}^n$

$$\sum_{i,j=1}^{n} a_{ij}(\mathbf{w})\frac{\partial^2 u}{\partial w_i \partial w_j} + \sum_{i=1}^{n} b_i(\mathbf{w})\frac{\partial u}{\partial w_i} + c(\mathbf{w})u = f(\mathbf{w}). \tag{4.40}$$

In typical applications, the variables w_i represent the coordinates of a point in space or time. Equations of the type (4.40) are classified according to the behaviour of the term of highest degree. If, at a point $\mathbf{w} \in \mathbb{R}^n$, the symmetric matrix a_{ij} has all eigenvalues of the same sign, the equation is called *elliptic* at the point $\mathbf{w}$. If the eigenvalues are non-vanishing but have different signs, the equation is called *hyperbolic* at $\mathbf{w}$. Finally, if at least one of the eigenvalues vanishes, the equation is called *parabolic* at $\mathbf{w}$. The same equation can be of different type at different points, although in concrete application to physics the equations are of the same type everywhere. Examples of elliptic equations are given by the Laplace equation,

$$\sum_{i=1}^{n} \frac{\partial^2 u}{\partial x_i^2} = 0, \tag{4.41}$$

and its inhomogeneous version, the Poisson equation,

$$\sum_{i=1}^{n} \frac{\partial^2 u}{\partial x_i^2} = f(\mathbf{x}), \tag{4.42}$$

which appear in electrostatics and Newtonian gravitational problems, as well as the time-independent Schrödinger equation of quantum mechanics

$$-\frac{\hbar^2}{2m}\sum_{i=1}^{n} \frac{\partial^2 u}{\partial x_i^2} + V(\mathbf{x})u = Eu. \tag{4.43}$$

For all these equations, $w_i = x_i$ are space variables, the matrix a_{ij} is proportional to the identity, and all its eigenvalues are equal. Examples of parabolic equations are given by the heat equation,

$$\frac{\partial u}{\partial t} = \kappa \sum_{i=1}^{n-1} \frac{\partial^2 u}{\partial x_i^2}, \tag{4.44}$$

which describes conduction phenomena, and the time-dependent Schrödinger equation

$$i\hbar\frac{\partial u}{\partial t} = -\frac{\hbar^2}{2m}\sum_{i=1}^{n-1} \frac{\partial^2 u}{\partial x_i^2} + V(\mathbf{x})u. \tag{4.45}$$

Here the n variables w_k are time t and $n-1$ spatial variables x_i. The time derivative appears only at first order, so that the $n \times n$ matrix a_{ij} is proportional to the matrix $\text{diag}(0, 1, \ldots, 1)$ and it has one zero eigenvalue. Finally, an example of hyperbolic equation is given by the wave equation,

$$\frac{\partial^2 u}{\partial t^2} - \sum_{i=1}^{n-1} \frac{\partial^2 u}{\partial x_i^2} = 0. \tag{4.46}$$

The n variables are time t and $n-1$ spatial variables x_i, and the $n \times n$ matrix a_{ij} is proportional to the matrix $\text{diag}(1, -1, \ldots, -1)$ with eigenvalues of different sign.

The Cauchy problem for equation (4.40) can be formulated by choosing a surface Σ in $\mathbb{R}^n$ that extends to infinity and looking for solutions that have a prescribed value for u and its derivative $\partial_n u$ in the direction orthogonal to Σ,

$$u|_\Sigma = u_0, \qquad \partial_n u|_\Sigma = u_1, \tag{4.47}$$

where u_0 and u_1 are functions defined on Σ. In typical applications, Σ is a surface of constant time $t = t_0$. We then specify the value of u and its time derivative at time t_0, $u(t_0, x_i) = u_0(x_i)$ and $\partial_t u(t_0, x_i) = u_1(x_i)$ and we look for a solution $u(t, x_i)$ at later times $t \geq t_0$. One typically also requires that the solution vanishes or has a nice behaviour for large x_i.

According to Hadamard's definition, a problem involving partial differential equations is *well-posed* if the solution exists, it is unique and it depends continuously on the initial conditions. It turns out that the Cauchy problems for the heat and the wave equations are well-posed if the initial surface Σ is space-like.[6] Also the Cauchy problem for the time-dependent Schrödinger equation is well-posed. These are the kind of Cauchy problems we usually encounter in physics. On the contrary, the Cauchy problem for the Laplace equation can be not well-posed (see Example 4.8).

In physics, the type of boundary conditions we impose on a partial differential equation depends on the system we want to describe. In most applications, Cauchy problems are formulated for parabolic and hyperbolic

---

[6]This concept should be familiar from special relativity. The simplest example of space-like surface is a constant time surface in $\mathbb{R}^n$.

equations. Elliptic equations instead are usually associated with boundary problems, which we discuss extensively in Chapter 5. In these kinds of problems conditions are specified at the boundary of a region Ω in $\mathbb{R}^n$. Since we have more than one variable we can also impose mixed boundary conditions, for example Cauchy conditions for some variables and boundary conditions for the others.

We now give few examples of Cauchy problems for second-order linear partial differential equations.

Example 4.6 (Heat equation). We want to study the variation in time of the temperature of an infinite metal wire. We describe the wire as an infinite line, which we parameterise with the variable x. At the time $t = 0$, the temperature in the wire is given by the function $u_0(x)$. The time evolution of the temperature is described by the heat equation

$$\frac{\partial u(x,t)}{\partial t} = \kappa \frac{\partial^2 u(x,t)}{\partial x^2} \tag{4.48}$$

with conditions $u(x,0) = u_0(x)$ and $\lim_{|x| \to \infty} u(x,t) = 0$. This is a Cauchy problem with initial surface Σ given by the line $t = 0$. Notice that we also impose vanishing conditions at spatial infinity.

In order to solve (4.48), we perform a Fourier transform in the variable x

$$U(p,t) = \frac{1}{\sqrt{2\pi}} \int_{-\infty}^{+\infty} dx \, e^{-ipx} u(x,t), \tag{4.49}$$

where we assumed that $u(x,t)$ is integrable on $\mathbb{R}$ as a function of x. Since the Fourier transform is linear and maps derivation into multiplication by the conjugate variable (see Appendix A) we find

$$\frac{\partial U(p,t)}{\partial t} = -\kappa p^2 U(p,t). \tag{4.50}$$

Note that the effect of the Fourier transform is to reduce the original partial differential equation to an ordinary one in the variable t. Moreover (4.50) can be easily integrated and the solution is

$$U(p,t) = U_0(p)e^{-\kappa p^2 t}. \tag{4.51}$$

Here $U(p,0) = U_0(p)$ is the Fourier transform of the function $u_0(x)$, as it can be easily seen from (4.49). Using the Fourier transform of a Gaussian,

(A.14), and the convolution theorem (see Appendix A), we find the solution $u(x, t)$ for $t \geq 0$

$$u(x, t) = \frac{1}{\sqrt{4\pi\kappa t}} \int_{-\infty}^{+\infty} dy \, e^{-\frac{(x-y)^2}{4kt}} u_0(y). \qquad (4.52)$$

The function

$$h(x - y, t) = \frac{1}{\sqrt{4\pi\kappa t}} e^{-\frac{(x-y)^2}{4kt}} \qquad (4.53)$$

is called *heat kernel* and is an example of Green function for the heat equation, as we will discuss in Section 6.4.2.

Example 4.7 (The two-dimensional wave equation). Consider the Cauchy problem for the two-dimensional wave equation

$$\frac{\partial^2 u}{\partial t^2} - \frac{\partial^2 u}{\partial x^2} = 0, \qquad \begin{cases} u(x, 0) = h(x), \\ \partial_t u(x, 0) = v(x). \end{cases} \qquad (4.54)$$

The initial surface Σ is given by $t = 0$. The problem is easily solved by the change of variables

$$\zeta_+ = x + t, \qquad \zeta_- = x - t. \qquad (4.55)$$

In the new variables the equation becomes

$$\partial_{\zeta_-} \partial_{\zeta_+} u = 0, \qquad (4.56)$$

and the general solution is immediately found

$$u(x, t) = f_+(\zeta_+) + f_-(\zeta_-) = f_+(x + t) + f_-(x - t), \qquad (4.57)$$

where $f_\pm$ are two arbitrary functions. The physical interpretation of the solution is simple. $f_+(x + t)$ has the same value for all $x = -t + c$, where c is a constant, and describes a wave of shape f_+ that propagates with constant velocity -1 in the negative direction of the real axis. $f_-(x - t)$ has the same value for all $x = t + c$ and describes a wave of shape f_- that propagates with constant velocity 1 in the positive direction of the real axis. The shapes of the two waves are arbitrary. Imposing the initial conditions

$$u(x, 0) = f_+(x) + f_-(x) = h(x), \qquad \partial_t u(x, 0) = f'_+(x) - f'_-(x) = v(x),$$

we determine $f_\pm$, up to a constant a,

$$f_+(x) = \frac{h(x)}{2} + \frac{1}{2} \int_a^x v(x')dx', \qquad f_-(x) = \frac{h(x)}{2} - \frac{1}{2} \int_a^x v(x')dx'.$$

The problem is well-posed because we have chosen a space-like surface. It is easy to find a surface Σ' where things go wrong. Suppose we give the value of u and its first normal derivative on the light-like surface $x = t$ ($\zeta_- = 0$). In other words, we try to find a solution that satisfies $u(\zeta_+, \zeta_- = 0) = \tilde{h}(\zeta_+)$ and $\partial_{\zeta_-} u(\zeta_+, \zeta_- = 0) = \tilde{v}(\zeta_+)$. Since the general solution of the equation is (4.57), we see immediately that the second condition, which requires $f'_-(0) = \tilde{v}(\zeta_+)$, is impossible to satisfy. The surfaces $x = \pm t$, called *characteristic surfaces*, are the curves where the Cauchy problem is not well-posed and play an important role in understanding front waves for the wave equation.

Example 4.8 (Hadamard's example). Consider the Cauchy problem for the two-dimensional Laplace equation

$$\frac{\partial^2 u}{\partial x^2} + \frac{\partial^2 u}{\partial y^2} = 0, \qquad \begin{cases} u(x,0) = 0, \\ \partial_y u(x,0) = \epsilon \sin\left(\dfrac{x}{\epsilon}\right). \end{cases} \tag{4.58}$$

The initial surface Σ is the real axis. One can easily see that

$$u(x,y) = \epsilon^2 \sin\left(\frac{x}{\epsilon}\right) \sinh\left(\frac{y}{\epsilon}\right), \tag{4.59}$$

is the only solution (see Exercise 5.10). The problem is not well-posed. Indeed, the solution is not continuous with respect to the initial conditions. When $\epsilon \to 0$, the initial condition becomes simply $u(x,0) = u_y(x,0) = 0$ and the corresponding Cauchy problem has the obvious unique solution $u(x,y) = 0$. However, (4.59) blows up for $\epsilon \to 0$ for any finite value of y.

4.4. Exercises

Exercise 4.1. Separation of variables. Find the solution of the Cauchy problems

$$\text{(a) } y' = xy^2 + x, \quad y(2) = 0; \qquad \text{(b) } y' = \left(\frac{y+2}{x-1}\right)^3, \quad y(0) = -1,$$

writing the equations in the form $f(y)dy = g(x)dx$ and integrating.

Exercise 4.2. Newton's equation for a conservative system. Given the equation $y'' = f(y)$,

(a) show that the energy $E = y'^2/2 + U(y)$, where $U'(y) = -f(y)$, is constant on any solution;

(b) use separation of variables on the result of point (a) to find the general solution of the equation;

(c) solve $y'' = -k \sin y$ (pendulum).

Exercise 4.3. Find the general solution of the first-order linear ordinary differential equation

$$a_1(x)y' + a_0(x)y = g(x).$$

[Hint: write the equation in the form $d(A(x)y)/dx = B(x)$ and integrate.]

Exercise 4.4. Find the solutions of (4.26) and (4.29) when the characteristic equation has coincident roots.

Exercise 4.5. Variation of arbitrary constants. Show that one can always find a solution of (4.34) of the form (4.35).

Exercise 4.6. Find the general solution of

$$xy'' + 2y' - \frac{2x}{(1+x)^2}y = \frac{2}{1+x}.$$

[Hint: perform a change of variables to eliminate the term y' and use the method of variation of arbitrary constants.]

Exercise 4.7. Bernoulli and Riccati equations.

(a) Find the general solution of the Bernoulli equation $y' + A(x)y = B(x)y^n$, with $n \geq 2$ integer. [Hint: set $u = y^{1-n}$.]

(b) Solve, in particular, $y' + y/x = x^3 y^2$.

(c) Show that the general solution of the Riccati equation $y' = p(x) + q(x)y + t(x)y^2$ can be written as $y_1 + u$, where y_1 is a particular solution and u solves a Bernoulli equation.

* Exercise 4.8. Use a Laplace transform with respect to time to solve the initial value problem (4.48) for the heat equation.

Exercise 4.9. Use a Fourier transform with respect to space to solve the initial value problem (4.54) for the two-dimensional wave equation.

Exercise 4.10. Use a Laplace transform with respect to time to solve the initial/boundary value problem for the heat equation

$$\frac{\partial u(x,t)}{\partial t} = \kappa \frac{\partial^2 u(x,t)}{\partial x^2}, \qquad \begin{cases} u(x,0) = 0, \\ u(0,t) = g(t), \end{cases}$$

in the region $t > 0$, $x > 0$, where $g(t)$ is a function with compact support. Assume that $u(x,t)$ is bounded.

5

Boundary Value Problems

In physics we often have to solve a differential equation in a region of space-time with specified conditions at the boundary. This is called a boundary value problem. One of the most famous examples, and one of the first to have been studied, is the Dirichlet problem arising in electrostatics and many other contexts. It consists in finding a solution of the Laplace equation in a domain $\Omega \subset \mathbb{R}^n$ that takes a prescribed value at the boundary $\partial\Omega$ of the domain,

$$\Delta u = \sum_{i=1}^{n} \frac{\partial^2 u}{\partial x_i^2} = 0, \quad u|_{\partial\Omega} = g. \tag{5.1}$$

We already encountered the Dirichlet problem in two dimensions in Section 1.3.1. The boundary conditions in (5.1) are called *Dirichlet*. Another commonly used set of boundary conditions are the *Neumann* ones, which require that the normal derivative[1] of u vanishes at the boundary

$$\partial_n u|_{\partial\Omega} = 0. \tag{5.2}$$

There is a plethora of other boundary problems obtained by changing the differential operator, considering inhomogeneous equations or changing the boundary conditions. In this chapter, we provide simple examples of boundary value problems in one and higher dimensions. As we will see in

---

[1]The normal derivative is the derivative in the direction orthogonal to the boundary and is denoted as $\partial_n u$.

Chapter 9, also the time-independent Schrödinger equation can be related to a boundary value problem.

For partial differential equations, boundary problems are often solved by the method of separation of variables. This allows to reduce the equation to an equivalent set of one-dimensional problems that, in many cases, can be solved by generalised Fourier analysis. Most of the special functions and orthogonal polynomials of mathematical physics arise in this way. We will discuss the most common ones. We will also discuss the relation between boundary problems and the spectral theory for differential operators.

A lot of activity in the field of partial differential equations has been devoted to study the existence, uniqueness and continuous dependence on the boundary data of the solutions of boundary value problems. This is usually done by formulating the problem in a suitable space of functions. The general theory is beyond the scope of this book.

5.1.   Boundary Value Problems in One Dimension

Consider a second-order linear ordinary differential equation depending on a parameter λ

$$y'' + P(x, \lambda)y' + Q(x, \lambda)y = 0 \tag{5.3}$$

to be solved in the compact interval $[x_0, x_1]$. We saw in the previous chapter that the Cauchy problem with $y(x_0) = y_0$ and $y'(x_0) = y_1$ has a unique solution in $[x_0, x_1]$, provided P and Q are continuous. The boundary value problem, where instead we impose vanishing conditions at the boundary of the interval,

$$y(x_0) = 0, \quad y(x_1) = 0, \tag{5.4}$$

behaves quite differently. It certainly admits the trivial solution $y(x) = 0$. The question is whether there are other, non-trivial, solutions. Call $y_i(x, \lambda)$ the two linearly independent solutions of (5.3). As we saw in Section 4.2, the general solution is the linear combination

$$y(x) = c_1 y_1(x, \lambda) + c_2 y_2(x, \lambda), \tag{5.5}$$

with arbitrary coefficients c_i. The boundary value problem (5.4) reduces to the homogeneous linear system

$$c_1 y_1(x_0, \lambda) + c_2 y_2(x_0, \lambda) = 0,$$
$$c_1 y_1(x_1, \lambda) + c_2 y_2(x_1, \lambda) = 0, \qquad (5.6)$$

in the variables c_1 and c_2 and admits a non-trivial solution only if the determinant of the matrix of coefficients, $\det y_i(x_j, \lambda)$, vanishes. This happens only for particular values of the parameter λ.

Of particular interest is the case of equations of the form

$$-y'' - P(x)y' - Q(x)y = \lambda y. \qquad (5.7)$$

The values of λ for which (5.7) admits a non-vanishing solution are called *eigenvalues* of the boundary problem and the corresponding solutions *eigenvectors* or *eigenfunctions*. The reason for this nomenclature is that λ can be interpreted as the eigenvalue of the differential operator $L = -d^2/dx^2 - P d/dx - Q$.

Example 5.1. Consider the boundary problem

$$y'' + \lambda y = 0, \quad y(0) = y(\pi) = 0 \qquad (5.8)$$

in the interval $[0, \pi]$. The general solution of the differential equation is $y(x) = C_1 \sin \sqrt{\lambda} x + C_2 \cos \sqrt{\lambda} x$. The boundary conditions impose $C_2 = 0$ and $C_1 \sin \sqrt{\lambda} \pi = 0$. For generic λ the only solution is for $C_1 = C_2 = 0$. However, when $\lambda = n^2$ with $n = 1, 2, \ldots$, we have $\sin \sqrt{\lambda} \pi = 0$. C_1 is then arbitrary and we find a solution of the form $y(x) = C_1 \sin nx$. The problem has a numerable set of eigenvalues $\lambda_n = n^2$ with corresponding eigenfunctions $y_n(x) = A_n \sin nx$. Notice that we have just solved the eigenvalue problem $Ly = \lambda y$ for the operator $L = -d^2/dx^2$ with domain $\mathcal{D}(L) = \{y_{ac}, y', y'' \in L^2[0, \pi] \,|\, y(0) = y(\pi) = 0\}$.[2]

---

[2]The subindex ac stands for absolutely continuous. This is a natural requirement when considering differentiable functions in Lebesgue theory (see Rudin (1987)). A useful characterisation is the following. A function f defined on an interval $I = [a, b]$ is absolutely continuous on I if the derivative $f'(x)$ exists almost everywhere in I and the fundamental theorem of calculus $\int_a^x dy \frac{df(y)}{dy} = f(x) - f(a)$ holds.

The above discussion can be generalised to higher order equations and more general boundary conditions. Consider a homogeneous linear differential equation of order n

$$\sum_{i=0}^{n} a_i(x, \lambda_a) y^{(i)}(x) = 0 \qquad (5.9)$$

in the interval $[x_0, x_1]$, depending on a set of parameters λ_a. A boundary problem consists in imposing n linear homogeneous conditions in y and its first $n-1$ derivatives at x_0 and x_1

$$\sum_{i=0}^{n-1} (\alpha_i^j y^{(i)}(x_0) + \beta_i^j y^{(i)}(x_1)) = 0, \quad j = 1, \dots, n. \qquad (5.10)$$

The general solution of the differential equation (5.9) is a linear combination of n linearly independent solutions $y(x) = \sum_{i=1}^{n} c_i y_i(x, \lambda_a)$ and it involves n arbitrary constants c_i. Imposing the boundary conditions (5.10), we find a system of n linear homogeneous equations for the n constants c_i. The system has, in general, only the trivial solution $c_i = 0$ corresponding to $y(x) = 0$. Only for particular values of the parameters λ_a we can find non-trivial solutions.

We should stress that we will only consider boundary conditions of homogeneous type, namely when the right-hand side of (5.4) and (5.10) are zero.

5.1.1. *Sturm–Liouville problems*

The *Sturm–Liouville problem* is a boundary value problem for second-order ordinary differential equations of the form[3]

$$\frac{d}{dx} \left[p(x) \frac{dy}{dx} \right] + q(x)y = -\lambda \omega(x)y, \qquad (5.12)$$

---

[3]There is no loss of generality here. Any linear second-order differential equation of the form

$$a_2(x)y'' + a_1(x)y' + a_0(x)y = -\lambda y, \qquad (5.11)$$

can be reduced to the form (5.12) by simple algebraic manipulations (see Exercise 5.1).

in the interval $a \leq x \leq b$. We assume that the real functions $p(x), q(x), w(x)$ are continuous on $[a, b]$, and $p(x) > 0, w(x) > 0$ at least on the open interval (a, b).

Equation (5.12) can be interpreted as the eigenvalue equation, $Ly = \lambda y$, for the differential operator

$$L = -\frac{1}{w(x)} \left(\frac{d}{dx} \left[p(x) \frac{d}{dx} \right] + q(x) \right) \tag{5.13}$$

in the Hilbert space $L_w^2[a, b]$ with scalar product

$$(f, g) = \int_a^b \overline{f(x)} g(x) w(x) dx. \tag{5.14}$$

The boundary conditions for the Sturm–Liouville problem are chosen in such a way that L is a self-adjoint operator. There are various types of boundary conditions of the form (5.10) satisfying this requirement. Among the most common ones are

$$\alpha_1 y(a) + \alpha_2 y'(a) = 0,$$
$$\beta_1 y(b) + \beta_2 y'(b) = 0, \tag{5.15}$$

where α_i and β_i are real numbers, and at least one of the α_i and one of the β_i are non-zero, or the periodic ones

$$y(a) = y(b), \quad y'(a) = y'(b). \tag{5.16}$$

It is simple to check that L, defined on the domain

$$\mathcal{D}(L) = \{f_{ac}, f', f'' \in L_w^2[a, b] \,|\, \alpha_1 f(a) + \alpha_2 f'(a) = \beta_1 f(b) + \beta_2 f'(b) = 0\},$$

is self-adjoint. We have indeed

$$(f, Lg) = -\int_a^b \bar{f} \left(\frac{d}{dx} (pg') + qg \right) dx = -\bar{f} pg'|_a^b - \int_a^b \left(-\bar{f}' pg' + \bar{f} qg \right) dx$$

$$= -p(\bar{f} g' - \bar{f}' g)|_a^b - \int_a^b \left(\frac{d}{dx} (p\bar{f}') + q\bar{f} \right) g \, dx = (Lf, g), \tag{5.17}$$

where the boundary terms cancel due to (5.15). Notice that the function w in the integration measure precisely cancels the factor $1/w$ in L.

Equation (5.17) only shows that L is symmetric.[4] One can easily see that L is also self-adjoint.[5] A similar computation shows that the operator L with domain

$$\mathcal{D}(L) = \{f_{ac}, f', f'' \in L^2_\omega[a, b] \mid y(b) = y(a), y'(b) = y'(a)\} \qquad (5.18)$$

is self-adjoint.

Boundary conditions of the form (5.15) and (5.16) assume that the interval $[a, b]$ is bounded and the functions $p(x)$ and $\omega(x)$ do not vanish at the boundary. When this is not the case, the solutions of the differential equation may diverge at one or both boundaries of the interval and (5.15)–(5.16) are not applicable. We need to specify different boundary conditions. These typically require that the solution is bounded or square-integrable on $[a, b]$. The latter is also the natural condition in quantum mechanics. As before, the interesting boundary conditions are those that make L self-adjoint.

In summary, a Sturm–Liouville problem is equivalent to the eigenvalue problem

$$Ly = \lambda y, \qquad (5.19)$$

for the self-adjoint operator L in (5.13). The general theory of self-adjoint operators[6] tells us that the eigenvalues λ are real and the corresponding eigenvectors y_λ are mutually orthogonal. The family of eigenvectors, after normalisation, are a basis in the Hilbert space $L^2_\omega[a, b]$. In general, the spectrum of L contains also a continuous part and the family of eigenvectors is a generalised basis for $L^2_\omega[a, b]$.

5.1.2. *Notable examples of Sturm–Liouville problems*

A Sturm–Liouville problem is *regular* if the interval $[a, b]$ is bounded, the functions p, p', q, ω are continuous and $p > 0, \omega > 0$ in the closed interval

---

[4]Recall that an operator L is symmetric if $(Lf, g) = (f, Lg)$ for all $f, g \in \mathcal{D}(L)$. This implies that $L^\dagger$ is an extension of L. L is self-adjoint if $\mathcal{D}(L^\dagger) = \mathcal{D}(L)$ (see Petrini *et al.*, 2017, Chapter 10).

[5]Consider for example the endpoint b. If f satisfies (5.15), $\bar{f}'(b) = -\beta_1/\beta_2 \bar{f}(b)$ and the boundary term at b becomes $-p(b)\bar{f}(b)(g'(b) + \beta_1/\beta_2 g(b))$ implying that also $g(x)$ should satisfy the condition (5.15) at b. A similar analysis can be done when $\beta_2 = 0$, and at the endpoint a.

[6]See Petrini *et al.* (2017) and Appendix A for a brief summary.

$[a, b]$. Regular Sturm–Liouville problems are interesting because the spectrum of the corresponding operator L is purely discrete.[7] It follows that the eigenfunctions of a regular Sturm–Liouville problem, when suitably normalised, are a complete orthonormal system in $L^2_\omega[a, b]$. One can show that for a regular Sturm–Liouville problem with boundary conditions of the form (5.15) the eigenvalues are always not degenerate, the eigenvectors can be chosen to be real and have a number of zeros that increases with n (see Exercise 5.4).

Example 5.2. Consider again Example 5.1. This is a regular Sturm–Liouville problem with boundary conditions of the form (5.15). The spectrum, $\lambda_n = n^2$ with $n = 1, 2, \ldots$, is purely discrete and the normalised eigenvectors $y_n(x) = \sqrt{2/\pi} \sin nx$ are a basis in $L^2[0, \pi]$.[8] Notice that $y_n(x)$ has precisely $n - 1$ zeros in the interval $(0, \pi)$, and their number increases with n.

Example 5.3. The Sturm–Liouville problem with periodic boundary conditions of the form (5.16)

$$y'' + \lambda y = 0, \qquad \begin{cases} y(0) = y(2\pi), \\ y'(0) = y'(2\pi) \end{cases} \tag{5.20}$$

is regular. The eigenvalues are given by $\lambda = m^2$ with $m \in \mathbb{Z}$ and the normalised eigenvectors $y_m(x) = e^{imx}/\sqrt{2\pi}$ are the Fourier basis in $L^2[0, 2\pi]$. Notice that the spectrum is degenerate: $y_{\pm m}$ for $m \neq 0$ correspond to the same eigenvalue $\lambda = m^2$.

A Sturm–Liouville problem that is not regular is called *singular*. Typical examples occur when the interval is not bounded or one or both functions $p(x)$ and $\omega(x)$ vanish at the boundary. The solutions of the differential

---

[7]This follows from the fact that L is an operator with compact resolvent (see Section 8.1 and Exercise 8.5).

[8]The functions $s_n = \sqrt{\frac{2}{L}} \sin \frac{\pi n x}{L}$, with $L = b - a$, form a basis in $L^2[a, b]$. It is not the standard Fourier series because the sines have period $2L$. However any function $f(x)$ defined on $[0, L]$ can be expanded in s_n. Indeed $f(x)$ can be extended to an odd function on $[-L, L]$ and expanded there in Fourier series. Since the function is odd the expansion contains only sines. Its restriction to $[0, L]$ gives the required expansion in the functions s_n.

equation can then be discontinuous or unbounded at the endpoints of the interval $[a, b]$. Most of the special functions and orthogonal polynomials of mathematical physics arise as eigenfunctions of some singular Sturm–Liouville problem. We discuss below the examples of the Hermite, Legendre and Laguerre polynomials, which are commonly used in physics. Properties of these orthogonal polynomials are discussed in details in Petrini *et al.* (2017), Section 6.4. In all cases below, the solutions of the Sturm–Liouville problems can be found by power series methods. We illustrate it explicitly for the Hermite equation, which is the simplest one. The case of Legendre equation is discussed in details in Chapter 7, where the general theory of power series expansion is also discussed.

Example 5.4 (Hermite polynomials). The *Hermite equation*

$$\frac{d}{dx}\left(e^{-x^2}\frac{dy}{dx}\right) + e^{-x^2}\lambda y = 0 \tag{5.21}$$

can be formulated as a Sturm–Liouville problem on the real axis $\mathbb{R}$ with $p(x) = w(x) = e^{-x^2}$ and $q(x) = 0$. The problem is singular because the interval $(-\infty, +\infty)$ is unbounded. The natural domain for the operator L is $\mathcal{D}(L) = \{f_{ac}, f', f'' \in L_w^2(\mathbb{R})\}$, and we look for solutions that are square-integrable at infinity with the measure $e^{-x^2}dx$. The equation can also be written as

$$y'' - 2xy' + \lambda y = 0, \tag{5.22}$$

and can be solved by power series methods. We assume that y can be expanded as

$$y(x) = \sum_{n=0}^{\infty} c_n x^n. \tag{5.23}$$

Substituting (5.23) into equation (5.22) we find

$$\sum_{n=0}^{\infty} c_n n(n-1)x^{n-2} - 2x\sum_{n=0}^{\infty} c_n n x^{n-1} + \lambda\sum_{n=0}^{\infty} c_n x^n$$

$$= \sum_{p=0}^{\infty}[(p+2)(p+1)c_{p+2} - (2p-\lambda)c_p]x^p = 0, \tag{5.24}$$

where we redefined the summation index to $n = p + 2$ in the first series and $n = p$ in the other two. Equation (5.24) can be satisfied for all x if and only if all coefficients of the power series in the second line are zero. This condition gives

$$c_{p+2} = \frac{2p - \lambda}{(p+1)(p+2)} c_p. \tag{5.25}$$

This is a recursion relation that allows to determine all the coefficients c_p in terms of c_0 and c_1. The constant c_0 determines inductively all the even coefficients c_{2p} and c_1 all the odd ones c_{2p+1}. Thus the general solution of (5.22) depends on two parameters c_0 and c_1

$$y(x) = \sum_{k=0}^{\infty} c_{2k} x^{2k} + \sum_{k=0}^{\infty} c_{2k+1} x^{2k+1} = c_0 y_+(x) + c_1 y_-(x). \tag{5.26}$$

The two linearly independent solutions $y_\pm$ are even and odd, respectively. We can actually extend the solutions to the complex plane. We can easily evaluate the radius of convergence of y_+ using (A.2)

$$R = \lim_{k \to \infty} \left| \frac{c_{2k}}{c_{2(k+1)}} \right| = \lim_{k \to \infty} \left| \frac{(2k+1)(2k+2)}{4k - \lambda} \right| = \infty, \tag{5.27}$$

and similarly for y_-. We see that the general solution (5.26) is an entire function on the complex plane. We still have to impose that the solutions are elements of $L_\omega^2(\mathbb{R})$. In order to see if $y \in L_\omega^2(\mathbb{R})$, we need to know how y grows at infinity. This can be done by estimating the behaviour of the power series. Consider y_+. For large x, the power series is dominated by the terms with large k. For these we have

$$c_{2k+2} = \frac{4k - \lambda}{(2k+1)(2k+2)} c_{2k} \sim \frac{1}{k} c_{2k}, \tag{5.28}$$

which is consistent with $c_{2k} \sim 1/k!$ for large k. We see that

$$y_+(x) \sim \sum_k \frac{1}{k!} x^{2k} = e^{x^2}. \tag{5.29}$$

We similarly find $y_- \sim e^{x^2}$. We are then in trouble. The norm of $y_\pm$ in $L_\omega^2(\mathbb{R})$

$$||y_\pm||^2 = \int_{\mathbb{R}} |y_\pm(x)|^2 e^{-x^2} dx \sim \int_{\mathbb{R}} e^{x^2} dx \tag{5.30}$$

is certainly not finite. It seems that there are no square-integrable solutions at all. However, we are assuming here that $y_\pm$ are power series with infinitely many non-zero coefficients and this is not always the case. From (5.25) we see that if $\lambda = 2n$ with $n \in \mathbb{N}$, the recursion stops at the nth step since $c_{n+2} = 0$ no matter what c_n is. By recursion, if the coefficient c_{n+2} is zero all the coefficients c_{n+2k} with $k > 1$ are also zero. The power series truncates to a polynomial of degree n and this is certainly normalisable with the measure $e^{-x^2} dx$. Thus, we can find a normalisable solution if and only if $\lambda = 2n$ for $n = 0, 1, 2, \ldots$. The corresponding eigenfunctions are the *Hermite polynomials*

$$H_n(x) = (-1)^n e^{x^2} \frac{d^n}{dx^n} e^{-x^2}. \tag{5.31}$$

The first polynomials are

$$H_0(x) = 1, \quad H_1(x) = 2x, \quad H_2(x) = 4x^2 - 2. \tag{5.32}$$

One can show that, after normalisation, the Hermite polynomials form an orthonormal basis in $L^2_\omega(\mathbb{R})$. Notice that the presence of $w(x)$ in the integration measure is necessary to ensure the normalisability of the polynomials. The Hermite equation arises in the study of the harmonic oscillator in quantum mechanics (see Example 9.7).

Example 5.5 (Legendre polynomials). The *Legendre equation*

$$\frac{d}{dx}\left((1 - x^2)\frac{dy}{dx}\right) + \lambda y = 0 \tag{5.33}$$

can be formulated as a Sturm–Liouville problem in the interval $[-1, 1]$ with $p(x) = 1 - x^2$, $q(x) = 0$ and $w(x) = 1$. The problem is singular because the function $p(x)$ vanishes at $x = \pm 1$. If we require that the solutions are bounded in $x = \pm 1$, one can show (see Example 7.6) that the eigenvalues are $\lambda_l = l(l + 1)$ with $l = 0, 1, 2, \ldots$. The eigenvectors are the *Legendre polynomials*

$$P_l(x) = \frac{1}{2^l l!} \frac{d^l}{dx^l} (x^2 - 1)^l. \tag{5.34}$$

The first polynomials are

$$P_0(x) = 1, \quad P_1(x) = x, \quad P_2(x) = \frac{1}{2}(3x^2 - 1).$$

One can show that, after normalisation, the Legendre polynomials form an orthonormal basis for $L^2[-1, 1]$. The Legendre polynomials are related to the spherical harmonics that we discuss in Section 5.3.

Example 5.6 (Laguerre polynomials). The *Laguerre equation*

$$\frac{d}{dx}\left(xe^{-x}\frac{dy}{dx}\right) + e^{-x}\lambda y = 0 \tag{5.35}$$

can be formulated as a Sturm–Liouville problem on the interval $[0, \infty)$ with $p(x) = xe^{-x}$, $\omega(x) = e^{-x}$ and $q(x) = 0$. The problem is singular because the interval $[0, \infty)$ is unbounded. We choose boundary conditions of the form (5.15) at $x = 0$, $y(0) = 0$, and we require square-integrability at infinity with the measure $e^{-x}dx$. Then, the domain of the operator L is $\mathcal{D}(L) = \{f_{ac}, f', f'' \in L^2_\omega[0, \infty) \mid f(0) = 0\}$. The eigenvalues are $\lambda_n = n$ with $n \in \mathbb{N}$ and the corresponding eigenfunctions are the *Laguerre polynomials*

$$L_n(x) = \frac{e^x}{n!}\frac{d^n}{dx^n}[e^{-x}x^n]. \tag{5.36}$$

The first polynomials are

$$L_0(x) = 1, \quad L_1(x) = 1 - x, \quad L_2(x) = \frac{1}{2}(x^2 - 4x + 2).$$

One can show that, after normalisation, the Laguerre polynomials form an orthonormal basis in $L^2_\omega[0, \infty)$. The presence of $\omega(x)$ in the integration measure is necessary to ensure the normalisability of the polynomials L_n. A generalisation of the Laguerre equation arises in the study of the hydrogen atom in quantum mechanics.

In the three examples above the solutions of the Sturm–Liouville problem form a complete orthonormal system. However, this is not the case for a general singular Sturm–Liouville problem, for which the spectrum is not necessarily discrete.

5.2.   Boundary Value Problems for Partial Differential Equations

In this section we give few examples of boundary value problems for partial differential equations in two dimensions. We use the method of separation of variables to reduce the problem to a set of ordinary differential equations. In the process we introduce integration constants and encounter Sturm–Liouville problems in one or more variables. The general solution is obtained by using a generalised Fourier expansion in the eigenvectors of the Sturm–Liouville problem.

Example 5.7 (Dirichlet problem on a disk). Consider the Dirichlet problem for the Laplace equation in two dimensions

$$\Delta u = \frac{\partial^2 u}{\partial x^2} + \frac{\partial^2 u}{\partial y^2} = 0, \qquad u|_{\partial\Omega} = g, \qquad (5.37)$$

on the unit disk $\Omega = \{x^2 + y^2 \leq 1\}$. Given the symmetry of the problem, it is useful to work in polar coordinates $x = \rho\cos\phi, y = \rho\sin\phi$ with $\rho \geq 0$ and $0 \leq \phi < 2\pi$ (see Figure 5.1(a)). In the new variables, the Laplace equation reads

$$\frac{\partial^2 u}{\partial\rho^2} + \frac{1}{\rho}\frac{\partial u}{\partial\rho} + \frac{1}{\rho^2}\frac{\partial^2 u}{\partial\phi^2} = 0. \qquad (5.38)$$

We look for particular solutions of product form $u(\rho, \phi) = R(\rho)\Phi(\phi)$. The equation becomes

$$\rho^2 R''(\rho)\Phi(\phi) + \rho R'(\rho)\Phi(\phi) + R(\rho)\Phi''(\phi) = 0 \qquad (5.39)$$

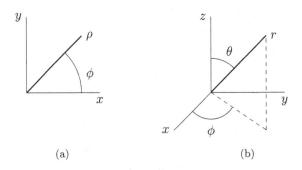

(a)   (b)

Fig. 5.1.   (a) polar coordinates in the plane; (b) spherical coordinates in the space.

and, by dividing by $R(\rho)\Phi(\phi)$ and reorganising the terms, we find

$$\frac{\rho^2 R''(\rho) + \rho R'(\rho)}{R(\rho)} = -\frac{\Phi''(\phi)}{\Phi(\phi)} = \lambda, \tag{5.40}$$

where we named λ the common value of the two resulting expressions. λ is simultaneously equal to an expression that only depends on the variable ρ and to one that only depends on the variable ϕ. This is possible only if λ is a constant. We thus obtain two ordinary differential equations

$$\rho^2 R''(\rho) + \rho R'(\rho) = \lambda R(\rho), \tag{5.41}$$

$$\Phi''(\phi) = -\lambda \Phi(\phi) \tag{5.42}$$

that depend on the real parameter λ. Consider first (5.42). Since ϕ is an angle, we require $\Phi(\phi)$ to be periodic of period 2π. Equation (5.42) is then equivalent to the regular Sturm–Liouville problem

$$\Phi''(\phi) + \lambda\Phi(\phi) = 0, \quad \begin{cases} \Phi(0) = \Phi(2\pi), \\ \Phi'(0) = \Phi'(2\pi), \end{cases} \tag{5.43}$$

which we solved in Example 5.3. The eigenvalues are $\lambda = m^2$ with $m \in \mathbb{Z}$ and the eigenvectors are $\Phi_m = A_m e^{im\phi}$, which form a basis for $L^2[0, 2\pi]$. Let us now move to (5.41). This is a special case of the Euler equation discussed in Example 4.5 and we know that the solutions are of the form $R(\rho) = \rho^\alpha$. Substituting it in (5.41), we find the characteristic equation $(\alpha(\alpha - 1) + \alpha - \lambda)\rho^\alpha = 0$ which is solved by $\alpha = \pm\sqrt{\lambda} = \pm m$. Since $\rho = 0$ corresponds to the centre of the disk, we can only accept solutions $R(\rho)$ that are regular in $\rho = 0$. These are $R(\rho) = \rho^{|m|}$ for any $m \in \mathbb{Z}$. Thus the solutions of the Laplace equation are of the form $\rho^{|m|} e^{im\phi}$ with $m \in \mathbb{Z}$. Since the equation is linear, a more general solution is obtained by taking arbitrary linear combinations

$$u(\rho, \phi) = \sum_{m \in \mathbb{Z}} c_m \rho^{|m|} e^{im\phi}, \tag{5.44}$$

where $c_{-m} = \bar{c}_m$ as u is real. Now we can impose the boundary condition $u(1, \phi) = g(\phi)$. By restricting (5.44) to $\rho = 1$ we see that the coefficients

c_m are just the Fourier coefficients of the function g

$$g(\phi) = \sum_{m \in \mathbb{Z}} c_m e^{im\phi}. \tag{5.45}$$

They can be explicitly computed as

$$c_m = \frac{1}{2\pi} \int_0^{2\pi} g(t)e^{-imt} dt. \tag{5.46}$$

Combining all information, we find the *Poisson solution* of the Dirichlet problem on the disk in the integral form

$$u(\rho, \phi) = \frac{1}{2\pi} \sum_{m \in \mathbb{Z}} \int_0^{2\pi} \rho^{|m|} e^{im(\phi - t)} g(t) dt = \frac{1}{2\pi} \int_0^{2\pi} P_\rho(\phi - t)g(t)dt, \tag{5.47}$$

where $P_\rho(\phi) = \sum_{m \in \mathbb{Z}} \rho^{|m|} e^{im\phi}$ is the Poisson kernel we have already encountered in Section 1.3.1.

Example 5.8 (Heat equation for a bar). We consider the problem of heat conduction in a bar of length L whose temperature at time $t = 0$ is given by the function $g(x)$, $x \in [0, L]$. We also assume that the endpoints are kept at zero temperature at all times. This corresponds to the following boundary problem:

$$\frac{\partial u}{\partial t} = \kappa \frac{\partial^2 u}{\partial x^2}, \quad \begin{cases} u(0, t) = u(L, t) = 0, \\ u(x, 0) = g(x). \end{cases} \tag{5.48}$$

Notice that we are imposing boundary conditions for the variable x and Cauchy conditions for the variable t. We look for solutions of product form $u(x, t) = X(x)T(t)$. Substituting it in the equation and dividing term by term by $X(x)T(t)$, we obtain

$$\frac{T'(t)}{T(t)} = \kappa \frac{X''(x)}{X(x)} = \lambda, \tag{5.49}$$

where, as in the previous example, λ, the common value of the two expressions, must be a constant. The equation for T is solved by $T(t) = Ae^{\lambda t}$, where A is a constant. The boundary conditions $u(0, t) = X(0)T(t) = 0$ and

$u(L, t) = X(L)T(t) = 0$ can be solved for all t only if $X(0) = X(L) = 0$. The equation for X is then the regular Sturm–Liouville problem

$$X''(x) - \frac{\lambda}{\kappa}X(x) = 0, \quad X(0) = X(L) = 0 \tag{5.50}$$

that we already encountered in Examples 5.1 and 5.2. It has eigenvalues $\lambda = -\kappa\pi^2 n^2/L^2$ with $n = 1, 2, \ldots$ and eigenvectors $X_n(x) = \sin(\pi n x/L)$ which, after normalisation, form a basis in $L^2[0, L]$. The general solution of the boundary problem is

$$u(x, t) = \sum_{n=1}^{\infty} c_n e^{-\kappa\pi^2 n^2 t/L^2} \sin\frac{\pi n x}{L}. \tag{5.51}$$

The remaining boundary condition $u(x, 0) = g(x)$ is solved by identifying c_n with the Fourier coefficients of g

$$g(x) = \sum_{n=1}^{\infty} c_n \sin\frac{\pi n x}{L}, \quad c_n = \frac{2}{L} \int_0^L g(x) \sin\frac{n\pi x}{L} dx. \tag{5.52}$$

Example 5.9 (The vibrating string). The eigenvalues of differential operators naturally appear in physics as the normal modes of vibrations of a system. Consider, for instance, a string with fixed endpoints, initial profile $f(x)$ and zero initial velocity. Its evolution is given by the following boundary problem for the wave equation

$$\frac{\partial^2 u}{\partial t^2} = v^2 \frac{\partial^2 u}{\partial x^2}, \quad \begin{cases} u(0, t) = u(L, t) = 0, \\ u(x, 0) = f(x), \\ \partial_t u(x, 0) = 0. \end{cases} \tag{5.53}$$

Notice that, since the equation is second-order in t, we have to impose two conditions at $t = 0$, one for the initial profile and one for the velocity. As before, we are imposing boundary conditions for the variable x and Cauchy conditions for the variable t. We can again separate variables by setting $u(x, t) = X(x)T(t)$. By substituting in the wave equation and dividing term by term by $X(x)T(t)$, we obtain

$$\frac{T''(t)}{T(t)} = v^2 \frac{X''(x)}{X(x)} = \lambda. \tag{5.54}$$

The boundary conditions $u(0,t) = X(0)T(t) = 0$ and $u(L,t) = X(L)$ $T(t) = 0$ can be solved for all t only if $X(0) = X(L) = 0$. The equation for X becomes a Sturm–Liouville problem on the segment $[0, L]$ with boundary conditions $X(0) = X(L) = 0$, which has solutions $X_n = \sin(n\pi x/L)$ with $\lambda = -n^2\pi^2 v^2/L^2$. These correspond to the *normal modes* of vibration of the string. The condition $\partial_t u(x,0) = T'(0)X(x) = 0$ can be solved for all x only if $T'(0) = 0$. The equation $T''(t) - \lambda T(t) = 0$ with $T'(0) = 0$ has the solutions $T(t) = \cos(n\pi vt/L)$. Then the general solution is

$$u(x,t) = \sum_{n=1}^{\infty} c_n \cos\frac{n\pi vt}{L} \sin\frac{\pi n x}{L}. \qquad (5.55)$$

The boundary condition $u(x,0) = f(x)$ is solved by identifying c_n with the Fourier coefficients of f

$$f(x) = \sum_{n=1}^{\infty} c_n \sin\frac{\pi n x}{L}, \quad c_n = \frac{2}{L}\int_0^L g(x)\sin\frac{n\pi x}{L}\,dx. \qquad (5.56)$$

The boundary problems that we have discussed are all well-posed according to Hadamard's definition (see Section 4.3). In general, boundary problems (Dirichlet, Neumann or mixed) for the Laplace equation are well-posed. Also mixed problems for the heat and wave equation, where we impose Cauchy conditions on time and boundary conditions on a spatial surface, are usually well-posed.

5.3. The Dirichlet Problem for the Sphere

In this section we solve the Dirichlet problem for the Laplace equation in three dimensions in a region bounded by a sphere. The problem can be solved by separation of variables and it involves the use of special functions, a generalisation of the Legendre polynomials. It also involves finding the eigenvalues and eigenvectors of the Laplacian operator on a sphere. The eigenfunctions are called spherical harmonics and are ubiquitous in physics, from electrostatics to quantum mechanics. They show up any time we study a spherically symmetric problem involving the Laplace equation.

In any problem with spherical symmetry, it is convenient to use spherical coordinates

$$x = r \sin \theta \cos \phi, \quad y = r \sin \theta \cos \phi, \quad z = r \cos \theta, \qquad (5.57)$$

where $r \geq 0$, $0 \leq \theta \leq \pi$ and $0 \leq \phi < 2\pi$ (see Figure 5.1(b)). The Laplace equation becomes

$$\Delta u = \frac{1}{r} \frac{\partial^2}{\partial r^2}(ru) + \frac{1}{r^2}\hat{\Delta} u = 0, \qquad (5.58)$$

where the operator

$$\hat{\Delta} u = \frac{1}{\sin \theta} \frac{\partial}{\partial \theta}\left(\sin \theta \frac{\partial u}{\partial \theta}\right) + \frac{1}{\sin^2 \theta} \frac{\partial^2 u}{\partial \phi^2} \qquad (5.59)$$

is the restriction of the Laplace operator to the unit sphere, $r = 1$, and only depends on the angular coordinates θ and ϕ. We try to solve the Laplace equation by separation of variables, using[9]

$$u(r, \theta, \phi) = \frac{U(r)}{r} Y(\theta, \phi). \qquad (5.60)$$

We get as usual two equations, one for the radial and one for the angular part

$$r^2 U''(r) - \lambda U(r) = 0, \qquad (5.61)$$

$$\hat{\Delta} Y(\theta, \phi) + \lambda Y(\theta, \phi) = 0. \qquad (5.62)$$

The radial equation is of Euler type (see Example 4.5) and it is easily solved. The two linearly independent solutions are given by

$$U(r) = r^{\alpha_i}, \qquad i = 1, 2, \qquad (5.63)$$

where α_i are the two solutions of the equation $\alpha(\alpha - 1) = \lambda$. The angular equation deserves a longer discussion.

---

[9]Given the form of the radial derivative in (5.58), we have isolated an explicit factor of $1/r$ in the expression. This is just a matter of convenience.

5.3.1. *Spectrum of the Laplace operator on the sphere*

We look for solutions $Y(\theta, \phi)$ of equation (5.62) that are regular on the unit sphere. We can further separate the variables θ and ϕ by writing $Y(\theta, \phi) = \Theta(\theta)\Phi(\phi)$. Substituting in (5.62), using (5.59) and dividing by $\Theta(\theta)\Phi(\phi)$, we find

$$\frac{\sin\theta}{\Theta(\theta)}\frac{d}{d\theta}\left(\sin\theta\frac{d\Theta(\theta)}{d\theta}\right) + \lambda\sin^2\theta = -\frac{1}{\Phi(\phi)}\frac{d^2\Phi(\phi)}{d\phi^2} = \mu, \qquad (5.64)$$

where μ is a new constant. The equation for ϕ, $\Phi''(\phi) = -\mu\Phi(\phi)$, is easily solved by $\Phi(\phi) = e^{\pm i\sqrt{\mu}\phi}$. Since ϕ is the angle of rotation around z, the function Y should go back to the original value when ϕ is shifted by 2π. This implies that $\mu = m^2$ with $m \in \mathbb{Z}$ and the general solution for Φ is

$$\Phi(\phi) = e^{im\phi}, \qquad m \in \mathbb{Z}. \qquad (5.65)$$

The equation for Θ becomes

$$\frac{1}{\sin\theta}\frac{d}{d\theta}\left(\sin\theta\frac{d\Theta(\theta)}{d\theta}\right) + \left(\lambda - \frac{m^2}{\sin^2\theta}\right)\Theta(\theta) = 0, \qquad (5.66)$$

or, in terms of the variable $x = \cos\theta \in [-1, 1]$,

$$\frac{d}{dx}\left((1 - x^2)\frac{d\Theta(x)}{dx}\right) + \left(\lambda - \frac{m^2}{1 - x^2}\right)\Theta(x) = 0, \qquad (5.67)$$

which is called the *associated Legendre equation*. It reduces to the Legendre equation (5.33) when $m = 0$. We want solutions of (5.67) in the interval $[-1, 1]$ that are regular at $x = \pm 1$, the North and South pole of the sphere. As for the Legendre equation, one can show that there are regular solutions only for $\lambda = l(l + 1)$ with $l = 0, 1, \ldots$, given by the *associated Legendre functions*

$$P_l^m(x) = \frac{(-1)^m}{2^l l!}(1 - x^2)^{m/2}\frac{d^{l+m}}{dx^{l+m}}(x^2 - 1)^l. \qquad (5.68)$$

The functions $P_l^m(x)$ reduce to $P_l(x)$ for $m = 0$ and are non-vanishing only for $|m| \le l$.[10]

---

[10] Indeed, for $m < -l$ the expression does not make sense. For $m > l$ the number of derivatives is greater than the degree of the polynomial $(x^2 - 1)^l$ and the result is zero.

The (suitably normalised) regular solutions of the angular equation in (5.62) are given by[11]

$$Y_{lm}(\theta, \phi) = \sqrt{\frac{2l+1}{4\pi} \frac{(l-m)!}{(l+m)!}} P_l^m(\cos\theta) e^{im\phi}, \qquad (5.69)$$

with $l = 0, 1, \ldots$ and $m = -l, -l+1, \ldots, l-1, l$. The functions Y_{lm} are called *spherical harmonics* and are the eigenfunctions of the Laplace operator on the sphere

$$\hat{\Delta} Y_{lm} = -l(l+1)Y_{lm}. \qquad (5.70)$$

With the normalisation in (5.69), they satisfy the orthogonality condition

$$\int_0^{2\pi} d\phi \int_0^{\pi} d\theta \sin\theta \, \overline{Y_{lm}(\theta, \phi)} \, Y_{l'm'}(\theta, \phi) = \delta_{ll'} \delta_{mm'}, \qquad (5.71)$$

and are an orthonormal basis for the Hilbert space of square-integrable functions on the sphere S^2

$$L^2(S^2) = \left\{ f(\theta, \phi) : S^2 \to \mathbb{C} \mid \int_0^{2\pi} d\phi \int_0^{\pi} d\theta \sin\theta |f(\theta, \phi)|^2 \leq \infty \right\}, \qquad (5.72)$$

where we use the natural measure $\sin\theta d\theta d\phi$ on S^2. They are related by complex conjugation as follows:

$$\overline{Y_{lm}(\theta, \phi)} = (-1)^m Y_{l,-m}(\theta, \phi). \qquad (5.73)$$

The first spherical harmonics are

$$Y_{2\pm2} = \sqrt{\frac{15}{32\pi}} \sin^2\theta e^{\pm 2i\phi}$$

$$Y_{1\pm1} = \mp\sqrt{\frac{3}{8\pi}} \sin\theta e^{\pm i\phi}$$

$$Y_{00} = \frac{1}{\sqrt{4\pi}}$$

$$Y_{2\pm1} = \mp\sqrt{\frac{15}{8\pi}} \sin\theta \cos\theta e^{\pm i\phi}.$$

$$Y_{10} = \sqrt{\frac{3}{4\pi}} \cos\theta$$

$$Y_{20} = \sqrt{\frac{5}{16\pi}} (3\cos^2\theta - 1)$$

The fact that the spectrum of the Laplacian on the sphere is purely discrete is part of a general result about elliptic operators: self-adjoint elliptic

---

[11]The spherical harmonics are sometimes defined with an extra $(-1)^m$ sign factor.

operators on a compact space have a discrete spectrum and their eigenvectors are C^∞ functions (see Rudin, 1991).

5.3.2. *The solution of the Dirichlet problem*

We are now ready to solve the Dirichlet problem on the unit sphere

$$\Delta u = 0, \quad u|_{r=1} = g(\theta, \phi).$$

Combining (5.60), (5.63), (5.70) and the linearity of the Laplace equation, we find the general solution

$$u = \sum_{l,m} c_{lm} r^l Y_{lm}(\theta, \phi) + \sum_{l,m} \frac{d_{lm}}{r^{l+1}} Y_{lm}(\theta, \phi), \tag{5.74}$$

where we used $\lambda = l(l+1)$ to determine the exponent $\alpha_i = -l, l+1$ in (5.63). The sum is over $l = 0, 1, \ldots$ and $m = -l, -l+1, \ldots, l-1, l$. The coefficients c_{lm} and d_{lm} are complex numbers subject to the only constraint that u is real. Since we want u to be regular at the centre of the sphere, $r = 0$, we set $d_{lm} = 0$. The boundary condition at $r = 1$ now imposes

$$\sum_{l,m} c_{lm} Y_{lm}(\theta, \phi) = g(\theta, \phi), \tag{5.75}$$

which can be solved for any regular function g since the spherical harmonics Y_{lm} are a basis for $L^2(S^2)$. The coefficients are explicitly given by

$$c_{lm} = \int_0^{2\pi} d\phi \int_0^\pi d\theta \sin\theta \overline{Y_{lm}(\theta, \phi)} g(\theta, \phi). \tag{5.76}$$

5.4. Exercises

Exercise 5.1. Show that (5.11) can be reduced to the Sturm–Liouville form (5.12) by the redefinitions

$$p(x) = e^{\int \frac{a_1}{a_2} dx}, \qquad q(x) = \frac{a_0}{a_2} e^{\int \frac{a_1}{a_2} dx}, \qquad w(x) = \frac{1}{a_2} e^{\int \frac{a_1}{a_2} dx}.$$

Exercise 5.2. Consider the Sturm–Liouville problem

$$-y'' + 4y = \lambda y, \qquad \begin{cases} y(0) = \alpha y(\pi), \\ \alpha y'(0) = y'(\pi) \end{cases} \tag{5.77}$$

with $\alpha \in [0, 1]$.

(a) Verify that $L = -d^2/dx^2 + 4$ is self-adjoint in $\mathcal{D}(L) = \{f_{ac}, f', f'' \in L^2[0, \pi]$
$\mid y(0) = \alpha y(\pi), \alpha y'(0) = y'(\pi)\}$.

(b) Find eigenvalues and eigenvectors.

(c) Solve the inhomogeneous equation $Ly = f$. [Hint: expand y and f in eigenvectors of L].

Exercise 5.3. Show that for an equation in Sturm–Liouville form, $(py')' + qy = -\lambda wy$, the product $p(x)W(x)$, where W is the Wronskian, is independent of x.

* **Exercise 5.4.** Consider the regular Sturm–Liouville problem (5.12) with boundary conditions (5.15). Show that

(a) the eigenvalues are not degenerate;

(b) the eigenfunctions can be chosen real;

(c) if $\lambda_2 > \lambda_1$, the eigenfunction $y_2(x)$ associated with λ_2 has at least one zero between two consecutive zeros of the eigenfunction $y_1(x)$ associated with λ_1.

Exercise 5.5. Solve the problem

$$\frac{\partial u}{\partial x} = 3\frac{\partial u}{\partial y} + 2u, \qquad u(x, 0) = \cosh x$$

with the method of separation of variables.

Exercise 5.6. Solve the Dirichlet problem for the two-dimensional Laplace equation on the square of side L, $\Omega = \{0 \leq x \leq L, 0 \leq y \leq L\}$, with boundary conditions $u(x, 0) = u(0, y) = u(L, y) = 0$ and $u(x, L) = 1$.

Exercise 5.7. Solve the Dirichlet problem for the two-dimensional Laplace equation on the semi-annulus, $\Omega = \{1 \leq \rho \leq 2, 0 \leq \phi \leq \pi\}$ with boundary conditions $u(\rho, 0) = u(\rho, \pi) = u(2, \phi) = 0$ and $u(1, \phi) = \sin 2\phi$.

Exercise 5.8. Solve the Dirichlet problem for the Laplace equation inside a unit sphere in $\mathbb{R}^3$ with boundary value $u(1, \theta, \phi) = 1 + \cos \theta$. Find the solution of the Laplace equation with the same boundary condition outside the sphere. [Hint: use the terms $1/r^{l+1}$ in (5.74).]

* **Exercise 5.9.** Consider two vectors $\mathbf{x}, \mathbf{x}' \in \mathbb{R}^3$.

(a) Show that $\frac{1}{\mathbf{x} - \mathbf{x}'}$ solve the Laplace equation as a function of $\mathbf{x}$ for $\mathbf{x} \neq \mathbf{x}'$.

(b) Show that $\frac{1}{\mathbf{x}-\mathbf{x}'} = \sum_{l=0}^{\infty} \frac{|\mathbf{x}'|^l}{|\mathbf{x}|^{l+1}} P_l(\cos\gamma)$ for $|\mathbf{x}| > |\mathbf{x}'|$, where γ is the angle between $\mathbf{x}$ and $\mathbf{x}'$. [Hint: choose the z-axis in the direction of $\mathbf{x}'$ and use the results of Section 5.3.2.]

(c) Show that the *generating function* of Legendre polynomials is given by
$$\frac{1}{\sqrt{1+t^2-2tx}} = \sum_{l=0}^{\infty} P_l(x)t^l.$$

Exercise 5.10. Find the solution of (4.58) using the method of separation of variables.

Exercise 5.11. Solve the problem of heat conduction in a bar of length $L = 1$ whose endpoints are kept at different temperature, $u(0,t) = 1$ and $u(1,t) = 2$, and whose initial temperature is $u(x,0) = x^2 + 1$.

Exercise 5.12. Solve the boundary problem for the heat equation
$$\frac{\partial u}{\partial t} = \kappa \frac{\partial^2 u}{\partial x^2}, \quad \begin{cases} u(0,t) = 0, \\ \partial_x u(L,t) = -cu(L,t), \\ u(x,0) = g(x). \end{cases}$$

Exercise 5.13. Solve the boundary problem for the wave equation
$$\frac{\partial^2 u}{\partial t^2} = v^2 \frac{\partial^2 u}{\partial x^2}, \quad \begin{cases} u(0,t) = u(L,t) = 0, \\ u(x,0) = 0, \\ \partial_t u(x,0) = \sin\frac{3\pi x}{L}. \end{cases}$$

Exercise 5.14. Find the eigenfunctions of the Laplacian $\Delta u = \lambda u$ on a square of side L with vanishing conditions at the boundary. Use this result to
(a) solve the heat equation $v_t = \kappa(v_{xx} + v_{yy})$ on the square with boundary conditions $v(0,y,t) = v(L,y,t) = v(x,0,t) = v(x,L,t) = 0$ and $v(x,y,0) = g(x,y)$;
(b) solve the Poisson equation $u_{xx} + u_{yy} = f(x,y)$ on the square with boundary conditions $u(0,y) = u(L,y) = u(x,0) = u(x,L) = 0$.

Exercise 5.15. Consider the unit vector $\mathbf{n} = (\sin\theta\cos\phi, \sin\theta\sin\phi, \cos\theta)$ in $\mathbb{R}^3$ and the vector space V_l of homogeneous polynomials of degree l
$$\sum_{i_1=1,\ldots,i_l=1}^{3} a_{i_1\cdots i_l} n_{i_1} \cdots n_{i_l},$$

where the complex coefficients $a_{i_1 \cdots i_l}$ are totally symmetric in the indices $i_1, \ldots, i_l$ and traceless: $\sum_{k=1}^{3} a_{kk i_3 \cdots i_l} = 0$. With an explicit computation for $l = 0, 1, 2$, verify the following statements (true for all l):

(a) V_l is a vector space of complex dimension $2l + 1$;

(b) the spherical harmonics Y_{lm} with $m = -l, -l+1, \ldots, l$ are a basis for V_l.

6

Green Functions

Green functions are fundamental tools in the study of differential operators. They provide solutions of inhomogeneous differential equations and are useful in the study of boundary value problems. In this chapter we give examples of Green functions for the most common differential equations and we briefly discuss their relation to the spectral theory of self-adjoint operators. In physics, Green functions appear in many different fields, from electromagnetism to quantum field theory, where they are also called *propagators*, and are used in many kinds of perturbation theory.

6.1. Fundamental Solutions and Green Functions

Consider a linear differential operator of order p in $\mathbb{R}^n$ with constant coefficients

$$L_{\mathbf{x}} = \sum_{i_1 + \cdots i_n \leq p} a_{i_1 \cdots i_n} \partial_{x_1}^{i_1} \cdots \partial_{x_n}^{i_n}. \tag{6.1}$$

A solution G of the equation

$$L_{\mathbf{x}} G(\mathbf{x}) = \delta(\mathbf{x}), \tag{6.2}$$

where $\delta(\mathbf{x}) = \prod_{i=1}^{d} \delta(x_i)$, is called *fundamental solution* for $L_{\mathbf{x}}$ in $\mathbb{R}^n$. The previous expression should be understood as an equation in the space of distributions and G itself is a distribution.[1] Fundamental solutions are

---

[1]With an abuse of notation we indicated a dependence on $\mathbf{x}$ of the distribution G.

important because they allow to determine a particular solution of the inhomogeneous equation

$$L_{\mathbf{x}}f(\mathbf{x}) = h(\mathbf{x}) \tag{6.3}$$

as the convolution

$$f(\mathbf{x}) = (G * h)(\mathbf{x}) = \int G(\mathbf{x} - \mathbf{y})h(\mathbf{y})d\mathbf{y}, \tag{6.4}$$

where all the integrals are over $\mathbb{R}^n$. Indeed, applying the operator $L_{\mathbf{x}}$ to the previous expression we obtain[2]

$$L_{\mathbf{x}}f(\mathbf{x}) = \int L_{\mathbf{x}}G(\mathbf{x} - \mathbf{y})h(\mathbf{y})d\mathbf{y} = \int \delta(\mathbf{x} - \mathbf{y})h(\mathbf{y})d\mathbf{y} = h(\mathbf{x}). \tag{6.5}$$

$G(\mathbf{x})$ is also called the fundamental solution of the differential equation (6.3). Fundamental solutions are not unique. If G is a fundamental solution for $L_{\mathbf{x}}$ and u is a solution of the homogeneous equation, $L_{\mathbf{x}}u = 0$, $\tilde{G} = G + u$ is again a fundamental solution.

From a mathematical point of view, fundamental solutions allow to invert the operator $L_{\mathbf{x}}$. Its inverse is the integral operator with kernel given by the fundamental solution

$$L_{\mathbf{x}}f(\mathbf{x}) = h(\mathbf{x}) \implies f(\mathbf{x}) = L_{\mathbf{x}}^{-1}h(\mathbf{x}) = \int G(\mathbf{x} - \mathbf{y})h(\mathbf{y})d\mathbf{y}. \tag{6.6}$$

Notice that, strictly speaking, $L_{\mathbf{x}}^{-1}$ exists and the previous formula makes sense only if $\ker L_{\mathbf{x}} = \{\mathbf{0}\}$. In this case, there are no non-trivial solutions u of the homogeneous equation and G is uniquely defined.

From a physical point of view, fundamental solutions give the response of a physical system described by the equation $L_{\mathbf{x}}f = h$ to the presence of a point-like source $h = \delta$. A standard example is the following.

Example 6.1 (Poisson equation). Consider the *Poisson equation* in $\mathbb{R}^3$

$$\Delta f(\mathbf{x}) = -4\pi\rho(\mathbf{x}) \tag{6.7}$$

---

[2]Notice that this manipulation works only when $L_{\mathbf{x}}$ has constant coefficients. Indeed only in this case the operator is translational invariant and $L_{\mathbf{x}}G(\mathbf{x} - \mathbf{y}) = L_{\mathbf{x}-\mathbf{y}}G(\mathbf{x} - \mathbf{y}) = \delta(\mathbf{x} - \mathbf{y})$.

describing the electrostatic potential $f(\mathbf{x})$ produced by a distribution of charges $\rho(\mathbf{x})$. We know from electrostatics that the potential generated by a point-like unit charge

$$\Delta G(\mathbf{x}) = -4\pi\delta(\mathbf{x}) \tag{6.8}$$

is given by Coulomb law, $G(\mathbf{x}) = 1/r$ with $r = |\mathbf{x}|$. We also know that the potential generated by a generic distribution of charges is obtained by superposition of many point-like ones

$$f(\mathbf{x}) = \int \frac{\rho(\mathbf{y})}{|\mathbf{x} - \mathbf{y}|}\mathrm{d}\mathbf{y}. \tag{6.9}$$

This is precisely (6.4) with $G(\mathbf{x} - \mathbf{y}) = 1/|\mathbf{x} - \mathbf{y}|$. We will verify in Section 6.4.1 that G is indeed a fundamental solution for the Laplace equation. It is the only fundamental solution that vanishes at infinity.

When the coefficients of $L_\mathbf{x}$ are not constant, we need to consider functions of two variables and generalise (6.2) as follows:

$$L_\mathbf{x}G(\mathbf{x}, \mathbf{y}) = \delta(\mathbf{x} - \mathbf{y}), \tag{6.10}$$

where $\mathbf{y}$ is a parameter, which physically corresponds to the position of the external point-like source. The solution of (6.3) is now

$$f(\mathbf{x}) = \int G(\mathbf{x}, \mathbf{y})h(\mathbf{y})\mathrm{d}\mathbf{y}. \tag{6.11}$$

Indeed, the same manipulations as in (6.5) give

$$L_\mathbf{x}f(\mathbf{x}) = \int L_\mathbf{x}G(\mathbf{x}, \mathbf{y})h(\mathbf{y})\mathrm{d}\mathbf{y} = \int \delta(\mathbf{x} - \mathbf{y})h(\mathbf{y})\mathrm{d}\mathbf{y} = h(\mathbf{x}). \tag{6.12}$$

Similarly to (6.2), the solution is not unique. If $G(\mathbf{x}, \mathbf{y})$ solves (6.10), also

$$G(\mathbf{x}, \mathbf{y}) + c(\mathbf{y})u(\mathbf{x}) \tag{6.13}$$

does, where $L_\mathbf{x}u(\mathbf{x}) = 0$ and $c(\mathbf{y})$ is an arbitrary function of $\mathbf{y}$. The case of operators with constant coefficients corresponds to a translational invariant system. In this case $G(\mathbf{x}, \mathbf{y}) = G(\mathbf{x} - \mathbf{y})$ becomes a function of the difference only and we recover (6.2).

In most physical problems, we also have to specify initial or boundary conditions. These translate into suitable conditions on the function $G(\mathbf{x}, \mathbf{y})$.

The conditions can be imposed at spatial infinity or at the boundary of a region $\Omega \subset \mathbb{R}^n$.[3] A solution of (6.10) that allows to solve a Cauchy or boundary problem for the operator $L_\mathbf{x}$ is called the *Green function* of the problem.[4] To understand how the method works, consider a simple example.

Example 6.2. Consider a boundary value problem for a second-order ordinary differential equation on a segment $[a, b]$

$$L_x f(x) = h(x), \quad \begin{cases} \alpha_1 f(a) + \alpha_2 f'(a) = 0, \\ \beta_1 f(b) + \beta_2 f'(b) = 0. \end{cases} \tag{6.14}$$

The solution can be written as $f(x) = \int_a^b G(x, y) h(y) \mathrm{d}y$ where the Green function $G(x, y)$ satisfies for all $x, y \in [a, b]$

$$L_x G(x, y) = \delta(x - y), \quad \begin{cases} \alpha_1 G(a, y) + \alpha_2 \partial_x G(a, y) = 0, \\ \beta_1 G(b, y) + \beta_2 \partial_x G(b, y) = 0. \end{cases} \tag{6.15}$$

We can use the arbitrariness (6.13) to choose the appropriate Green function for the boundary problem we are considering. Different boundary conditions for L_x correspond to different Green functions.

6.2. Linear Ordinary Differential Equations

We start our analysis of fundamental solutions and Green functions with ordinary linear differential equations. We consider, for simplicity, equations of order two. The results can be easily extended to higher order equations.

6.2.1. *Fundamental solution for equations with constant coefficients*

Consider the fundamental solution $G(x)$ of an ordinary linear second-order differential equation with constant coefficients

$$a_2 G''(x) + a_1 G'(x) + a_0 G(x) = \delta(x). \tag{6.16}$$

---

[3] In the latter case, all previous integrals are taken over Ω.
[4] As customary in physics, most often we will not distinguish between fundamental solutions and Green functions.

For $x \neq 0$, the delta function vanishes and $G(x)$ must reduce to a solution of the homogeneous equation $a_2 u'' + a_1 u' + a_0 u = 0$. Then we can write

$$G(x) = \begin{cases} \mu_1 u_1(x) + \mu_2 u_2(x), & x > 0, \\ \nu_1 u_1(x) + \nu_2 u_2(x), & x < 0, \end{cases} \qquad (6.17)$$

where $u_1(x)$ and $u_2(x)$ are two linearly independent solutions of the homogeneous equation. We need to solve (6.16) also in a neighbourhood of $x = 0$. In the theory of distributions delta functions arise when we take the derivative of a function with a jump discontinuity. We can then solve (6.16) imposing that $G(x)$ is continuous while $G'(x)$ has a jump discontinuity at $x = 0$. The magnitude of the jump can be easily determined by integrating (6.16) over an infinitesimal interval $[-\epsilon, \epsilon]$

$$(a_2 G'(x) + a_1 G(x))|_{-\epsilon}^{\epsilon} + \int_{-\epsilon}^{\epsilon} a_0 G(x) \mathrm{d}x = 1. \qquad (6.18)$$

Since G is continuous in $x = 0$, the last two terms on the left-hand side do not contribute in the limit $\epsilon \to 0$ and the equation above reduces to

$$a_2(G'(0^+) - G'(0^-)) = 1, \qquad (6.19)$$

so that the jump is $\Delta G'(0) = 1/a_2$. Consider again (6.17). Imposing that $G(x)$ is continuous and $G'(x)$ has a jump of $1/a_2$ at $x = 0$, we find

$$c_1 u_1(0) + c_2 u_2(0) = 0, \quad c_1 u_1'(0) + c_2 u_2'(0) = \frac{1}{a_2}, \qquad (6.20)$$

where $c_i = \mu_i - \nu_i$. This is an inhomogeneous linear system for c_1 and c_2 that always admits a unique solution since the determinant of the coefficients is $W(0) \neq 0$.[5] Any function of the form (6.17) with $c_i = \mu_i - \nu_i$ satisfying (6.20) is a fundamental solution of the differential equation. $G(x)$ is obviously not unique. We can always shift μ_i and ν_i by a common constant without changing the value of c_i. This ambiguity corresponds to the possibility of adding to $G(x)$ any solution of the homogeneous equation.

---

[5] Recall from Section 4.2 that, if $u_1(x)$ and $u_2(x)$ are linearly independent, their Wronskian is different from zero.

Example 6.3 (The classical harmonic oscillator). Consider the fundamental solution for the harmonic oscillator

$$G''(x) + \omega^2 G(x) = \delta(x). \tag{6.21}$$

Two linearly independent solutions of the homogeneous equation associated with (6.21) are $u_1(x) = e^{i\omega x}$ and $u_2(x) = e^{-i\omega x}$ with Wronskian $W(x) = -2i\omega$. The system (6.20) gives $c_1 + c_2 = 0$ and $i\omega(c_1 - c_2) = 1$, which are solved by $c_1 = -c_2 = 1/2i\omega$. We find the general solution

$$G(x) = \theta(x)\frac{\sin \omega x}{\omega} + \nu_1 e^{i\omega x} + \nu_2 e^{-i\omega x}. \tag{6.22}$$

The solution of the inhomogeneous equation

$$f''(x) + \omega^2 f(x) = g(x), \tag{6.23}$$

is then given by (6.4)

$$f(x) = \int G(x-y)g(y)dy$$

$$= \int \theta(x-y)\frac{\sin \omega(x-y)}{\omega}g(y)dy + C_1 e^{i\omega x} + C_2 e^{-i\omega x},$$

where $C_1 = \nu_1 \int e^{-i\omega y} g(y)dy$ and $C_2 = \nu_2 \int e^{i\omega y} g(y)dy$. The same result can be obtained with the method of variation of arbitrary constants (see Exercise 6.1).

6.2.2. *Green functions for boundary and initial problems*

Consider now the boundary problem for a linear second-order equation

$$a_2(x)f''(x) + a_1(x)f'(x) + a_0(x)f(x) = h(x), \quad \begin{cases} \alpha_1 f(a) + \alpha_2 f'(a) = 0, \\ \beta_1 f(b) + \beta_2 f'(b) = 0, \end{cases}$$
$$\tag{6.24}$$

where $a_i(x)$ are continuous functions, $a_2(x) \neq 0$, α_i and β_i are real and at least one α_i and one β_i are non-zero. As already discussed in Example 6.2, we need to find a Green function $G(x, y)$ that satisfies

$$a_2(x)\partial_x^2 G(x, y) + a_1(x)\partial_x G(x, y) + a_0(x)G(x, y) = \delta(x-y), \tag{6.25}$$

and the boundary conditions

$$\alpha_1 G(a, y) + \alpha_2 \partial_x G(a, y) = 0,$$
$$\beta_1 G(b, y) + \beta_2 \partial_x G(b, y) = 0. \tag{6.26}$$

The solution of (6.24) is then given by $f(x) = \int_a^b G(x, y)h(y)\mathrm{d}y$.

Let us fix y and consider $G(x, y)$ as a function of x. For $x \neq y$, $G(x, y)$ must solve the homogeneous equation $a_2(x)u'' + a_1(x)u' + a_0(x)u = 0$ with boundary conditions (6.26). Then it must be of the form

$$G(x, y) = \begin{cases} a(y)u(x), & a \leq x < y, \\ b(y)\tilde{u}(x), & y < x \leq b, \end{cases} \tag{6.27}$$

where $u(x)$ and $\tilde{u}(x)$ are two solutions of the homogeneous equation satisfying $\alpha_1 u(a) + \alpha_2 u'(a) = 0$ and $\beta_1 \tilde{u}(b) + \beta_2 \tilde{u}'(b) = 0$, respectively. Following the same logic as in Section 6.2.1, we impose that, for $x = y$, $G(x, y)$ is continuous and $\partial_x G$ has a jump of $1/a_2(y)$,

$$a(y)u(y) - b(y)\tilde{u}(y) = 0, \quad a(y)u'(y) - b(y)\tilde{u}'(y) = -\frac{1}{a_2(y)}. \tag{6.28}$$

The Green function is

$$G(x, y) = \begin{cases} C(y)\tilde{u}(y)u(x), & a \leq x < y, \\ C(y)u(y)\tilde{u}(x), & y < x \leq b, \end{cases} \tag{6.29}$$

where $C^{-1}(y) = a_2(y)W(y) = a_2(y)(u(y)\tilde{u}'(y) - u'(y)\tilde{u}(y))$.

Let us discuss under which conditions the Green function of the boundary problem (6.24) exists and is unique. By the theorem of existence and uniqueness for ordinary differential equations, u and $\tilde{u}$ exist and are unique up to a multiplicative constant.[6] However, something special happens when the homogeneous problem, (6.24) with $h(x) = 0$, has a non-trivial solution. In this case there is a solution that simultaneously satisfies the boundary conditions in $x = a$ and $x = b$. This means that u and $\tilde{u}$ are actually proportional and $C(y)$ diverges. Then the Green function does not exist. Thus we see that the Green function exists only if the homogeneous equation has

---

[6]The function $u(x)$ can be seen as the solution of the Cauchy problem for the homogeneous equation with conditions $u(a) = -\alpha_2 C$ and $u'(a) = \alpha_1 C$, where C is an arbitrary constant. The same holds for $\tilde{u}$.

no non-trivial solution. In this case, the multiplicative constants in u and $\tilde{u}$ cancel in (6.29) and $G(x, y)$ is also unique.

We can rephrase this argument in the language of operators. Equation (6.24) is equivalent to the equation $L_x f = h$ where L_x is the operator

$$L_x = a_2(x) \frac{d^2}{dx^2} + a_1(x) \frac{d}{dx} + a_0(x) \tag{6.30}$$

with domain

$$\mathcal{D}(L_x) = \{f_{ac}, f', f'' \in L^2[a, b] | \alpha_1 f(a) + \alpha_2 f'(a) = \beta_1 f(b) + \beta_2 f'(b) = 0\}.$$

When $\ker L_x$ is trivial we can invert the operator and write $f = L_x^{-1} h = \int_a^b G(x, y) h(y) dy$. The solution is also unique. When $\ker L_x$ is non-trivial, we need more sophisticated methods to see whether (6.24) has a solution and to compute it (see Exercise 6.2 and Section 8.1.1).

When the operator L_x is of Sturm–Liouville type, (5.12), the function $C(y)$ reduces to a constant (see Exercise 5.3) and $G(y, x)$ is symmetric. This reflects the self-adjointness of L_x (see Section 8.1).

We can also use the Green function method to solve the Cauchy problem

$$a_2(x) f''(x) + a_1(x) f'(x) + a_0(x) f(x) = h(x),$$

$$f(x_0) = f'(x_0) = 0. \tag{6.31}$$

We write the solution as

$$f(x) = \int_{x_0}^{+\infty} G(x, y) h(y) dy, \tag{6.32}$$

where $G(x, y)$ satisfies (6.25). The initial conditions $f(x_0) = f'(x_0) = 0$ are satisfied if $G(x, y) = 0$ for $x < y$. Such a G is called a *retarded* Green function. Since G satisfies (6.25) and vanishes for $x < y$, it must be of the form

$$G(x, y) = \theta(x - y)(\alpha_1(y) u_1(x) + \alpha_2(y) u_2(x)), \tag{6.33}$$

where $u_1(x)$ and $u_2(x)$ are two linearly independent solutions of the homogeneous equation. As before, we require that $G(x, y)$ is continuous in $x = y$ and that its derivative with respect to x has a jump discontinuity in $x = y$.

Continuity in $x = y$ requires $\alpha_1(y) = u_2(y)k(y)$ and $\alpha_2(y) = -u_1(y)k(y)$ where $k(y)$ is an arbitrary function of y. Imposing that $\partial_x G$ has a jump discontinuity $1/a_2(y)$ for $x = y$,

$$\lim_{x \to y^+} \partial_x G(x, y) - \lim_{x \to y^-} \partial_x G(x, y) = -W(y)k(y) = \frac{1}{a_2(y)},$$

where $W(y) = u_1(y)u_2'(y) - u_2(y)u_1'(y)$ is the Wronskian, we find $k(y) = -1/(a_2(y)W(y))$. In conclusion

$$G(x, y) = -\theta(x - y)\frac{u_1(x)u_2(y) - u_2(x)u_1(y)}{a_2(y)W(y)}. \tag{6.34}$$

Notice that, since $G(x, y) = 0$ for $x < y$, the integral in the solution is restricted to the interval $[x_0, x]$

$$f(x) = \int_{x_0}^{x} G(x, y)h(y)\mathrm{d}y. \tag{6.35}$$

6.3. The Fourier Transform Method

Consider a linear differential operator of order k in $\mathbb{R}^n$ with constant coefficients

$$L_{\mathbf{x}} = \sum_{i_1 + \cdots + i_n \leq k} a_{i_1 \cdots i_n} \partial_{x_1}^{i_1} \cdots \partial_{x_n}^{i_n}. \tag{6.36}$$

We now show, using the Fourier transform, that $L_{\mathbf{x}}$ always admits a fundamental solution that is a tempered distribution. By introducing the polynomial L in n variables

$$L(\mathbf{y}) = \sum_{i_1 + \cdots + i_n \leq k} a_{i_1 \cdots i_n} y_1^{i_1} \cdots y_n^{i_n}, \tag{6.37}$$

where $\mathbf{y} = (y_1, \ldots, y_n)$, the operator can be compactly written as $L_{\mathbf{x}} = L(\nabla)$ with $\nabla = (\partial_{x_1}, \ldots, \partial_{x_n})$. We can easily solve the equation

$$L_{\mathbf{x}}G(\mathbf{x}) = L(\nabla)G(\mathbf{x}) = \delta(\mathbf{x}) \tag{6.38}$$

in Fourier space. The Fourier transform maps ∇ into $i\mathbf{p}$ and we find

$$L(i\mathbf{p})\hat{G}(\mathbf{p}) = (2\pi)^{-n/2}, \tag{6.39}$$

where $\hat{G}(\mathbf{p})$ is the Fourier transform of $G(\mathbf{x})$. This is an algebraic equation with solution $\hat{G}(\mathbf{p}) = (2\pi)^{-n/2}/L(i\mathbf{p})$. The Green function is then given by the inverse Fourier transform

$$G(\mathbf{x}) = \int \frac{d\mathbf{p}}{(2\pi)^{n/2}} \hat{G}(\mathbf{p}) e^{i(\mathbf{p},\mathbf{x})} = \int \frac{d\mathbf{p}}{(2\pi)^n} \frac{e^{i(\mathbf{p},\mathbf{x})}}{L(i\mathbf{p})}, \qquad (6.40)$$

where $(\mathbf{p},\mathbf{x})$ denotes the scalar product on $\mathbb{R}^n$. Notice, however, that the integral (6.40) is ill defined when the polynomial $L(i\mathbf{p})$ has zeros for $\mathbf{p} \in \mathbb{R}^n$. This is because we have been too cavalier in solving the algebraic equation (6.39) in the space of distributions. $1/L(i\mathbf{p})$ is not locally integrable near the simple zeros of $L(i\mathbf{p})$ and it is not a well-defined distribution. The general solution of (6.39) in the space of distributions is given by

$$\hat{G}(\mathbf{p}) = \frac{1}{(2\pi)^{n/2}} P \frac{1}{L(i\mathbf{p})} + \sum c_i \delta(\mathbf{p} - \mathbf{p}_i), \qquad (6.41)$$

where $\mathbf{p}_i$ are the real zeros of $L(i\mathbf{p})$, P is the principal value symbol and c_i are arbitrary constants (see Appendix A and Exercise 6.7). Equation (6.41) is now a tempered distribution and so is its inverse Fourier transform

$$G(\mathbf{x}) = P \int \frac{d\mathbf{p}}{(2\pi)^n} \frac{e^{i(\mathbf{p},\mathbf{x})}}{L(i\mathbf{p})} + \sum_i \frac{c_i}{(2\pi)^{n/2}} e^{i(\mathbf{p}_i,\mathbf{x})}. \qquad (6.42)$$

The constants c_i reflect the possibility of adding to $G(\mathbf{x})$ a solution of the homogeneous equation $L_{\mathbf{x}} u = 0$. Mathematically, they can be interpreted as different regularisations of the zeros of $L(i\mathbf{p})$ in the integral (6.40) and, physically, they correspond to different choices of boundary conditions for the physical problem. This is clearly shown in the example below.

Example 6.4. Consider again the fundamental solution for the harmonic oscillator of Example 6.3

$$G''(x) + \omega^2 G(x) = \delta(x). \qquad (6.43)$$

The Fourier transform of (6.43) gives $(-p^2 + \omega^2)\hat{G}(p) = (2\pi)^{-1/2}$ whose general distributional solution is

$$\hat{G}(p) = \frac{1}{\sqrt{2\pi}} P \frac{1}{(\omega^2 - p^2)} + c_+ \delta(p + \omega) + c_- \delta(p - \omega). \qquad (6.44)$$

The fundamental solutions are then

$$G(x) = \int \frac{dp}{\sqrt{2\pi}} \hat{G}(p) e^{ipx} = P \int \frac{dp}{2\pi} \frac{e^{ipx}}{\omega^2 - p^2} + \frac{c_+}{\sqrt{2\pi}} e^{-i\omega x} + \frac{c_-}{\sqrt{2\pi}} e^{i\omega x}.$$

By evaluating the integral in the expression above, we find again (6.22). As expected, the arbitrariness in $G(x)$ is given by the freedom of adding the solutions $e^{\pm i\omega x}$ of the homogeneous equation.

For particular values of $c_\pm$ we can write the solutions in a different form. This will allow to interpret the different choices of $c_\pm$ as different regularisations of the integral (6.40). When $c_+ = (\mp i/\omega)\sqrt{\pi/8}$ and $c_- = (\pm i/\omega)\sqrt{\pi/8}$, using the identities (A.7), we can write (6.44) as

$$\hat{G}(p) = -\frac{1}{\sqrt{2\pi}} \frac{1}{(p + \omega \pm i0)(p - \omega \pm i0)}. \tag{6.45}$$

These choices of $c_\pm$ lead to four Green functions

$$G(x) = -\lim_{\epsilon \to 0} \int_{-\infty}^{+\infty} \frac{dp}{2\pi} \frac{e^{ipx}}{(p + \omega \pm i\epsilon)(p - \omega \pm i\epsilon)}, \tag{6.46}$$

which differ for the positions of the poles of $\hat{G}(p)$, as in Figure 6.1. The integrals in (6.46) are easily computed using the residue theorem. Consider first the case in Figure 6.1(a). Using Jordan's lemma, for $x > 0$ we close the contour with a semi-circle in the upper half-plane. Since the contour contains no poles, the integral is zero. For $x < 0$ we close the contour with a semi-circle in the lower half-plane. We pick the contribution of the two simple poles $\pm\omega - i\epsilon$, with a minus sign because the contour is oriented

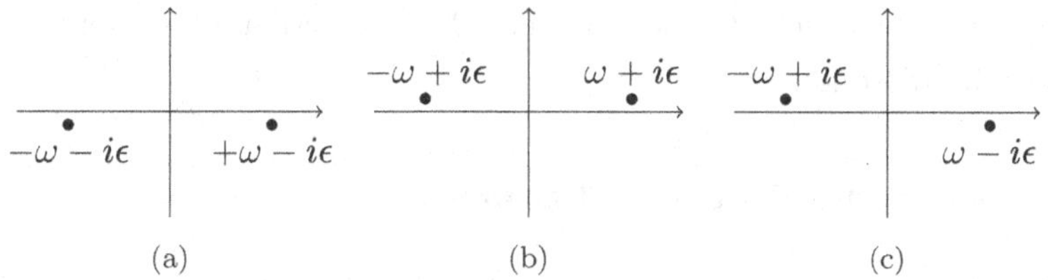

(a)   (b)   (c)

Fig. 6.1. Pole prescription for (a) advanced Green function; (b) retarded Green function; (c) causal Green function. We omit a fourth case, which is obtained from (c) by a reflection along the real axis and has similar properties.

clock-wise. The final result is

$$G_A(x) = i\theta(-x) \lim_{\epsilon \to 0} \left(\frac{e^{-ix\omega + \epsilon x}}{-2\omega} + \frac{e^{ix\omega + \epsilon x}}{2\omega} \right) = -\theta(-x) \frac{\sin \omega x}{\omega}. \quad (6.47)$$

This is non-zero only for $x < 0$ and is called the *advanced* Green function. The case in Figure 6.1(b) is analogous. Now the two poles contribute only when $x > 0$, with a plus sign, and we find

$$G_R(x) = \theta(x) \frac{\sin \omega x}{\omega}. \quad (6.48)$$

This is non-zero only for $x > 0$ and is called the *retarded* Green function. The generalisation to higher dimensions of the advanced and retarded Green functions is used in the solution of the wave equation (see Example 6.4.3). In the case in Figure 6.1(c), the pole $-\omega + i\epsilon$ contributes for $x > 0$ and $\omega - i\epsilon$ contributes for $x < 0$. We find

$$G_F(x) = i\theta(x) \frac{e^{-i\omega x}}{2\omega} + i\theta(-x) \frac{e^{i\omega x}}{2\omega}. \quad (6.49)$$

This is the *causal* or *Feynman* Green function. Its higher dimensional analogue is used in quantum field theory where it is also called *propagator*. In quantum field theory, x is a time variable and p an energy, and G_F propagates positive energy states forward in time and negative energy states (anti-particles) backward in time. The Feynman Green function can also be written as

$$G_F(x) = -\lim_{\epsilon \to 0} \int_{-\infty}^{+\infty} \frac{dp}{2\pi} \frac{e^{ipx}}{(p^2 - \omega^2 + i\epsilon)}, \quad (6.50)$$

since, for infinitesimal ϵ, the zeros of the denominator move in the complex plane as in Figure 6.1(c). The fourth case where the poles are at $-\omega - i\epsilon$ and $\omega + i\epsilon$ is analogous to the Feynman Green function with time running in opposite direction.

6.4.   Linear Partial Differential Equations

In this section we give examples of fundamental solutions and Green functions for some partial differential equations commonly used in mathematical physics. We also apply them to the solution of boundary problems.

6.4.1. *Laplace equation*

A fundamental solution of the Laplace operator in dimension d

$$\Delta G = \delta \qquad (6.51)$$

is given by

$$G(\mathbf{x}) = \begin{cases} -\dfrac{r^{2-d}}{(d-2)\Omega_d}, & d \geq 3, \\[2mm] \dfrac{1}{2\pi} \ln r, & d = 2, \end{cases} \qquad (6.52)$$

where $\Omega_d = 2\pi^{d/2}/\Gamma(d/2)$ is the surface area of the unit sphere in $\mathbb{R}^d$ and $r = |\mathbf{x}|$. In particular for $d = 3$ we have $G(\mathbf{x}) = -1/(4\pi r)$. In what follows, we give different derivations of this result.

Example 6.5 (Gauss law). Consider, for simplicity, the case $d = 3$. We want to show that

$$\Delta \frac{1}{r} = -4\pi\delta(\mathbf{x}), \qquad (6.53)$$

where $\delta(\mathbf{x}) = \delta(x)\delta(y)\delta(z)$. This is a distributional equation in $\mathcal{D}'(\mathbb{R}^3)$. Notice[7] that $1/r$ is locally integrable in $\mathbb{R}^3$ and therefore it is a well-defined element of $\mathcal{D}'(\mathbb{R}^3)$. The same is true for $\Delta(1/r)$, since distributions are infinitely differentiable. It follows easily from (5.58) that $\Delta(1/r) = 0$ for $r \neq 0$.[8] Integrate now $\Delta(1/r)$ over of a sphere of radius R centred in the origin. Using the divergence theorem,[9] we find

$$\int_{r \leq R} \Delta \frac{1}{r} d\mathbf{x} = \int_{r \leq R} \nabla^2 \frac{1}{r} d\mathbf{x} = \int_{r=R} (\nabla \frac{1}{r}, d\sigma) = -\frac{4\pi R^2}{R^2} = -4\pi, \quad (6.55)$$

---

[7]In dimension d, $1/r^\alpha$ is integrable in $r = 0$ if $\alpha < d$. This follows from the fact that the integration measure in spherical coordinates behaves like $r^{d-1}dr$.

[8]The reader is suggested to check this result also by a direct computation in Cartesian coordinates (see Exercise 5.9).

[9]The divergence theorem for a vector field $\mathbf{F}(\mathbf{x})$ in $S \subset \mathbb{R}^n$ states that

$$\int_S \operatorname{div} \mathbf{F} \, d\mathbf{x} = \int_{\partial S} (\mathbf{F}, d\sigma), \qquad (6.54)$$

where $\operatorname{div} \mathbf{F} = (\nabla, \mathbf{F}) = \sum_{i=1}^n \partial_i F_i$ is the divergence of $\mathbf{F}$ and $d\sigma$ is the infinitesimal element of area orthogonal to the boundary ∂S and pointing outwards. In computing the integral we used the three-dimensional gradient in spherical coordinates, $\nabla f = \partial_r f \mathbf{e}_r + r^{-1}\partial_\theta f \mathbf{e}_\theta + (r\sin\theta)^{-1}\partial_\phi f \mathbf{e}_\phi$ with $\mathbf{e}_r$, $\mathbf{e}_\theta$, $\mathbf{e}_\phi$ the unit vectors in the directions r, θ, ϕ, and the fact that for the sphere $d\sigma = r^2 \sin\theta d\theta d\phi \mathbf{e}_r$. We also used the fact that $\Delta = \nabla^2$.

where $d\sigma$ is the infinitesimal element of area orthogonal to the sphere and pointing outwards. Thus $\Delta(1/r)$ is a distribution with support in $r = 0$, and whose integral over a region containing the origin is -4π. We conclude that $\Delta(1/r) = -4\pi\delta$.

We can also give a more formal proof of the distributional equation (6.53) by applying it to a test function $\phi \in \mathcal{D}(\mathbb{R}^3)$,

$$\langle \Delta\frac{1}{r}, \phi \rangle = \langle \frac{1}{r}, \Delta\phi \rangle = \int_{\mathbb{R}^3} \frac{\Delta\phi}{r} dx = \lim_{\epsilon \to 0} \int_{r \geq \epsilon} \frac{\Delta\phi}{r} dx, \qquad (6.56)$$

where, in the first step, we used the definition of derivative of a distribution, and, in the last one, the fact that the integral is convergent around $r = 0$. We can now integrate by parts twice

$$\int_{r \geq \epsilon} \frac{\Delta\phi}{r} dx = -\int_{r \geq \epsilon} \left(\nabla\frac{1}{r}, \nabla\phi \right) dx - \int_{r=\epsilon} \frac{1}{r}\frac{\partial\phi}{\partial r} r^2 d\Omega$$
$$= \int_{r \geq \epsilon} \phi\Delta\frac{1}{r} dx + \int_{r=\epsilon} \phi\frac{\partial}{\partial r}\frac{1}{r} r^2 d\Omega - \int_{r=\epsilon} \frac{1}{r}\frac{\partial\phi}{\partial r} r^2 d\Omega, \qquad (6.57)$$

where $d\Omega = \sin\theta d\theta d\phi$ is the measure on the unit sphere. The first term in the last line vanishes since $\Delta(1/r) = 0$ for $r \neq 0$. We evaluate the other two using the mean value theorem. They reduce to $-4\pi\epsilon\partial_r\phi(\mathbf{x}_1) - 4\pi\phi(\mathbf{x}_2)$ where $\mathbf{x}_i$ are points on the sphere. Taking $\epsilon \to 0$, we also have $\mathbf{x}_i \to \mathbf{0}$ and we obtain the result (6.53)

$$\langle \Delta\frac{1}{r}, \phi \rangle = -4\pi\phi(0) = -4\pi\langle \delta, \phi \rangle. \qquad (6.58)$$

In electrostatics, $\frac{1}{r}$ is the potential generated by a unit charge placed in the origin, $\mathbf{E} = -\nabla\frac{1}{r}$ is the corresponding electric field and (6.55) amounts to Gauss law: the flux of the electric field $\int(\mathbf{E}, d\sigma)$ through a surface is equal to 4π times the total charge contained inside the surface.

The expression for the fundamental solution (6.52) for $d \neq 3$ is obtained in a similar way.

Example 6.6 (Fourier transform method). Introduce   the   Fourier transform of G

$$\hat{G}(\mathbf{p}) = \int \frac{dx}{(2\pi)^{d/2}} G(\mathbf{x}) e^{-i(\mathbf{p},\mathbf{x})}. \qquad (6.59)$$

Taking the Fourier transform of (6.51) and using $\Delta \to -\mathbf{p}^2$ we find

$$-\mathbf{p}^2 \hat{G}(\mathbf{p}) = \frac{1}{(2\pi)^{d/2}}. \tag{6.60}$$

We can invert this equation and obtain $\hat{G}(\mathbf{p}) = -1/((2\pi)^{d/2}\mathbf{p}^2)$. Notice that, contrary to Example 6.4, $\hat{G}(\mathbf{p})$ is a well-defined tempered distribution for $d \geq 3$, since it is locally integrable in $\mathbb{R}^d$ and vanishes at infinity. We now evaluate $G(\mathbf{x})$ for $d = 3$. Since $\hat{G}(\mathbf{p})$ is spherically symmetric, it is convenient to use spherical coordinates in Fourier space. We choose an adapted system of coordinates (p, θ, ϕ), defined such that the z-axis points in the direction specified by the vector $\mathbf{x}$. In this way, θ is the relative angle between $\mathbf{p}$ and $\mathbf{x}$. The Green function is thus given by

$$G(\mathbf{x}) = -\int \frac{d\mathbf{p}}{(2\pi)^3} \frac{e^{i(\mathbf{p},\mathbf{x})}}{\mathbf{p}^2} = -\frac{1}{(2\pi)^3} \int_0^{+\infty} dp \int_0^{2\pi} d\phi \int_0^{\pi} d\theta \sin\theta \, e^{ipr\cos\theta}$$

$$= -\frac{1}{2\pi^2} \int_0^{+\infty} dp \frac{\sin pr}{pr} = -\frac{1}{2\pi^2 r} \int_0^{+\infty} dy \frac{\sin y}{y} = -\frac{1}{4\pi r}, \tag{6.61}$$

where the last integral can be computed using the residue theorem. Notice that we have been able to perform the integral over θ with elementary methods. This is a feature of three dimensions. In $d \neq 3$ the integration requires the use of special functions.

The Green function method can also be used to solve boundary problems for the Laplace equation and, in general, for partial differential equations. In this case (6.11) must be modified by the addition of suitable boundary terms. Let us discuss an example where the solution can be found in closed form.

Example 6.7 (The inhomogeneous Dirichlet problem in $\mathbb{R}^n$). The solution of the boundary problem

$$\Delta f = h, \qquad f|_{\partial\Omega} = g, \tag{6.62}$$

is given by

$$f(\mathbf{x}) = \int_\Omega G(\mathbf{y}, \mathbf{x}) h(\mathbf{y}) d\mathbf{y} + \int_{\partial\Omega} g(\mathbf{y})(\nabla_y G(\mathbf{y}, \mathbf{x}), d\sigma_y), \tag{6.63}$$

where $d\sigma_y$ is the infinitesimal element of area orthogonal to the boundary $\partial\Omega$, and $G(\mathbf{x},\mathbf{y})$ is a Green function satisfying[10]

$$\Delta_y G(\mathbf{y},\mathbf{x}) = \delta(\mathbf{x}-\mathbf{y}), \qquad \mathbf{x},\mathbf{y} \in \Omega,$$

$$G(\mathbf{y},\mathbf{x}) = 0, \qquad \mathbf{y} \in \partial\Omega, \mathbf{x} \in \Omega. \qquad (6.64)$$

This can be checked using the identity[11]

$$\int_\Omega (u\Delta v - v\Delta u)d\mathbf{y} = \int_{\partial\Omega} (u\nabla v - v\nabla u, d\sigma_y), \qquad (6.65)$$

with $u(\mathbf{y}) = f(\mathbf{y})$ and $v(\mathbf{y}) = G(\mathbf{y},\mathbf{x})$. Indeed, from (6.62) and (6.64) we have $\Delta_y u(\mathbf{y}) = h(\mathbf{y})$ and $\Delta_y v(\mathbf{y}) = \delta(\mathbf{x}-\mathbf{y})$. Then we find

$$u(\mathbf{x}) - \int_\Omega G(\mathbf{y},\mathbf{x})h(\mathbf{y})d\mathbf{y} = \int_{\partial\Omega} (f(\mathbf{y})\nabla_y G(\mathbf{y},\mathbf{x}) - G(\mathbf{y},\mathbf{x})\nabla_y f(\mathbf{y}), d\sigma_y)$$

from which (6.63) follows since $G(\mathbf{y},\mathbf{x}) = 0$ and $f(\mathbf{y}) = g(\mathbf{y})$ for $\mathbf{y} \in \partial\Omega$. To solve the problem completely, we need to find solutions of (6.64), which in general is a difficult task. It can be done explicitly for certain domains Ω with simple geometries (see Exercise 6.8).

6.4.2. *Heat equation*

Using methods similar to those applied to the Laplace equation, one can prove that the fundamental solution of the heat equation in d spatial dimensions

$$\left(\frac{\partial}{\partial t} - \kappa\Delta\right) G(\mathbf{x},t) = \delta(t)\delta(\mathbf{x}) \qquad (6.66)$$

is given by

$$G(\mathbf{x},t) = \theta(t)h(\mathbf{x},t), \qquad h(\mathbf{x},t) = \frac{e^{-\mathbf{x}^2/4\kappa t}}{(4\kappa\pi t)^{d/2}}. \qquad (6.67)$$

---

[10]The positions of $\mathbf{x}$ and $\mathbf{y}$ in (6.63) and (6.64) might seem strange, but one can show that $G(\mathbf{x},\mathbf{y}) = G(\mathbf{y},\mathbf{x})$. This follows from the fact that the Laplace operator is self-adjoint.
[11]This is just the divergence theorem (6.54) with $\mathbf{F} = u\nabla v - v\nabla u$, since $u\Delta v - v\Delta u = \nabla\cdot(u\nabla v - v\nabla u)$.

The function $h(\mathbf{x}, t)$ is called the *heat kernel*. We have already seen it in Example 4.6 for $d = 1$. Notice that the very same function enters in the solution of the Cauchy problem for the homogeneous heat equation and in the Green function that allows to solve the inhomogeneous equation. This is a consequence of Duhamel's principle (see Exercise 6.9).

6.4.3. *Wave equation*

The most common Green functions of the wave equation in d spatial dimension

$$\left(\frac{\partial^2}{\partial t^2} - c^2 \Delta\right) G(\mathbf{x}, t) = \delta(t)\delta(\mathbf{x}), \tag{6.68}$$

are the advanced Green function $G_A(\mathbf{x}, t)$, which vanishes identically for $t > 0$, and the retarded one $G_R(\mathbf{x}, t)$, which vanishes for $t < 0$. Their physical meaning is the following. Consider the scattering problem

$$\left(\frac{\partial^2}{\partial t^2} - c^2 \Delta\right) \psi(\mathbf{x}, t) = f(\mathbf{x}, t), \tag{6.69}$$

where the source $f(\mathbf{x}, t)$ is non-zero only for a small amount of time near $t = 0$. The solution can be written in two different ways

$$\psi(\mathbf{x}, t) = \psi_{\text{in}}(\mathbf{x}, t) + \int G_R(\mathbf{x} - \mathbf{x}', t - t') f(\mathbf{x}', t') \mathrm{d}\mathbf{x}' \mathrm{d}t', \tag{6.70}$$

$$\psi(\mathbf{x}, t) = \psi_{\text{out}}(\mathbf{x}, t) + \int G_A(\mathbf{x} - \mathbf{x}', t - t') f(\mathbf{x}', t') \mathrm{d}\mathbf{x}' \mathrm{d}t', \tag{6.71}$$

where ψ_{in} and ψ_{out} are solutions of the homogeneous wave equation. Since $G_R(\mathbf{x} - \mathbf{x}', t - t')$ vanishes for $t < t'$ and $f(\mathbf{x}', t')$ is supported around $t' = 0$, the solution ψ in (6.70) reduces to ψ_{in} for $t \ll 0$. Then (6.70) represents a scattering process where an incoming wave ψ_{in} travels freely for $t \ll 0$ and it is scattered by the source at $t \sim 0$. On the other hand, (6.71) determines the wave at early times for a given outcoming wave ψ_{out}. We see that G_R and G_A propagate the effect of the source in the future and in the past, respectively.

The explicit form of the Green functions for the wave equation depends on the dimension d. For example,

$$G_R(\mathbf{x}, t) = \begin{cases} \dfrac{1}{2c}\theta(ct - |x|), & d = 1, \\[3mm] \dfrac{\theta(ct - |\mathbf{x}|)}{2\pi c\sqrt{c^2t^2 - |\mathbf{x}|^2}}, & d = 2, \\[3mm] \dfrac{\delta(ct - |\mathbf{x}|)}{4\pi c|\mathbf{x}|}, & d = 3. \end{cases} \tag{6.72}$$

Notice that G_R is supported inside the future light-cone $ct > r$. For $d = 3$ it is actually non-vanishing only on the future light-cone. This is in agreement with the theory of special relativity that states that a point A can receive a light signal from the point B only if it lives on the future light-cone centred in B. The form of the advanced Green function G_A is similar with t replaced by $-t$. The function G_A is supported inside the past light-cone. Equation (6.72) can be derived with the same methods we used for the Laplace equation.

Example 6.8. Consider the wave equation for $d = 3$. By Fourier transforming (6.68) in time

$$G(\mathbf{x}, t) = \int \frac{d\omega}{\sqrt{2\pi}} e^{it\omega} \hat{G}(\mathbf{x}, \omega), \tag{6.73}$$

we find

$$\left(\Delta + \frac{\omega^2}{c^2}\right)\hat{G}(\mathbf{x}, \omega) = -\frac{1}{\sqrt{2\pi}c^2}\delta(\mathbf{x}). \tag{6.74}$$

It is simple to see that $\hat{G}(\mathbf{x}, \omega) = e^{\pm i\omega r/c}/(\sqrt{32\pi^3}c^2 r)$ are solutions of (6.74) (see Exercise 6.10). The fundamental solutions are then

$$G(\mathbf{x}, t) = \frac{1}{8\pi^2 c^2 r}\int d\omega e^{\pm i\omega r/c + i\omega t} = \frac{\delta(t \pm r/c)}{4\pi r c^2} = \frac{\delta(ct \pm r)}{4\pi r c}, \tag{6.75}$$

where we used (A.13) and $\delta(\lambda x) = \frac{\delta(x)}{|\lambda|}$. The minus sign gives the retarded Green function (6.72), while the plus sign gives the advanced one.

6.4.4. *Cauchy–Riemann operator*

Consider the Cauchy–Riemann operator

$$\frac{\partial}{\partial \bar{z}} = \frac{1}{2}\left(\frac{\partial}{\partial x} + i\frac{\partial}{\partial y}\right). \tag{6.76}$$

Recall that a function f is holomorphic if $\partial_{\bar{z}} f = 0$. The fundamental solution for the Cauchy–Riemann operator satisfies

$$\partial_{\bar{z}} G(z, \bar{z}) = \delta(z), \tag{6.77}$$

where $\delta(z) = \delta(x)\delta(y)$, and is given by $G(z, \bar{z}) = 1/(\pi z)$. Notice that G is locally integrable in $\mathbb{R}^2$ and is a well-defined distribution in $\mathcal{D}'(\mathbb{R}^2)$. Notice also that G is holomorphic for $z \neq 0$. Nevertheless its derivative with respect to $\bar{z}$ is non zero in distributional sense. This is due to the singularity in $z = 0$. We can prove that

$$\frac{\partial}{\partial \bar{z}}\frac{1}{\pi z} = \delta(z) \tag{6.78}$$

as follows. Since $1/z$ is holomorphic for $z \neq 0$, the equation is certainly true for $z \neq 0$. Define the regularised function

$$f_\epsilon(z, \bar{z}) = \begin{cases} \dfrac{1}{z}, & |z| \geq \epsilon, \\[2mm] \dfrac{\bar{z}}{\epsilon^2}, & |z| < \epsilon. \end{cases}$$

Since f_ϵ is continuous[12] and locally integrable, it is an element of $\mathcal{D}'(\mathbb{R}^2)$. Moreover one can easily check that, for $\epsilon \to 0$, f_ϵ tends to $1/z$ in $\mathcal{D}'(\mathbb{R}^2)$.[13] Since, for distributions, limit and derivative commute, we have

$$\frac{\partial}{\partial \bar{z}}\frac{1}{\pi z} = \lim_{\epsilon \to 0}\frac{\partial}{\partial \bar{z}}\frac{f_\epsilon(z, \bar{z})}{\pi} = \lim_{\epsilon \to 0}\frac{\theta(\epsilon - |z|)}{\pi \epsilon^2} = \delta(z), \tag{6.79}$$

where the last identity is easily proved by applying it to a test function.

---

[12]On $|z| = \epsilon$, write $z = \epsilon e^{i\phi}$. Then $1/z = \bar{z}/\epsilon^2 = e^{-i\phi}/\epsilon$.

[13]Recall that a sequence of distributions $\{T_n\}$ converges to the distribution T if $\langle T_n, \phi \rangle \to \langle T, \phi \rangle$ (see Petrini *et al.* (2017) and Appendix A for other properties of distributions).

6.5.   Green Functions and Linear Response

A very common problem in physics consists in determining the response of a system, described by some variable $x(t)$, to a time-dependent external force $f(t)$. Linear response theory deals with the case where the external force $f(t)$ can be treated as a small perturbation. The assumption of the theory is that, at leading order in $f(t)$, the change in the system is proportional to the perturbation and is given by

$$x(t) = \int_{-\infty}^{+\infty} \mathrm{d}\tau\, G(t - \tau) f(\tau), \tag{6.80}$$

where $G(t)$ is called *linear response function* and it is often denoted by $\chi(t)$. We already saw an explicit example of $G(t)$ for the problem of the driven harmonic oscillator in Example 2.6.[14] In more general cases, $G(t)$ can be obtained by linearising the equations governing the system around the solution describing the unperturbed system and it can be identified with an appropriate Green function of the linearised system. In writing (6.80), we assumed that the unperturbed system is time independent so that $G(t)$ is a function of only one variable. Linear response theory can be applied to a variety of situations in classical and quantum statistical physics involving out of equilibrium physics and transport phenomena. In this section, we derive general properties of the linear response function $G(t)$ that follow from basic physical principles.

Causality requires that the motion of the system at a time t can only depend on the values of the force $f(t)$ at previous times. This in turn implies that any response function must satisfy

$$G(t) = 0 \qquad \forall t < 0. \tag{6.81}$$

Comparing with Example 6.4 and Section 6.4.3, we see that G is a retarded Green function of the linearised system. Thus, from now on, we will

---

[14]The integration in (2.35) is between 0 and t. The lower limit in (2.35) is due to the fact that the external force in Example 2.6 vanishes for $t < 0$, the upper limit is due to causality as we discuss below.

call it G_R. Since $G_R(t) = 0$ for $t < 0$ we can also write $x(t) = \int_{-\infty}^{t} d\tau G_R(t - \tau) f(\tau)$. It is convenient to work in Fourier space

$$G_R(t) = \frac{1}{\sqrt{2\pi}} \int_{-\infty}^{+\infty} d\omega e^{i\omega t} \hat{G}_R(\omega), \tag{6.82}$$

where ω is interpreted as a frequency. The Fourier transform of (6.80) takes the simple form (see Appendix A)

$$\hat{x}(\omega) = \sqrt{2\pi} \hat{G}_R(\omega) \hat{f}(\omega), \tag{6.83}$$

and shows that the response of the system is local in frequency space, namely the system oscillates at the frequency ω of the perturbation. Since G_R is a real function, its Fourier transform satisfies $\hat{G}_R(-\omega) = \overline{\hat{G}_R(\omega)}$ and therefore the real and imaginary parts have definite parity

$$\operatorname{Re} \hat{G}_R(-\omega) = \operatorname{Re} \hat{G}_R(\omega), \qquad \operatorname{Im} \hat{G}_R(-\omega) = -\operatorname{Im} \hat{G}_R(\omega). \tag{6.84}$$

Causality imposes strong conditions on the structure of $\hat{G}_R(\omega)$. Consider the analytic continuation of $\hat{G}_R(\omega)$ to the complex plane and assume that it vanishes at infinity. Then we can evaluate the Fourier integral (6.82) using the residue theorem. By Jordan's lemma, for $t < 0$, we close the contour in the lower half-plane. Since $G_R(t)$ is zero for $t < 0$, we see that the function $\hat{G}_R(\omega)$ must have no poles in the lower half-plane or, in other words, must be analytic in the lower half-plane. This allows to derive relations between the real and imaginary part of $\hat{G}_R(\omega)$, called *Kramers–Kronig relations*

$$\operatorname{Re} \hat{G}_R(\omega) = -\frac{1}{\pi} P \int_{-\infty}^{+\infty} d\omega' \frac{\operatorname{Im} \hat{G}_R(\omega')}{\omega' - \omega}, \tag{6.85}$$

$$\operatorname{Im} \hat{G}_R(\omega) = \frac{1}{\pi} P \int_{-\infty}^{+\infty} d\omega' \frac{\operatorname{Re} \hat{G}_R(\omega')}{\omega' - \omega}. \tag{6.86}$$

Proof. To prove (6.85) and (6.86) we compute the integral

$$P \int_{-\infty}^{+\infty} d\omega' \frac{\hat{G}_R(\omega')}{\omega' - \omega} \tag{6.87}$$

using the residue theorem. We consider a contour in lower half-plane $\Gamma_R = [-R, \omega - \epsilon] \cup \gamma_\epsilon \cup [\omega + \epsilon, R] \cup \gamma_R$, where γ_R and γ_ϵ are two semi-circles centred in

ω of radius R and ϵ, respectively, and $[-R, -\epsilon]$ and $[\epsilon, R]$ are two segments on the real axis. Since the function $\hat{G}_R(\omega')$ has no singularities in the lower half-plane, the residue theorem gives zero. We now explicitly compute the integral over Γ_R in the limit $R \to \infty$ and $\epsilon \to 0$. Since $\hat{G}_R(\omega)$ goes to zero when $|\omega| \to \infty$, by Jordan's lemma the integral over γ_R vanishes and the rest of the integral gives

$$P \int_{-\infty}^{+\infty} d\omega' \frac{\hat{G}_R(\omega')}{\omega' - \omega} = -\lim_{\epsilon \to 0} \int_{\gamma_\epsilon} d\omega' \frac{\hat{G}_R(\omega')}{\omega' - \omega}$$

$$= \lim_{\epsilon \to 0} \int_0^{-\pi} d\theta i \hat{G}_R(\omega + \epsilon e^{i\theta}) = -i\pi \hat{G}_R(\omega),$$

where, in the last step, we parameterised the circle γ_ϵ as $\omega' = \omega + \epsilon e^{i\theta}$. Taking the real and imaginary parts of the above equation gives (6.85) and (6.86). $\square$

The real and imaginary parts of $\hat{G}_R(\omega)$ have very different physical interpretations. We can write

$$\mathrm{Re}\, \hat{G}_R(\omega) = \frac{1}{2}[\hat{G}_R(\omega) + \overline{\hat{G}_R(\omega)}] = \frac{1}{\sqrt{8\pi}} \int_{-\infty}^{+\infty} dt e^{-i\omega t}(G_R(t) + G_R(-t)),$$

where we used the fact that $G_R(t)$ is real and we changed variable, $t \to -t$, in the second part of the integral. We see that the real part of $\hat{G}_R(\omega)$ only depends on the combination $G_R(t) + G_R(-t)$, which is invariant under time reversal. On the contrary, the imaginary part

$$\mathrm{Im}\, \hat{G}_R(\omega) = -\frac{i}{\sqrt{8\pi}} \int_{-\infty}^{+\infty} dt e^{-i\omega t}(G_R(t) - G_R(-t)) \qquad (6.88)$$

is sensitive to the part of $G_R(t)$ that is not invariant under time reversal and therefore is related to the dissipative properties of the system. For this reason $\mathrm{Im}\, \hat{G}_R(\omega)$ is also called the *dissipative*, or *absorptive*, part of the Green function.

The relation between $\mathrm{Im}\, \hat{G}_R$ and the dissipative properties of the system is even more apparent if we compute the energy absorbed by the system under the action of an external force $f(t)$

$$\frac{dE}{dt} = f(t)\dot{x}(t) = \int_{-\infty}^{+\infty} d\omega \int_{-\infty}^{+\infty} d\omega' \frac{i\omega}{\sqrt{2\pi}} \hat{G}_R(\omega) \hat{f}(\omega) \hat{f}(\omega') e^{i(\omega+\omega')t}, \qquad (6.89)$$

where we substituted the Fourier integrals of f and the time derivative $\dot{x}$, and used (6.83). Integrating the previous equation over time, we find the total energy absorbed by the system

$$
\begin{aligned}
\Delta E &= \frac{1}{\sqrt{2\pi}} \int_{-\infty}^{+\infty} dt \int_{-\infty}^{+\infty} d\omega \int_{-\infty}^{+\infty} d\omega' (i\omega) \hat{G}_R(\omega) \hat{f}(\omega) \hat{f}(\omega') e^{i(\omega+\omega')t} \\
&= \sqrt{2\pi} \int_{-\infty}^{+\infty} d\omega \int_{-\infty}^{+\infty} d\omega' (i\omega) \hat{G}_R(\omega) \hat{f}(\omega) \hat{f}(\omega') \delta(\omega + \omega') \\
&= i\sqrt{2\pi} \int_{-\infty}^{+\infty} d\omega\, \omega |\hat{f}(\omega)|^2 \hat{G}_R(\omega),
\end{aligned}
\tag{6.90}
$$

where we used (A.13) and $\hat{f}(-\omega) = \overline{\hat{f}(\omega)}$, which follows from the reality of f. Using also $|\hat{f}(\omega)|^2 = |\hat{f}(-\omega)|^2$ and (6.84), we see that the absorbed energy is proportional to the dissipative part of the retarded Green function

$$
\Delta E = -2\sqrt{2\pi} \int_{0}^{+\infty} d\omega\, \omega |\hat{f}(\omega)|^2 \operatorname{Im} \hat{G}_R(\omega).
\tag{6.91}
$$

Example 6.9. Consider the driven harmonic oscillator of Example 2.6

$$
\ddot{x}(t) + \eta \dot{x}(t) + \omega_0^2 x(t) = f(t),
\tag{6.92}
$$

where ω_0 is the natural frequency of the system and we set $m = 1$, for simplicity. By taking the Fourier transform of (6.92) and comparing with (6.83), we find the fundamental solution

$$
\hat{G}(\omega) = \frac{1}{\sqrt{2\pi}} \frac{1}{-\omega^2 + i\eta\omega + \omega_0^2},
\tag{6.93}
$$

whose real and imaginary parts are

$$
\operatorname{Re} \hat{G}_R(\omega) = \frac{1}{\sqrt{2\pi}} \frac{\omega_0^2 - \omega^2}{(\omega_0^2 - \omega^2)^2 + \eta^2 \omega^2},
$$

$$
\operatorname{Im} \hat{G}_R(\omega) = -\frac{1}{\sqrt{2\pi}} \frac{\eta\omega}{(\omega_0^2 - \omega^2)^2 + \eta^2 \omega^2}.
\tag{6.94}
$$

Suppose that the system is acted upon by an oscillatory external force $f(t) = f_0 \sin \tilde{\omega} t$. To compute the energy absorbed by the oscillator we need the Fourier transform of f

$$\hat{f}(\omega) = -if_0 \sqrt{\frac{\pi}{2}} [\delta(\omega - \tilde{\omega}) - \delta(\omega + \tilde{\omega})], \qquad (6.95)$$

where we used (A.13). Plugging the above expression in (6.89) we find

$$\frac{dE}{dt} = \sqrt{2\pi} f_0^2 \tilde{\omega} \left[\frac{1}{2} \sin(2\tilde{\omega}t) \operatorname{Re} G_R(\tilde{\omega}) - \sin^2(\tilde{\omega}t) \operatorname{Im} G_R(\tilde{\omega}) \right], \qquad (6.96)$$

which, averaged over a period, gives

$$\Delta E = \frac{\tilde{\omega}}{2\pi} \int_0^{2\pi/\tilde{\omega}} dt \frac{dE}{dt} = -\sqrt{\frac{\pi}{2}} f_0^2 \tilde{\omega} \operatorname{Im} G_R(\tilde{\omega}). \qquad (6.97)$$

In statistical physics and quantum mechanics, the linear response functions can be written in terms of the correlation functions of the theory and relate the dissipation properties of a system to its fluctuations around equilibrium (see, for instance, Landau and Lifshitz, 2013).

6.6.   Green Functions and the Spectral Theorem

From the general theory of linear operators in Hilbert spaces, we know that the spectrum $\sigma(L)$ of a self-adjoint operator is the disjoint union of two parts: the discrete spectrum consisting of the complex numbers λ such that $L - \lambda \mathbb{I}$ is not invertible, and the continuous spectrum consisting of the complex numbers λ such that $(L - \lambda \mathbb{I})^{-1}$ exists but it not continuous. The complement in $\mathbb{C}$ of the spectrum, $\rho(L) = \mathbb{C} - \sigma(L)$, is called the *resolvent set*. By definition, when λ belongs to $\rho(L)$, the operator $L - \lambda \mathbb{I}$ is invertible and continuous. Its inverse $(L - \lambda \mathbb{I})^{-1}$ is called the *resolvent operator* and can be expressed in terms of a Green function $G(x, y; \lambda)$ depending on the parameter λ. As we will see, $G(x, y; \lambda)$ has singularities in correspondence to the elements of the spectrum $\lambda \in \sigma(L)$.

For simplicity, we focus on differential operators in one variable. The generalisation to more variables is straightforward. Consider the equation

$$L_x f - \lambda f = h, \qquad (6.98)$$

where L_x is a self-adjoint differential operator in $L^2[a, b]$ with domain $\mathcal{D}(L_x)$ and λ is a complex parameter. If λ is not an eigenvalue of L_x, the operator

$L_x - \lambda\mathbb{I}$ is invertible. The inverse can written in terms of a Green function as

$$f(x) = (L_x - \lambda\mathbb{I})^{-1}h(x) = \int_a^b dy G(x, y; \lambda)h(y). \qquad (6.99)$$

We are interested in the operators discussed in Section 6.2.2. The Green function $G(x, y; \lambda)$ satisfies

$$(L_x - \lambda)G(x, y; \lambda) = \delta(x - y), \qquad (6.100)$$

and belongs to the domain of L_x as a function of x for all y. Since $L_x - \lambda\mathbb{I}$ is invertible, the homogeneous equation $(L_x - \lambda)u = 0$ has no non-trivial solutions and $G(x, y; \lambda)$ is unique.

We want to write $G(x, y; \lambda)$ in terms of the eigenvectors of L_x. When the spectrum of L_x is discrete, we can write a very simple expression for $G(x, y; \lambda)$:

$$G(x, y; \lambda) = \sum_{n=1}^{\infty} \frac{\overline{u_n(y)}u_n(x)}{\lambda_n - \lambda}, \qquad (6.101)$$

where λ_n and u_n are the eigenvalues and eigenvectors of L_x, $L_x u_n = \lambda_n u_n$.

Proof. By the spectral theorem we can choose the eigenvectors u_n of L to be a basis in $L^2[a, b]$. Since the function $G_{y,\lambda}(x) = G(x, y; \lambda)$ is an element of $L^2[a, b]$ for all y, we can expand it as $G_{y,\lambda}(x) = \sum_n c_n(y, \lambda)u_n(x)$ where $c_n(y, \lambda) = (u_n, G_{y,\lambda})$. Multiplying (6.100) by $\overline{u_n(x)}$ and integrating over x, we obtain

$$\overline{u_n(y)} = (u_n, (L_x - \lambda\mathbb{I})G_{y,\lambda})$$

$$= (L_x u_n, G_{y,\lambda}) - \lambda(u_n, G_{y,\lambda}) = (\lambda_n - \lambda)(u_n, G_{y,\lambda})$$

where we used the fact that L_x is self-adjoint and that its eigenvalues λ_n are real. Then we have $c_n(y, \lambda) = (u_n, G_{y,\lambda}) = \overline{u_n(y)}/(\lambda_n - \lambda)$ and (6.101) follows. $\qquad \square$

By applying the operator $(L_x - \lambda)$ to (6.101) we find

$$(L_x - \lambda)G(x, y; \lambda) = \sum_{n=1}^{\infty} \frac{\overline{u_n(y)}}{\lambda_n - \lambda}(L_x - \lambda)u_n(x) = \sum_{n=1}^{\infty} \overline{u_n(y)}u_n(x), \quad (6.102)$$

since $(L_x - \lambda)u_n = (\lambda_n - \lambda)u_n$. This equation, combined with (6.100), gives

$$\sum_{n=1}^{\infty} \overline{u_n(y)}u_n(x) = \delta(x - y). \tag{6.103}$$

In quantum mechanics books, this equation is often presented as the condition of completeness of the orthonormal system u_n and it is in fact equivalent to it.

Equation (6.101) tells us that $G(x, y; \lambda)$ is a holomorphic function of λ in the resolvent set $\rho(L_x) = \mathbb{C} - \sigma(L_x)$. The function $G(x, y; \lambda)$ is actually holomorphic on $\mathbb{C}$ except for simple poles at the eigenvalues $\lambda = \lambda_n$ of L_x. Knowing the Green function $G(x, y; \lambda)$ we can find the eigenvalues of L_x by looking at the poles of $G(x, y; \lambda)$ and the eigenvectors by evaluating the residue of $G(x, y; \lambda)$ at these poles. Indeed the residue is $-\overline{u_n(y)}u_n(x)$.

Example 6.10. Consider the operator $L_x = -d^2/dx^2$ in $L^2[0, 1]$ with boundary conditions $u(0) = u(1) = 0$. It is easy to check that L_x is self-adjoint with eigenvalues $n^2\pi^2$ and eigenfunctions $u_n = A_n \sin n\pi x$ with $n = 1, 2, \ldots$. Using the method of Section 6.2.2, we can compute the Green function $G(x, y; \lambda)$ of $L_x - \lambda\mathbb{I}$. The functions u and $\tilde{u}$ in (6.29) are the solutions of the homogeneous equation $-f'' - \lambda f = 0$ that satisfy the conditions $u(0) = 0$ and $\tilde{u}(1) = 0$, respectively, and are given by $u(x) = \sin(\sqrt{\lambda}x)$ and $\tilde{u}(x) = \sin(\sqrt{\lambda}(x - 1))$. They are defined up to a multiplicative constant. The Wronskian is constant: $W = (u\tilde{u}' - \tilde{u}u') = \sqrt{\lambda}\sin\sqrt{\lambda}$. The Green function is then given by

$$G(x, y; \lambda) = \begin{cases} -\dfrac{\sin(\sqrt{\lambda}x)\sin(\sqrt{\lambda}(y - 1))}{\sqrt{\lambda}\sin\sqrt{\lambda}}, & 0 < x < y, \\[4mm] -\dfrac{\sin(\sqrt{\lambda}y)\sin(\sqrt{\lambda}(x - 1))}{\sqrt{\lambda}\sin\sqrt{\lambda}}, & y < x < 1. \end{cases} \tag{6.104}$$

At first sight, $G(x, y; \lambda)$ seems to have a branch cut singularity at $\lambda = 0$. However, by expanding the sine in Taylor series around zero, one easily see that $G(x, y; \lambda)$ is regular in $\lambda = 0$. The function $G(x, y; \lambda)$ is then holomorphic except at the eigenvalues $\lambda_n = n^2\pi^2$, with $n \neq 0$, where it has simple poles. The residue at each λ_n is proportional to $\sin n\pi x \sin n\pi y$, in agreement with (6.101). This example illustrates quite explicitly how the singularities in the Green function arise. As a function of λ, $G(x, y; \lambda)$ can have singularities only when the Wronskian vanishes. This happens for those values of λ such that u and $\tilde{u}$ are linearly dependent. For such values we can find a solution of the homogeneous equation that satisfies the boundary conditions simultaneously at both endpoints. Therefore λ is an eigenvalue.

For an arbitrary self-adjoint operator L_x whose spectrum is not only discrete, one can show that $G(x, y; \lambda)$ is still holomorphic in $\rho(L_x) = \mathbb{C} - \sigma(L_x)$. The function $G(x, y; \lambda)$ has poles at the eigenvalues of L_x and cuts along the continuous spectrum of L_x, as the following example shows.

Example 6.11. Consider the self-adjoint operator $L_x = -d^2/dx^2$ in $L^2[0, \infty]$ with boundary condition $u(0) = 0$. It is easy to see that L_x has no proper eigenvectors and a family of generalised eigenvectors $u_p = A \sin px$ for $p \in [0, \infty)$. The spectrum is therefore purely continuous: $\sigma(L_x) = [0, \infty)$. We compute the Green function using again (6.29). Consider $\lambda \notin [0, \infty)$. The two linearly independent solutions of the homogeneous equations $-u'' - \lambda u = 0$ are $e^{\pm i\sqrt{\lambda}x}$. The solution that satisfies the condition $u(0) = 0$ is $u(x) = \sin(\sqrt{\lambda}x)$. For $\tilde{u}$ there are no explicit boundary conditions at infinity, but we need to choose a solution that belongs to the Hilbert space, namely that is square-integrable at infinity. This is $\tilde{u}(x) = e^{i\sqrt{\lambda}x}$. Indeed, since $\lambda \notin [0, \infty)$, we can define $\lambda = \rho e^{i\phi}$ with $\phi \in (0, 2\pi)$. Then $\sqrt{\lambda} = \sqrt{\rho}e^{i\phi/2} = \mu_r + i\mu_i$ with $\mu_i = \operatorname{Im}\sqrt{\lambda} > 0$. Since $|\tilde{u}| = e^{-\mu_i x}$, $\tilde{u}$ is square-integrable at infinity. The Wronskian is $W = (u\tilde{u}' - \tilde{u}u') = -\sqrt{\lambda}$ and the Green function reads

$$G(x, y; \lambda) = \begin{cases} \dfrac{1}{\sqrt{\lambda}} \sin(\sqrt{\lambda}x)e^{i\sqrt{\lambda}y}, & 0 < x < y, \\[2mm] \dfrac{1}{\sqrt{\lambda}} \sin(\sqrt{\lambda}y)e^{i\sqrt{\lambda}x}, & y < x. \end{cases} \tag{6.105}$$

We see that the Green function, as expected, is holomorphic in $\mathbb{C} - [0, \infty)$ and has a branch cut in correspondence of the continuous spectrum of L_x.

6.7. Exercises

Exercise 6.1. Solve (6.23) with the method of variation of arbitrary constants.

Exercise 6.2. Fredholm alternative. Consider a Sturm–Liouville operator (5.13) with discrete spectrum. By expanding f and h in eigenfunctions of L, show that the equation $Lf = h$
(a) has a unique solution if $\ker L = \{0\}$;
(b) for $\ker L \neq \{0\}$, can be solved only if $h \in (\ker L)^{\perp}$ and the solution is not unique.

Exercise 6.3. Find the solution of the boundary problem

$$xf'' + 2f' + xf = 1, \qquad \begin{cases} f(\pi/2) = 0, \\ f(\pi) = 0, \end{cases}$$

using the Green function method. Discuss also the case where the condition in $\pi/2$ is replaced by $f(\pi/2) + \frac{\pi}{2} f'(\pi/2) = 0$.

Exercise 6.4. Find the canonical form of the differential operator

$$L = \frac{1}{1+t^2} \frac{d^2}{dt^2} - \frac{4t}{(1+t^2)^2} \frac{d}{dt} - \frac{4t^4 + 2t^2 + 6}{(1+t^2)^3},$$

defined on the domain $\mathcal{D}(L) = \{y_{\mathrm{ac}}, y', y'' \in L^2(\mathbb{R}) | y(-\infty) = y(+\infty) = 0\}$ and compute its Green function.

Exercise 6.5. Solve the Cauchy problem

$$y'(x) + 5x^4 y(x) = x^4, \quad y(0) = 0,$$

using both the variation of arbitrary constants and the Green function method.

Exercise 6.6. Solve the boundary problem

$$x^2 f'' - 3xf' + 3f = 2x^2, \qquad \begin{cases} f(1) = 0, \\ f'(1) = 0, \end{cases}$$

using the Green function method.

Exercise 6.7. Show that $T = P\frac{1}{x} + c\delta(x)$, where c is an arbitrary constant, is the general solution of $xT = 1$ in the space of distributions $\mathcal{D}'(\mathbb{R})$.

* Exercise 6.8. Let $G_L(x)$ be the fundamental solution of the Laplace equation (6.52). Show that the Green function (6.64) for the Dirichlet problem
(a) in half-space $\Omega = \{\mathbf{x} = (x_1, \ldots, x_d), x_d \geq 0\}$ is $G(\mathbf{y}, \mathbf{x}) = G_L(\mathbf{y} - \mathbf{x}) - G_L(\mathbf{y} - \mathbf{x}')$ where $\mathbf{x}' = (x_1, \ldots, -x_d)$;
(b) in the sphere $\Omega = \{\mathbf{x} = (x_1, \ldots, x_n), \sum_{i=1}^{d} x_i^2 \leq 1\}$ is $G(\mathbf{y}, \mathbf{x}) = G_L(\mathbf{y} - \mathbf{x}) - G_L(|\mathbf{x}|(\mathbf{y} - \mathbf{x}'))$ where $\mathbf{x}' = \mathbf{x}/|\mathbf{x}|^2$.

Exercise 6.9. *Duhamel's principle.* Let A be a linear operator on the Hilbert space $\mathcal{H}$. Define the evolution operator $U(t)$ in $\mathcal{H}$ as the solution of the equation $U'(t) + AU(t) = 0$ with initial condition $U(0) = \mathbb{I}$.
(a) Show that the homogeneous Cauchy problem $\mathbf{u}'(t) + A\mathbf{u}(t) = 0$ in $\mathcal{H}$ with initial condition $\mathbf{u}(0) = \mathbf{u}_0$ has solution $\mathbf{u}(t) = U(t)\mathbf{u}_0$.
(b) Show that the inhomogeneous differential equation $\mathbf{u}'(t) + A\mathbf{u}(t) = \mathbf{f}(t)$ in $\mathcal{H}$ with initial condition $\mathbf{u}(0) = \mathbf{u}_0$ has solution $\mathbf{u}(t) = \int_0^t U(t - \tau)\mathbf{f}(\tau)d\tau + U(t)\mathbf{u}_0$.
(c) Show that the inhomogeneous heat equation $\partial_t u(t, \mathbf{x}) - \kappa \Delta u(t, \mathbf{x}) = f(t, \mathbf{x})$ with initial condition $u(0, \mathbf{x}) = u_0(x)$ has solution

$$u(t, \mathbf{x}) = \int_0^t \int_{\mathbb{R}^d} h(t - \tau, \mathbf{x} - \mathbf{y})f(\tau, \mathbf{y})d\tau d\mathbf{y} + \int_{\mathbb{R}^d} h(t, \mathbf{x} - \mathbf{y})u_0(\mathbf{y})d\mathbf{y},$$

where $h(t, \mathbf{x})$ is the heat kernel.
(d) Show that (6.67) is a fundamental solution of the heat equation.

Exercise 6.10. Prove that $G_\pm(r) = -e^{\pm imr}/(4\pi r)$ are fundamental solutions of the *Helmholtz operator* $\Delta + m^2$ in three dimensions.

7

Power Series Methods

There are no general methods to find explicit solutions of ordinary differential equations with non-constant coefficients. However, we saw in Example 5.4 that a way to determine the solutions of the Hermite equation is to represent them as power series. In this chapter we will discuss in a more general way how to use power series to determine the local behaviour of the solution of an ordinary differential equation around a given point. As in the previous chapters, we will mostly focus on second-order equations, since most of the equations encountered in physics are of this type. The results can be extended to higher-order equations.

7.1. Ordinary and Singular Points of an Ordinary Differential Equation

Consider a homogeneous linear differential equation of order n in the complex variable z

$$\frac{d^n y(z)}{dz^n} + \sum_{k=0}^{n-1} a_k(z) \frac{d^k y(z)}{dz^k} = 0, \tag{7.1}$$

where all the functions $a_k(z)$ are defined in a neighbourhood U of a point z_0. We have the following classification depending on the behaviour of the coefficients $a_k(z)$ in z_0.

- The point z_0 is said to be an *ordinary point* for (7.1) if all the coefficients $a_k(z)$ are holomorphic in z_0.

- The point z_0 is said to be a *regular singular* or *Fuchsian point*[1] for (7.1) if, for any given k, the function $a_k(z)$ has a pole of order $n - k$ at most.
- The point z_0 is said to be an *irregular* or *essential* singularity for (7.1) in all other cases.

In the discussion above we assumed that $z_0 \in \mathbb{C}$. The same analysis can be carried out for $z = \infty$. Making the change of variable $w = 1/z$, we say that $z = \infty$ is an ordinary, Fuchsian or irregular point for (7.1) if $w = 0$ is an ordinary, Fuchsian or irregular point for the transformed equation, respectively.

Example 7.1. The Legendre equation of Example 5.5

$$(1 - z^2)y'' - 2zy' + \lambda y = 0 \tag{7.2}$$

has three Fuchsian points at $z = \pm 1$ and $z = \infty$. Indeed, by writing

$$y'' - \frac{2z}{1 - z^2}y' + \frac{\lambda}{1 - z^2}y = 0 \tag{7.3}$$

we see immediately that both functions $a_0(z)$ and $a_1(z)$ have simple poles in $z = \pm 1$. To study the behaviour in $z = \infty$, we use the variable $w = 1/z$. The equation becomes[2]

$$\frac{d^2y}{dw^2} + \frac{2w}{w^2 - 1}\frac{dy}{dw} + \frac{\lambda}{w^2(w^2 - 1)}y = 0, \tag{7.5}$$

from which we see that $w = 0$ is a double pole in the coefficient of the y term.

Example 7.2. The equation for the harmonic oscillator

$$y''(z) + \omega^2 y(z) = 0 \tag{7.6}$$

has only an essential singularity in $z = \infty$.

---

[1]Even if standard, we find the name regular singular point a bit misleading. For this reason, in the rest of the book we will rather use Fuchsian for this kind of singularities.
[2]Under the change of coordinate $z = 1/w$

$$\frac{dy(z)}{dz} = -w^2\frac{dy}{dw}, \quad \frac{d^2y(z)}{dz^2} = w^4\frac{d^2y}{dw^2} + 2w^3\frac{dy}{dw}. \tag{7.4}$$

As we will discuss below, it is always possible to find the general solution of equation (7.1) as a convergent series expansion around a point z_0 if this is an ordinary or Fuchsian point. For essential singularities the situation is more complicated. It is still possible to write the solution around the singular point as a series expansion, but this makes sense mostly as an asymptotic series.

7.2. Series Solutions for Second-Order Linear Ordinary Differential Equations

In the rest of this chapter we will show how to find series solutions for second-order ordinary homogeneous linear differential equations. These can always be set in normal form

$$y''(z) + p(z)y' + q(z)y(z) = 0. \tag{7.7}$$

7.2.1. Solutions around ordinary points

We want to determine the solution of equation (7.7) in a neighbourhood $U(z_0)$ of an ordinary point z_0. In this case $p(z)$ and $q(z)$ are holomorphic in a region Ω containing $U(z_0)$. This implies that the general solution of (7.7) around z_0 is always given by a Taylor series expansion

$$y(x) = y_1(x) + y_2(x) = \sum_{n=0}^{\infty} c_n^1 (z - z_0)^n + \sum_{n=0}^{\infty} c_n^2 (z - z_0)^n, \tag{7.8}$$

where $y_1(x)$ and $y_2(x)$ are two independent power series that converge in the neighbourhood $U(z_0)$.

To see this, we take, for simplicity, $z_0 = 0$ and we use the holomorphicity of $p(z)$ and $q(z)$ to express them as power series around $z = 0$

$$p(z) = \sum_{n=0}^{\infty} p_n z^n, \quad q(z) = \sum_{n=0}^{\infty} q_n z^n. \tag{7.9}$$

It is then natural to look for a solution that also has a power series expansion

$$y(z) = \sum_{n=0}^{\infty} c_n z^n. \tag{7.10}$$

Plugging (7.9) and (7.10) in (7.7) one finds

$$\sum_{n=2}^{\infty} n(n-1)c_n z^{n-2}$$

$$+ \left(\sum_{k=0}^{\infty} p_k z^k\right)\left(\sum_{n=1}^{\infty} nc_n z^{n-1}\right) + \left(\sum_{k=0}^{\infty} q_k z^k\right)\left(\sum_{n=0}^{\infty} c_n z^n\right) = 0.$$

Collecting all the terms with the same power z^n gives an infinite set of equations[3]

$$(n+2)(n+1)c_{n+2} + \sum_{r+k-1=n} kc_k p_r + \sum_{k+r=n} c_k q_r = 0, \qquad (7.12)$$

which can be solved recursively for c_{n+2}. In particular the equation for $n=0$ gives $c_2 = -(c_1 p_0 + c_0 q_0)/2$ in terms of the two arbitrary parameters c_1 and c_0. Notice that all c_n are linear in c_0 and c_1. Setting $c_1 = 0$ in the ansatz (7.10) we find the first independent solutions $y_1(z)$ in (7.8). Similarly $y_2(z)$ is obtained with $c_0 = 0$.

One can show that the two series converge on $U(z_0)$ and that, using analytic continuation, they can be extended to the whole domain Ω of holomorphicity of $p(z)$ and $q(z)$.

Example 7.3. The Hermite equation

$$y'' - 2zy' + \lambda y = 0 \qquad (7.13)$$

has no singular points on $\mathbb{C}$ and an essential singularity in $z = \infty$. As shown in Example 5.4 the power series around $z = 0$ is given by

$$y(z) = y_1(z) + y_2(z) = \sum_{k=0}^{\infty} c_{2k} z^{2k} + \sum_{k=0}^{\infty} c_{2k+1} z^{2k+1}, \qquad (7.14)$$

where the coefficients c_k are determined by the recursion relation (5.25)

$$c_{2k} = \frac{1}{(2k)!} \prod_{r=0}^{k-1} (4r - \lambda)c_0, \qquad c_{2k+1} = \frac{1}{(2k+1)!} \prod_{r=0}^{k-1} [2(2r+1) - \lambda]c_1.$$

---

[3]To obtain the expansions below we use the Cauchy product for series. Given two power series $\sum_k a_k z^k$ and $\sum_k b_k z^k$, their product is

$$\left(\sum_{i=0}^{\infty} a_i z^i\right)\left(\sum_{j=0}^{\infty} b_j z^j\right) = \sum_{k=0}^{\infty} c_k z^k, \qquad c_k = \sum_{r=0}^{k} a_r b_{k-r}. \qquad (7.11)$$

The two solutions $y_1(x)$ and $y_2(x)$ contain even and odd powers of z. As we saw in Example 5.4, when $\lambda = 2n$, one of the two series, y_1 or y_2 depending on whether n is even or odd, truncates and gives the Hermite polynomials.

7.2.2. Solutions around fuchsian points

When z_0 is a Fuchsian singular point, (7.7) can be written as

$$y''(z) + \frac{A(z)}{z - z_0} y'(z) + \frac{B(z)}{(z - z_0)^2} y(z) = 0, \tag{7.15}$$

where $A(z)$ and $B(z)$ are holomorphic in z_0 and, therefore, can be expanded in a power series around it

$$A(z) = \sum_{n=0}^{\infty} a_n (z - z_0)^n, \quad B(z) = \sum_{n=0}^{\infty} b_n (z - z_0)^n. \tag{7.16}$$

In this case, it is not possible to find solutions of the type (7.10). An ansatz of this kind would generically give the trivial solution with all coefficients c_k equal to zero. In order to have a non-trivial solution one needs an ansatz of the form

$$y(z) = (z - z_0)^\alpha \sum_{n=0}^{\infty} c_n (z - z_0)^n, \tag{7.17}$$

with $c_0 \neq 0$, where the extra term in front of the power series compensates the negative powers in (7.15).[4] Series of the form (7.17) are called *Frobenius series* and this way of solving the differential equation is called *Frobenius method*. Plugging (7.17) and (7.16) into (7.15), and collecting terms with the same power of $z - z_0$, we obtain a set of equations for the coefficients c_n

$$[(\alpha + n)(\alpha + n - 1) + (\alpha + n)a_0 + b_0]c_n + \sum_{r=0}^{n-1} [(\alpha + r)a_{n-r} + b_{n-r}]c_r = 0. \tag{7.18}$$

For $n = 0$, (7.18) gives the *indicial equation*

$$\alpha(\alpha - 1) + \alpha a_0 + b_0 = 0, \tag{7.19}$$

---

[4]The prefactor $(z-z_0)^\alpha$ in the ansatz gives the leading behaviour of the solution around z_0 and can be motivated as follows. We approximate (7.15) near $z = z_0$ by keeping only the most singular terms and we obtain the Euler equation $(z-z_0)^2 y'' + a_0(z-z_0)y' + b_0 y = 0$, whose solution is of the form $(z - z_0)^\alpha$ where α satisfies $\alpha(\alpha - 1) + \alpha a_0 + b_0 = 0$ (see Example 4.5).

which determines two possible values, α_1 and α_2, of the *indicial exponent* α. For each choice α_1 and α_2 one has to solve (7.18) to determine the coefficients c_n. There are two possible cases, depending on the relation between α_1 and α_2. Let us call α_1 the largest of the two roots.[5] Then we have the following cases.

- For $\alpha_1 - \alpha_2 \neq \mathbb{N}$, the two solutions are of the form

$$y_i(z) = (z - z_0)^{\alpha_i} \sum_{n=0}^{\infty} c_n^i (z - z_0)^n, \qquad i = 1, 2. \qquad (7.20)$$

- For $\alpha_1 - \alpha_2 = p \in \mathbb{N}$, the two solutions are of the form

$$y_1(z) = (z - z_0)^{\alpha_1} \sum_{n=0}^{\infty} c_n (z - z_0)^n, \qquad (7.21)$$

$$y_2(z) = C\, y_1(z) \ln(z - z_0) + (z - z_0)^{\alpha_2} \sum_{n=0}^{\infty} d_n (z - z_0)^n, \qquad (7.22)$$

with C constant. In some particular cases $C = 0$ and the solutions are still of the form (7.20).[6]

Proof. For $\alpha_1 - \alpha_2 \neq \mathbb{N}$ the recurrence relation (7.18) can be solved for both α_1 and α_2, and the corresponding solutions are (7.20). For $\alpha_1 - \alpha_2 = p \in \mathbb{N}$ the recurrence relations (7.18) can be solved for α_1 and the corresponding solution is (7.21). However, it is easy to show that for $\alpha = \alpha_2$ the coefficient of c_n for $n = p$ in (7.18) vanishes. In some particular cases also the other terms in (7.18) vanish for $n = p$. Then the coefficient c_p is an arbitrary constant that can be used in (7.18) to determine c_{p+1}, c_{p+2}, etc. The resulting solution is still of the form (7.20). In all other cases the equation has no solution and we need another method to determine the second independent solution. A way to derive the expression for $y_2(z)$ is to consider the Wronskian $W = y_1 y_2' - y_1' y_2$. As discussed in Section 4.2 we find the integral expression (4.33)

$$y_2(z) = y_1(z) \int_a^z du \frac{W(u)}{y_1^2(u)} = C y_1(z) \int_a^z du \frac{e^{-\int_a^u dv\, p(v)}}{y_1^2(u)}. \qquad (7.23)$$

Using (7.16) and (7.21) we can write $p(z) = a_0/(z - z_0) + g(z)$ and $y_1(z) = (z - z_0)^{\alpha_1} h(z)$, where $g(z)$ and $h(z)$ are holomorphic in the neighbourhood of z_0

---

[5]When the roots are complex we choose α_1 to be the one with largest real part.
[6]See, for instance, Example 7.9 and Exercises 7.3 and 7.4.

and $h(z_0) \neq 0$. We then have

$$y_2(z) = y_1(z) \int_a^z du \frac{\tilde{C} e^{-\int_a^u dv g(v)}}{(u - z_0)^{a_0 + 2\alpha_1} h(u)^2}. \qquad (7.24)$$

We now use the Taylor expansion around z_0, $\tilde{C} e^{-\int_a^u dv g(v)} / h(u)^2 = \sum_{k=0} \tilde{d}_k (u - z_0)^k$. Using repeatedly $\alpha_1 - \alpha_2 = p$ and the fact that, from (7.19), $a_0 - 1 = -\alpha_1 - \alpha_2$,[7] we obtain

$$y_2(z) = y_1(z) \left[\sum_{k=0}^{\infty} \tilde{d}_k \int_a^z du (u - z_0)^{k-1-\alpha_1+\alpha_2} \right]$$

$$= y_1(z) \left[\tilde{d}_p \ln(z - z_0) + \sum_{k \neq p} \frac{\tilde{d}_k}{k - p} (z - z_0)^{k-p} + const \right]$$

$$= \tilde{d}_p \ln(z - z_0) y_1(z) + (z - z_0)^{\alpha_2} \sum_{k=0}^{\infty} d_k (u - z_0)^k, \qquad (7.25)$$

where the coefficients d_k can be obtained multiplying term by term the two series $\sum_{n=0}^{\infty} c_n (z - z_0)^n$ and $\sum_{k \neq p} \frac{\tilde{d}_k}{k-p} (z - z_0)^k + const(z - z_0)^p$. $\qquad \square$

The series (7.20)–(7.22) have a non-zero radius of convergence given by the distance between z_0 and the nearest other singularity of the equation. Differently from the case of an ordinary point, in general, at least one of the solutions is multivalued. Indeed (7.22) contains a logarithmic term, while in (7.20) at least one of the α_i is not an integer.

Example 7.4 (The hypergeometric equation). The *hypergeometric equation*

$$z(1 - z)y'' + [c - (a + b + 1)z]y' - aby = 0, \qquad (7.26)$$

where a, b and c are complex parameters, has three Fuchsian points in $z = 0, 1$ and $z = \infty$. One can show that the indicial exponents associated with the points $z = 0, 1$ and $z = \infty$ are

$$z = 0: \quad \alpha_1^{(0)} = 0, \quad \alpha_2^{(0)} = 1 - c, \qquad (7.27)$$

---

[7] For an algebraic equation $d_2 x^2 + d_1 x + d_0 = d_2(x - x_1)(x - x_2) = 0$ the two roots x_1 and x_2 satisfy $x_1 + x_2 = -d_1/d_2$ and $x_1 x_2 = d_0/d_2$.

$$z = 1: \quad \alpha_1^{(1)} = 0, \quad \alpha_2^{(1)} = c - a - b, \tag{7.28}$$

$$z = \infty: \quad \alpha_1^{(\infty)} = a, \quad \alpha_2^{(\infty)} = b. \tag{7.29}$$

Let us consider first the point $z = 0$. It is easy to verify that, in a neighbourhood of $z = 0$, the hypergeometric equation can be put in the form (7.15) with $a_0 = c$, $b_0 = 0$, $a_k = c - a - b - 1$ and $b_k = -ab$ for $k \geq 1$. The indicial equation is then $\alpha(\alpha - 1 + c) = 0$ and it has solutions $\alpha_1^{(0)} = 0$ and $\alpha_2^{(0)} = 1 - c$. Consider first $\alpha_1^{(0)} = 0$. Subtracting the recurrence relations (7.18) for n and $n - 1$ we obtain

$$n(n - 1 + c)c_n - (n - 1 + a)(n - 1 + b)c_{n-1} = 0, \tag{7.30}$$

which is solved by

$$c_n = \frac{c_0 \, \prod_{j=0}^{n-1}(a + j) \prod_{k=0}^{n-1}(b + k)}{n! \, \prod_{j=0}^{n-1}(c + j)} = \frac{c_0}{n!} \frac{\Gamma(c)}{\Gamma(a)\Gamma(b)} \frac{\Gamma(a + n)\Gamma(b + n)}{\Gamma(c + n)},$$

where[8] (see Appendix A)

$$\prod_{j=0}^{n-1}(a + j) = \frac{\Gamma(a + n)}{\Gamma(a)}. \tag{7.31}$$

Then the first independent solution is given in terms of the *hypergeometric series*[9]

$$y_1^{(0)}(z) = F(a, b, c; z) = \frac{\Gamma(c)}{\Gamma(a)\Gamma(b)} \sum_{n=0}^{\infty} \frac{\Gamma(a + n)\Gamma(b + n)}{\Gamma(c + n)} \frac{z^n}{n!} \tag{7.32}$$

$$= 1 + \sum_{n=1}^{\infty} \frac{a(a + 1)\ldots(a + n - 1)b(b + 1)\ldots(b + n - 1)}{c(c + 1)\ldots(c + n - 1)} \frac{z^n}{n!},$$

where we set $c_0 = 1$. The function $F(a, b, c; z)$ is symmetric in a and b. The series (7.32) is not defined for $c = -m$ with $m \in \mathbb{N}$.[10] It converges for

---

[8]The expression $(a)_n = \Gamma(a + n)/\Gamma(a)$ is also called *Pochhammer symbol*.

[9]Another commonly used way to denote the hypergeometric series is $_2F_1(a, b, c; z)$.

[10]For $c = -m$ negative integer, a solution of the equation is obtained by setting $c_0 = 1/\Gamma(c)$ and taking the limit $c \to -m$.

$|z| < 1$ for any value of a, b and $c \neq -m$, as one can verify by taking the limit

$$R = \lim_{n \to \infty} \left| \frac{a_n}{a_{n+1}} \right| = \lim_{n \to \infty} \left| \frac{c+n}{(a+n)(b+n)}(n+1) \right| = 1, \qquad (7.33)$$

with $a_n = \frac{a(a+1)...(a+n-1)b(b+1)...(b+n-1)}{c(c+1)...(c+n-1)n!}$.

The recurrence relation associated with the second root $\alpha_2^{(0)}$ is derived in a similar way

$$(1 - c + n)nc_n - (n - c + a)(n - c + b)c_{n-1} = 0. \qquad (7.34)$$

Notice that this relation can also be obtained from the equation for $\alpha_1^{(0)}$ by the replacement $a \to a - c + 1$, $b \to b - c + 1$ and $c \to 2 - c$. Hence the second solution is given by

$$y_2^{(0)}(z) = z^{1-c}F(a - c + 1, b - c + 1, 2 - c; z). \qquad (7.35)$$

The full solution around $z = 0$, when $\alpha_1^{(0)} - \alpha_2^{(0)} = c - 1 \neq \mathbb{N}$, is

$$y(z) = AF(a, b, c; z) + Bz^{1-c}F(a - c + 1, b - c + 1, 2 - c; z). \qquad (7.36)$$

When $c - 1 \in \mathbb{N}$, we still have the solution y_1, while y_2 can be obtained as in (7.22).

The same analysis holds for the other singular points. For $\alpha_1^{(i)} - \alpha_2^{(i)} \neq \mathbb{N}$, in a neighbourhood of $z = 1$, we find

$$y_1^{(1)}(z) = F(a, b, 1 + a + b - c; 1 - z),$$

$$y_2^{(1)}(z) = (1 - z)^{c-a-b}F(c - a, c - b, 1 + c - a - b; 1 - z), \qquad (7.37)$$

while, in a neighbourhood of $z = \infty$,

$$y_1^{(\infty)}(z) = \left(\frac{1}{z} \right)^a F\left(a, 1 + a - c, 1 + a - b; \frac{1}{z} \right),$$

$$y_2^{(\infty)}(z) = \left(\frac{1}{z} \right)^b F\left(b, 1 + b - c, 1 + b - a; \frac{1}{z} \right). \qquad (7.38)$$

The series (7.37) and (7.38) converge for $|z-1| < 1$ and $|z| > 1$, respectively.

The solutions (7.36), (7.37) and (7.38) are not linearly independent, the space of solutions of the hypergeometric equation (7.26) being two-dimensional. Indeed, each pair of solutions can be expressed as a linear combination of any other two on the intersection of their domains of definition. For instance one can show that, in the intersection of the circles $|z| < 1$ and $|z - 1| < 1$,

$$y_1^{(0)}(z) = \frac{\Gamma(c-a-b)\Gamma(c)}{\Gamma(c-a)\Gamma(c-b)} y_1^{(1)}(z) + \frac{\Gamma(a+b-c)\Gamma(c)}{\Gamma(a)\Gamma(b)} y_2^{(1)}(z), \qquad (7.39)$$

$$y_2^{(0)}(z) = \frac{\Gamma(c-a-b)\Gamma(2-c)}{\Gamma(1-a)\Gamma(1-b)} y_1^{(1)}(z) + \frac{\Gamma(a+b-c)\Gamma(2-c)}{\Gamma(a+1-c)\Gamma(b+1-c)} y_2^{(1)}(z).$$

Similar relations connect $y_i^{(1)}(z)$ and $y_i^{(\infty)}(z)$.

The above relations among the different solutions can be used to analytically continue $y_1^{(0)}(z)$ to a function defined also outside the unit disk. This is called *hypergeometric function* and we will still denote it by $F(a, b, c; z)$. Since $y_2^{(1)}(z)$ has a branch point singularity in $z = 1$ due to the prefactor $(1 - z)^{c-a-b}$, (7.39) implies, in general, that $F(a, b, c; z)$ has a singularity in $z = 1$. Indeed, one can see that, for generic values of a, b and c, $F(a, b, c; z)$ is holomorphic on $\mathbb{C} - [1, \infty)$ and has a branch cut on $[1, +\infty)$.

For specific values of the parameters a, b, c, the series (7.32) truncates and the hypergeometric function reduces to a polynomial. This happens, for instance, when $a = -p$ or $b = -p'$ with $p, p' \in \mathbb{N}$. Most of the commonly used orthogonal polynomials arise in this way. For example, the Legendre polynomials can be obtained as $P_l(x) = F(l+1, -l, 1; (1-x)/2)$ (see Exercise 7.5). For other values of the parameters, the hypergeometric series reduces to elementary functions (see Exercise 7.6). For example, for $c = b$, using (A.17), we find

$$F(a, b, b; z) = \sum_{n=0}^{\infty} \frac{\Gamma(a+n)}{\Gamma(a)} \frac{z^n}{n!} = (1-z)^{-a}.$$

Example 7.5 (Equations with three Fuchsian points). The hypergeometric equation is an example of a larger class of equations that are characterised by having as only singularities three Fuchsian points, z_1, z_2 and z_3. The general form of the equation is completely determined by the requirement of having only Fuchsian singularities. When the three singular points are at finite values, the equation is

$$y'' + \frac{2z^2 + a_1 z + a_0}{(z - z_1)(z - z_2)(z - z_3)} y' + \frac{b_2 z^2 + b_1 z + b_0}{(z - z_1)^2 (z - z_2)^2 (z - z_3)^2} y = 0,$$

(7.40)

while, when $z_3 = \infty$, it reduces to

$$y'' + \frac{a_1 z + a_0}{(z - z_1)(z - z_2)} y' + \frac{b_2 z^2 + b_1 z + b_0}{(z - z_1)^2 (z - z_2)^2} y = 0. \tag{7.41}$$

An interesting property of (7.40) and (7.41), is that, performing the linear fractional transformations

$$z \to \frac{(z - z_1)(z_2 - z_3)}{(z - z_3)(z_2 - z_1)}, \tag{7.42}$$

$$z \to \frac{z - z_1}{z_2 - z_1}, \tag{7.43}$$

respectively, the singular points z_1, z_2 and z_3 move to 0, 1 and ∞ while remaining Fuchsian, as one can see by a direct computation. In the new variables the equation is again of the form (7.41) with $z_1 = 0$, $z_2 = 1$ and new coefficients a_i and b_i. Moreover, the redefinition

$$\tilde{y}(z) = z^p (1 - z)^\sigma y(z) \tag{7.44}$$

can be used to change the coefficients a_i and b_i without changing the form of the equation. In this way, it is always possible to transform any equation with three Fuchsian points into the hypergeometric one (7.26).

Example 7.6. The Legendre equation

$$y'' - \frac{2z}{1 - z^2} y' + \frac{\nu(\nu + 1)}{1 - z^2} y = 0 \tag{7.45}$$

has three Fuchsian points in $z = \pm 1$ and $z = \infty$. With the transformation

$$w = -\frac{1}{2}(z - 1) \tag{7.46}$$

the singular points $z = -1$, $z = 1$ and $z = \infty$ are mapped into $w = 1$, $w = 0$ and $w = \infty$, respectively, and (7.45) becomes the hypergeometric equation (7.26) with $a = -\nu$, $b = \nu + 1$ and $c = 1$

$$y'' + \frac{1 - 2w}{w(1 - w)}y' + \frac{\nu(\nu + 1)}{w(1 - w)}y = 0. \tag{7.47}$$

We can use this information to study the regularity of the solutions. We stated in Example 5.5 that only for $\nu = l \in \mathbb{N}$ we can have bounded solutions in the closed interval $[-1, 1]$. Now we can see why. The change of variable $w = -\frac{1}{2}(z - 1)$ maps the interval $[-1, 1]$ into the interval $[0, 1]$. As in $w = 0$ the indicial exponents are equal, $\alpha_1^{(0)} = \alpha_2^{(0)} = 0$, one of the linearly independent solutions is given by (7.32), $F(-\nu, \nu + 1, 1, w)$, and is regular, while the other has a logarithmic singularity. The same happens near $w = 1$. We can impose that the solution is regular in $w = 0$ by choosing $y(z) = F(-\nu, \nu + 1, 1, w)$. However, as discussed in Example 7.4, we know that for generic values of the parameters the hypergeometric function is singular in $w = 1$. In this particular case, since $\alpha_1^{(1)} = \alpha_2^{(1)} = 0$, $F(-\nu, \nu + 1, 1, w)$ behaves like $\ln(1 - w)$ near $w = 1$ and is not bounded. Thus, for generic ν, it is impossible to find a solution that is regular in both points $w = 0$ and $w = 1$. The usual way out is to remember that for negative integer values of the parameters a and b the hypergeometric function truncates to a polynomial, which is obviously bounded. This happens when $\nu = l = 0, 1, 2 \ldots$ and the resulting solutions are the Legendre polynomials $P_l(x) = F(l + 1, -l, 1; (1 - x)/2)$ (see Exercise 7.5).

Example 7.7 (Confluent hypergeometric equation). The confluent hypergeometric equation is a limit of the equation with three Fuchsian singularities where two of them merge. Due to the merging, an essential singularity appears. We can see this using the hypergeometric equation (7.26) and performing an appropriate rescaling. As discussed in Example 7.4, the three Fuchsian points of the hypergeometric equation are $z = (0, 1, \infty)$. By rescaling the variable z one can map the singularity $z = 1$ to the point $z = b$, thus obtaining an equation with three Fuchsian points $(0, b, \infty)$. Letting $b \to \infty$, two of the singularities coalesce in $z = \infty$. The resulting equation (see Exercise 7.7)

$$zy'' + (c - z)y' - ay = 0, \qquad a, c \in \mathbb{C}, \tag{7.48}$$

is the *confluent hypergeometric* equation. It is easy to check that (7.48) has two singularities, a Fuchsian one in $z = 0$ and an essential one in $z = \infty$. Let us consider the solutions around the Fuchsian singularity $z = 0$. We will discuss the behaviour near $z = \infty$ in Section 7.2.3. The indicial exponents are $\alpha_1^{(0)} = 0$ and $\alpha_2^{(0)} = 1 - c$, and the two corresponding solutions are obtained plugging the ansatz (7.17) in (7.48). We find[11] (see Exercise 7.7)

$$y_1^{(0)}(z) = \Phi(a, c; z) = \frac{\Gamma(c)}{\Gamma(a)} \sum_{n=0}^{\infty} \frac{\Gamma(a + n)}{\Gamma(c + n)} \frac{z^n}{n!}, \qquad (7.49)$$

$$y_2^{(0)}(z) = z^{1-c}\Phi(a - c + 1, 2 - c, z), \qquad (7.50)$$

where $\Phi(a, c; z)$ is called *confluent hypergeometric function*.[12] Notice that $\Phi(a, c; z)$ is an entire function, as it can be seen from its radius of convergence

$$R = \frac{1}{\lim_{n \to \infty} |c_n|^{1/n}} \sim \lim_{n \to \infty} (n!)^{1/n} = \infty. \qquad (7.51)$$

As for the hypergeometric function, $\Phi(a, c; z)$ reduces to an elementary function or a polynomial for specific values of the parameters a and c. In particular, for $a = -n$ with $n \in \mathbb{N}$ and $c = 1$ we find the Laguerre polynomials, $L_n(x) = \Phi(-n, 1; x)$. For $c = m + 1$ we obtain the *associated Laguerre polynomials* $L_n^m(x) = \frac{(n+m)!}{n!m!}\Phi(-n, m + 1; x)$, which enter, for example, in the solution of the Schrödinger equation for the hydrogen atom (see Example 9.9).

Example 7.8 (Integral representation of hypergeometric functions). The hypergeometric and confluent hypergeometric functions admit various integral representations. Here we focus on the Euler's integral representation. Consider, for simplicity, the confluent hypergeometric function (see Exercise 7.8 for the case of the hypergeometric). We look for

---

[11] These are the solutions for non-integer c. When $c \in \mathbb{N}$ the solutions are of the form (7.21) and (7.22).
[12] The confluent hypergeometric is also called *Kummer function* and is sometimes denoted by $M(a, c, z)$. Another common notation is $_1F_1(a, c; z)$.

solutions of (7.48) of the form

$$y(z) = \int_\gamma dt \, e^{-zt} U(t), \tag{7.52}$$

where γ is a curve in the complex plane to be determined. Substituting (7.52) into the confluent hypergeometric equation (7.48), we obtain

$$\int_\gamma dt \, e^{-zt} (zt^2 U(t) - (c - z)tU(t) - aU(t)) = 0. \tag{7.53}$$

After integrating by parts the terms in z, we find

$$e^{-zt}(t^2 + t)U(t)|_{\partial\gamma} = \int_\gamma dt \, e^{-zt}((t^2 + t)U'(t) + (2t - ct + 1 - a)U(t)),$$

where $\partial\gamma$ is the boundary of γ. A solution is obtained by separately setting to zero the boundary terms and the integrand. The vanishing of the integrand gives a first-order differential equation for $U(t)$ with solution

$$U(t) = Ct^{a-1}(1 + t)^{c-a-1}, \tag{7.54}$$

where C is a constant. With U given in (7.54) the boundary terms vanish for various choices of the curve γ, and each of these gives a different solution of the confluent hypergeometric equation. For instance, the confluent hypergeometric function $\Phi(a, c; z)$ in (7.49) is obtained by choosing γ to be the segment $[-1, 0]$.[13] It is easy to see that the boundary term $e^{-zt}(t^2 + t)U(t)|_0^1$ vanishes for $\operatorname{Re} c > \operatorname{Re} a > 0$. Plugging (7.54) in (7.52) and making the change of variable $t \to -t$, we obtain the *Euler's integral representation* of $\Phi(a, c; z)$

$$\Phi(a, c; z) = \frac{\Gamma(c)}{\Gamma(a)\Gamma(c - a)} \int_0^1 dt \, e^{zt} t^{a-1}(1 - t)^{c-a-1}. \tag{7.55}$$

The determination of the multivalued functions in the integrand is specified by demanding $\arg(t) = \arg(1 - t) = 0$ on the segment $[0, 1]$. The integral representation (7.55) is valid for all z and $\operatorname{Re} c > \operatorname{Re} a > 0$. We can verify that the integral in (7.55) is precisely the hypergeometric function $\Phi(a, c; z)$

---

[13] We will see another example of contour in Section 7.2.3.

by expanding e^{tz} in power series around $z = 0$ and using the results of Exercise 3.5(a),

$$\sum_{n=0}^{\infty} \frac{z^n}{n!} \frac{\Gamma(c)}{\Gamma(a)\Gamma(c-a)} \int_0^1 dt\, t^{a-1+n}(1-t)^{c-a-1} = \frac{\Gamma(c)}{\Gamma(a)} \sum_{n=0}^{\infty} \frac{z^n}{n!} \frac{\Gamma(a+n)}{\Gamma(c+n)}.$$

Notice that the integral representation (7.55) gives the useful relation

$$\Phi(a, c; z) = e^z \Phi(c - a, c; -z). \tag{7.56}$$

The restriction $\operatorname{Re} c > \operatorname{Re} a > 0$ can be relaxed by considering a more complicated closed curve γ that surrounds the points $t = 0$ and $t = 1$.[14]

Example 7.9 (Bessel equation). By an appropriate rescaling of the variable z in the confluent hypergeometric equation (7.48) (see Exercise 7.9) we obtain *Bessel equation*

$$zy''(z) + y'(z) + \left(z - \frac{\nu^2}{z}\right) y(z) = 0, \tag{7.57}$$

where $\nu \in \mathbb{C}$. As for the confluent hypergeometric, the singularities are a Fuchsian point in $z = 0$ and an essential singular point in $z = \infty$. The solution around $z = 0$ is found substituting the ansatz (7.20) in the equation, which becomes

$$\sum_{n=0}^{\infty} [(\alpha + n)^2 - \nu^2] c_n z^n + \sum_{n=0}^{\infty} c_n z^{n+2} = 0. \tag{7.58}$$

It is easy to see that the constant term gives the indicial exponents $\alpha = \pm\nu$ with c_0 arbitrary, while the term proportional to z gives $c_1 = 0$. For $n \geq 2$ the form of the solution depends on whether ν is an integer or not. When $\nu \notin \mathbb{Z}$, the recurrence relation gives

$$c_{2n} = (-1)^n \prod_{l=1}^{n} \frac{c_0}{2l(2l \pm 2\nu)} \tag{7.59}$$

---

[14]One can use, for instance, the *Pochhammer contour* $P = \gamma_0 + \gamma_1 - \gamma_0 - \gamma_1$ where γ_0 and γ_1 are closed curved around $t = 0$ and $t = 1$, respectively (see Bateman *et al.*, 1953). The integral over P is not zero, because the integrand in (7.55) is multivalued and transforms non-trivially after a turn around $t = 0$ or $t = 1$.

for $n \geq 1$ and $c_{2n+1} = 0$. Setting $c_0 = \frac{1}{2^{\pm\nu}\Gamma(1\pm\nu)}$, we obtain the *Bessel function of first kind* of order ν

$$J_{\pm\nu} = \sum_{n=0}^{\infty} \frac{(-1)^n}{n!\,\Gamma(\pm\nu + n + 1)} \left(\frac{z}{2}\right)^{\pm\nu+2n}. \tag{7.60}$$

They have a branch point in $z = 0$. For $\nu = p/2$ with p odd[15] the Bessel functions can be expressed in terms of trigonometric functions. For instance, using the identities for the Gamma function given in Appendix A, for $\nu = 1/2$ we find

$$J_{\frac{1}{2}} = \sqrt{\frac{2z}{\pi}} \sum_{n=0}^{\infty} \frac{(-1)^n}{(2n+1)!} z^{2n} = \sqrt{\frac{2}{\pi z}}\, \sin z,$$

$$J_{-\frac{1}{2}} = \sqrt{\frac{2}{\pi z}} \sum_{n=0}^{\infty} \frac{(-1)^n}{(2n)!} z^{2n} = \sqrt{\frac{2}{\pi z}}\, \cos z. \tag{7.61}$$

For $\nu \notin \mathbb{Z}$ the general solution is

$$y(z) = a_1 J_\nu + a_2 J_{-\nu}, \tag{7.62}$$

where a_1 and a_2 are arbitrary constants. When the parameter ν is an integer one can show (see Exercise 7.9) that the two solutions $J_{\pm n}$ are not independent: $J_{-n} = (-1)^n J_n$. However, one can define a second set of solutions, called the *Bessel functions of second kind*,

$$Y_\nu(z) = \frac{J_\nu(z) \cos \nu\pi - J_{-\nu}(z)}{\sin \nu\pi}, \tag{7.63}$$

that are linearly independent from J_ν for any non-integer value of ν and have a non-trivial limit for $\nu = n \in \mathbb{N}$ (see Exercise 7.10)

$$Y_n(z) = \frac{2}{\pi} \left[\ln\frac{z}{2} + \gamma\right] J_n(z) - \frac{1}{\pi} \sum_{k=0}^{n-1} \frac{(n-k-1)!}{k!} \left(\frac{z}{2}\right)^{2k-n}$$

$$- \frac{1}{\pi} \sum_{k=0}^{\infty} \frac{(-1)^k}{k!(n+k)!}[h(k+n) + h(k)] \left(\frac{z}{2}\right)^{2k+n}, \tag{7.64}$$

---

[15] In this case $\alpha_1 - \alpha_2 = p \in \mathbb{N}$. Thus, if we take p positive, the solution for $\alpha_2 = -p/2$ should not be given by a Frobenius series. However, the recurrence relation for $n \geq 2$ becomes $n(n-p)c_n = -c_{n-2}$. Since p is odd and n even, $p-n$ is always different from zero and the recurrence relation can still be solved for c_n. So the second solution is again given by $J_{-\nu}$.

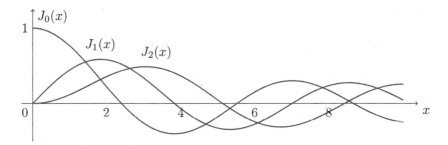

Fig. 7.1. Restriction to the real axis of the Bessel functions of first kind for $n = 0, 1, 2$.

where $\gamma = 0.577215$ is the Euler constant and $h(n) = \sum_{k=1}^{n} 1/k$. Thus, for $\nu = n$, the general solution is given by $y(z) = a_1 J_n(z) + a_2 Y_n(z)$. The functions J_n and Y_n have an oscillatory behaviour and an infinite number of zeros as shown in Figure 7.1.

The Bessel functions $J_{\pm\nu}$ can also be recovered as special cases of the confluent hypergeometric function (see Exercise 7.9)

$$J_\nu(z) = e^{-iz} \left(\frac{z}{2}\right)^\nu \Gamma^{-1}(\nu + 1) \Phi \left(\nu + \frac{1}{2}, 2\nu + 1; 2iz\right), \qquad (7.65)$$

and have an integral representation given by

$$J_\nu(z) = \left(\frac{z}{2}\right)^\nu \frac{2}{\Gamma(\nu + 1/2)\Gamma(1/2)} \int_0^{\pi/2} \cos(z \cos \theta)(\sin \theta)^{2\nu} d\theta, \qquad (7.66)$$

valid for Re $\nu > -1/2$, which we already used in Exercise 3.8.

7.2.3. *Solutions around an essential singular point*

There is no general theory for determining the solutions of equations of the type (7.7) near an essential singular point. However, in some cases it is still possible to formally write the solution in the neighbourhood of the singular point as a series, but, differently from the previous cases, the series is not in general convergent. It is enough to consider the case where the essential singularity is in $z = \infty$, as we can move the singularity to any point in $\mathbb{C}$ by a linear fractional transformation. We assume, for simplicity, that (7.7) takes the form

$$y''(z) + \left(p_0 + \frac{p_1}{z} + \frac{p_2}{z^2} + \cdots\right)y' + \left(q_0 + \frac{q_1}{z} + \frac{q_2}{z^2} + \cdots\right)y(z) = 0, \quad (7.67)$$

where p_0 or q_0 or q_1 are non-zero (see Exercise 7.11). This includes many of the equations relevant for physics.

To solve the equation we generalise the ansatz (7.17) to

$$y(z) = z^\alpha e^{\lambda z} \sum_{n=0}^{\infty} \frac{c_n}{z^n}, \quad (7.68)$$

and plug it into (7.67) to determine a set of relations among the coefficients c_n and the exponents α and λ. The first three terms are

$$(\lambda^2 + p_0\lambda + q_0)c_0 = 0,$$

$$(\lambda^2 + p_0\lambda + q_0)c_1 + [(p_0 + 2\lambda)\alpha + p_1\lambda + q_1]c_0 = 0, \quad (7.69)$$

$$(\lambda^2 + p_0\lambda + q_0)c_2 + [(p_0 + 2\lambda)(\alpha - 1) + p_1\lambda + q_1]c_1$$

$$+ [\alpha(\alpha - 1) + p_2\lambda + p_1\alpha + q_2]c_0 = 0.$$

Assuming $c_0 \neq 0$, the first equation in (7.69),

$$\lambda^2 + p_0\lambda + q_0 = 0, \quad (7.70)$$

determines two possible values of λ, while the second equation gives the exponent α

$$\alpha = -\frac{p_1\lambda + q_1}{p_0 + 2\lambda}. \quad (7.71)$$

The remaining equations can then be solved for the coefficients c_n with $n \geq 1$. When the two solutions λ_1 and λ_2 are different,[16] (7.71) gives two different values of α and the equation has two independent solutions of the form (7.68). Notice however that the series (7.68) is in general divergent

---

[16] When $\lambda_1 = \lambda_2$ the method fails since $p_0 + 2\lambda = 0$ and (7.71) gives a singular value for α. However, by writing $y(z) = e^{\lambda z}u(z)$ one obtains an equation for the unknown $u(z)$ that, after the change of variable $z = t^2$, is either an equation with a Fuchsian point at $t = \infty$ or is again of the form (7.67) with $\lambda_1 \neq \lambda_2$.

for any value of z (see Exercise 7.12), but it makes sense as an asymptotic series as discussed in Chapter 3. The expression (7.68) gives the leading behaviour of the solution for large z. The general solution of (7.67) is of the form

$$y(z) = a_1 z^{\alpha_1} e^{\lambda_1 z} \sum_{k=0}^{\infty} \frac{c_k^1}{z^k} + a_2 z^{\alpha_2} e^{\lambda_2 z} \sum_{k=0}^{\infty} \frac{c_k^2}{z^k} \tag{7.72}$$

with coefficients a_1 and a_2 that can take different values in different sectors of the complex plane. This is a manifestation of Stokes phenomenon.

Example 7.10. Bessel equation (7.57) has an essential singular point in $z = \infty$ with $p(z) = 1/z$ and $q(z) = 1 - \nu^2/z^2$. From (7.70) and (7.71), we immediately see that $\alpha = -1/2$, $\lambda = \pm i$ and one can check (see Exercise 7.13) that the general solution is given by

$$y(z) = a_1 \frac{e^{iz}}{\sqrt{2\pi z}} \sum_{k=0}^{\infty} \frac{c_k^1}{z^k} + a_2 \frac{e^{-iz}}{\sqrt{2\pi z}} \sum_{k=0}^{\infty} \frac{c_k^2}{z^k} \tag{7.73}$$

with coefficients $c_0^1 = c_0^2 = 1$ and, for $k \geq 1$,

$$c_k^1 = \frac{1}{k!} \left(\frac{i}{8}\right)^k \prod_{l=1}^{k} [4\nu^2 - (2l-1)^2], \quad c_k^2 = \frac{1}{k!} \left(-\frac{i}{8}\right)^k \prod_{l=1}^{k} [4\nu^2 - (2l-1)^2].$$

Notice that $c_k^2 = (-1)^k c_k^1$. With an appropriate choice of the constants a_1 and a_2, one finds the Bessel function of first kind

$$J_\nu(z) = \frac{e^{i(z - \nu\pi/2 - \pi/4)}}{\sqrt{2\pi z}} \sum_{k=0}^{\infty} \frac{c_k^1}{z^k} + \frac{e^{-i(z - \nu\pi/2 - \pi/4)}}{\sqrt{2\pi z}} \sum_{k=0}^{\infty} \frac{c_k^2}{z^k}. \tag{7.74}$$

The expansion (7.74) is valid in the sector $|\arg z| < \pi$. The anti-Stokes line is the real axis $z = x$ where the two terms are of equal weight. In particular, on the real positive axis, $x > 0$, the leading behaviour is

$$J_\nu \sim \sqrt{\frac{2}{\pi z}} \cos\left(z - \nu\frac{\pi}{2} - \frac{\pi}{4}\right) + O(z^{-3/2}). \tag{7.75}$$

From (7.60) it is easy to see that $J_\nu(z) = e^{i\pi\nu} J_\nu(ze^{-i\pi})$. We can use this relation to derive the asymptotic expansion of $J_{\pm\nu}$ in the sector $0 < \arg z < 2\pi$[17]

$$J_\nu = \frac{e^{i(z+3\pi\nu/2+3\pi/4)}}{\sqrt{2\pi z}} \sum_{k=0}^{\infty} \frac{c_k^1}{z^k} + \frac{e^{-i(z-\nu\pi/2-\pi/4)}}{\sqrt{2\pi z}} \sum_{k=0}^{\infty} \frac{c_k^2}{z^k}. \qquad (7.76)$$

We see explicitly the Stokes phenomenon: the coefficients a_i are different in different sectors. For $-\pi < \arg z < \pi$ we have $a_1 = e^{-i\nu\pi/2-i\pi/4}$ and $a_2 = e^{i\nu\pi/2+i\pi/4}$, while, for $0 < \arg z < 2\pi$, we have $a_1 = e^{i3\nu\pi/2+i3\pi/4}$ and $a_2 = e^{i\nu\pi/2+i\pi/4}$.

Example 7.11. Consider again the confluent hypergeometric equation (7.48) of Example 7.7. We want to study the solutions in a neighbourhood of the essential singularity $z = \infty$. The coefficients of the equation are of the form (7.67) with $p_0 = -1, p_1 = c, q_0 = 0, q_1 = -a$ and all the others zero. Solving (7.70) gives $\lambda = 0$ and $\lambda = 1$. Then the other equations can be solved recursively for α and the coefficients c_n. Setting $c_0^i = 1$, we find

$$\alpha = -a, \qquad c_n^1 = \frac{(-1)^n}{n!} \frac{\Gamma(a+n)\Gamma(a-c+n+1)}{\Gamma(a)\Gamma(a-c+1)} \qquad (7.77)$$

for $\lambda = 0$, and for $\lambda = 1$

$$\alpha = a - c, \qquad c_n^2 = \frac{1}{n!} \frac{\Gamma(-a+n+1)\Gamma(c-a+n)}{\Gamma(-a+1)\Gamma(c-a)}. \qquad (7.78)$$

Thus the solution can be written as

$$y(z) = a_1 z^{-a} \sum_{n=0}^{\infty} \frac{c_n^1}{z^n} + a_2 e^z z^{a-c} \sum_{n=0}^{\infty} \frac{c_n^2}{z^n}. \qquad (7.79)$$

These expansions are only asymptotic. To understand better their properties we can use the integral representation of Example 7.8. We focus for simplicity on the first series in (7.79) and show that there exists a solution $\Psi(a, c; z)$[18] of the confluent hypergeometric equation that reduces asymptotically to it in an appropriate sector. We use an integral representation

---

[17]When the argument of z varies in $(0, 2\pi)$ we have $-\pi < \arg(ze^{-i\pi}) < \pi$.
[18]This solution is called *Tricomi function* and is also denoted with $U(a, c; z)$.

for $\Psi(a, c; z)$ of the form (7.52) and (7.54) where γ is the semi-axis $[0, \infty)$

$$\Psi(a, c; z) = \frac{1}{\Gamma(a)} \int_0^{+\infty} dt \; e^{-zt} t^{a-1} (1+t)^{c-a-1}. \tag{7.80}$$

Here we take $\arg t = \arg(1 + t) = 0$ along the contour of integration. With this choice of contour the boundary terms discussed in Example 7.8 vanish when $\mathrm{Re}\, a > 0$ and z belongs to the sector $-\pi/2 < \arg z < \pi/2$. The integral representation (7.80) can be extended to $\mathbb{C} - \mathbb{R}^-$ by an appropriate rotation of the integration contour (see Example 3.6 for a similar procedure). We can use (7.80) to derive an asymptotic expansion of $\Psi(a, c; z)$ around $z = \infty$. By substituting the binomial expansion around $t = 0$ (see (A.17))

$$(1+t)^{c-a-1} = \sum_{n=0}^{\infty} \frac{\Gamma(a - c + 1 + n)}{\Gamma(a - c + 1)\Gamma(n + 1)} (-t)^n \tag{7.81}$$

in (7.80) and integrating term by term using (A.15), we find

$$\Psi(a, c; z) = \sum_{n=0}^{\infty} \frac{\Gamma(a + 1 - c + n)}{\Gamma(a + 1 - c)} \frac{\Gamma(n + a)}{\Gamma(a)} \frac{(-1)^n}{n!} \frac{1}{z^{a+n}}. \tag{7.82}$$

We see that $\Psi(a, c; z)$ precisely reproduces the first series in (7.79). The binomial expansion is only convergent for $t < 1$ while, on the contour of integration, t becomes arbitrarily large. As a consequence, the expansion (7.82) is not convergent and is an asymptotic series.[19]

Notice that the choice of γ is different from the one used in Example 7.8 to give an integral representation for $\Phi(a, c; z)$. Then $\Psi(a, c; z)$ and $\Phi(a, c; z)$ are linearly independent. One can show that $\Psi(a, c; z)$ is related to $\Phi(a, c; z)$ by

$$\Psi(a, c; z) = \frac{\Gamma(1 - c)}{\Gamma(a + 1 - c)} \Phi(a, c; z) + \frac{\Gamma(c - 1)}{\Gamma(a)} z^{1-c} \Phi(a - c + 1, 2 - c; z).$$

---

[19] One can explicitly check that (7.82) is an asymptotic series for $\Psi(a, c; z)$ by evaluating the remainder as in Example 3.1. Notice that the particular case $a = c = 1$ corresponds to the asymptotic expansion of the function $Ei(x)$ discussed in Example 3.1

$$E_1(x) = \int_x^{+\infty} du \frac{e^{-u}}{u} = e^{-x} \Psi(1, 1; x) = e^{-x} \sum_{n=0}^{\infty} \frac{(-1)^n n!}{x^{n+1}}, \tag{7.83}$$

as it can be easily checked by performing the change of variable $u = x(t + 1)$.

7.3. Exercises

Exercise 7.1. Find the values of E such that

$$y'' + e^{-x}(E - 4e^{-x})y = 0$$

has a bounded solution for $x \to \pm\infty$.

Exercise 7.2. Find the singular points of the differential equation

$$2z(1 - z)u''(z) + (2 - 3z)u'(z) + u(z) = 0.$$

Consider a solution that is regular around $z = 1$ and compute the first three coefficients of its series expansion in $z = 1$.

Exercise 7.3. Study the singular points of the differential equation

$$u''(z) + \frac{2}{z}u'(z) - \frac{2}{z^2(z - 1)^2}u(z) = 0,$$

and find its solutions by power series.

Exercise 7.4. Consider the differential equation

$$(1 + 2z)z^2u''(z) + 2z^2u'(z) - \lambda u(z) = 0.$$

(a) Study the singular points and find the values of λ such that at least one of the two independent solutions around $z = 0$ is regular and single-valued.
(b) Determine the solution around $z = 0$ in the case $\lambda = 0$.

Exercise 7.5. Determine the polynomial $F(l + 1, -l, 1; (1 - x)/2)$ from the power series expansion of the hypergeometric function for small values of $l \in \mathbb{N}$ and compare it to the Legendre polynomials (5.34).

Exercise 7.6. Verify the equalities below

(a) $F(1, 1, 2; -z) = z^{-1}\ln(1 + z)$,
(b) $F(1/2, 1, 3/2; -z^2) = z^{-1}\arctan z$,

(c) $\Phi(a, a; z) = e^z$,
(d) $\Phi(1, 2; z) = z^{-1}(e^z - 1)$,
(e) $\Phi(2, 1; z) = (1 + z)e^z$.

Exercise 7.7. Consider the hypergeometric equation (7.26).
(a) Show that performing the rescaling $z \to z/b$ and then sending b to ∞, (7.26) is mapped to the confluent hypergeometric equation.

(b) Applying the same limiting procedure as in part (a) to the solutions (7.32) and (7.35) of the hypergeometric equation, derive the solutions (7.49) and (7.50).

(c) Use the methods of Section 7.2.2 to derive again (7.49) and (7.50).

* **Exercise 7.8.** (a) Prove the *Euler's integral representation* of the hypergeometric function

$$F(a, b, c; z) = \frac{\Gamma(c)}{\Gamma(a)\Gamma(c-a)} \int_0^1 dt\, t^{a-1}(1-t)^{c-a-1}(1-zt)^{-b},$$

which is valid for $z \in \mathbb{C} - [1, \infty)$ and $\operatorname{Re} c > \operatorname{Re} a > 0$. In this formula arg $t = \arg(1-t) = 0$ and $(1-zt)^{-a}$ has the principal determination along $[0, 1]$.

(b) Derive (7.55) from part (a) by using $\Phi(a, c; z) = \lim_{b\to\infty} F(a, b, c; z/b)$.

Exercise 7.9. (a) Recover Bessel equation, (7.57), by the change of variable $y = e^{z/2}z^{-\nu}u$ and an appropriate rescaling of the coordinate z in the confluent hypergeometric equation, (7.48).

(b) Verify that for $\nu = n$ integer, $J_{-n} = (-1)^n J_n$.

* **Exercise 7.10.** Derive the expression (7.64) for the Bessel function of second kind $Y_n(z)$ for $n \in \mathbb{N}$.

Exercise 7.11. Show that, with p and q given by (7.67), the equation (7.7) has an essential singularity at $z = \infty$.

Exercise 7.12. Consider (7.7) where only the coefficients p_0, p_1 and q_2 in (7.67) are non-zero. Determine λ, α and the coefficients c_n of the expansion (7.68) near $z = \infty$ and show that the corresponding series diverges for any value of z.

Exercise 7.13. Verify that the general solution of Bessel equation around $z = \infty$ has the form (7.73).

Exercise 7.14. Consider the vibrations of a circular membrane $x^2 + y^2 \leq a^2$ whose displacement is described by the wave equation

$$u_{tt} - v^2 \Delta u = 0,$$

and whose boundary is fixed, $u(x, y, t) = 0$ for $x^2 + y^2 = a^2$. Find the general solution of the boundary problem.

* **Exercise 7.15.** Show that the Airy equation $y''(z) - zy(z) = 0$ (see Example 3.10) reduces to a Bessel equation with $\nu = 1/3$ with the change of variables $y(z) =$

$(-z)^{1/2}\tilde{y}(\frac{2}{3}(-z)^{3/2})$. Using this result and comparing the asymptotic behaviours,
(a) show that the asymptotic behaviour of solutions of the Airy equation for large z is of the form $(c_0 = 1)$

$$y(z) \sim a_1 z^{-1/4} e^{-2z^{3/2}/3} \sum_{n=0}^{\infty} \frac{c_n}{z^{3n/2}} + a_2 z^{-1/4} e^{2z^{3/2}/3} \sum_{n=0}^{\infty} (-1)^n \frac{c_n}{z^{3n/2}};$$

(b) find the constant a_1 and a_2 for $Ai(z)$ and $Bi(z)$ and show that they are different in different sectors (Stokes phenomenon);

(c) show that, for positive z, $Ai(-z) = \sqrt{\frac{z}{9}}(J_{1/3}(\frac{2}{3}z^{3/2}) + J_{-1/3}(\frac{2}{3}z^{3/2}))$ and $Bi(-z) = \sqrt{\frac{z}{3}}(J_{1/3}(\frac{2}{3}z^{3/2}) - J_{-1/3}(\frac{2}{3}z^{3/2}))$.

PART III
Hilbert Spaces

Introduction

This part is devoted to applications of the theory of linear operators in Hilbert spaces.[1] Two of the most common ones are the study of integral equations and the formalism of quantum mechanics. In both cases the spectral theorem for self-adjoint operators plays a central role.

The spectral properties of a particular class of operators, called compact operators, are remarkably simple. They have purely discrete spectrum. Compact operators are useful to study solutions of integral equations. These are equations of the type

$$\lambda(x)\phi(x) + f(x) = \int_a^b \mathrm{d}y \ k(x,y)\phi(y),$$

where $\lambda(x)$, $f(x)$ and $k(x,y)$ are known functions and $x \in (a,b)$. The integral equation can be associated to an eigenvalue problem in a Hilbert space. In this book we will focus on a specific class of integral equations, the Fredholm equations of second type, that have $\lambda(x) = const$. We will discuss their resolution methods and how they are related to compact operators.

Another notable application of Hilbert spaces and the spectral theorem is quantum mechanics. Indeed, the states of a quantum mechanical system are given by functions in $L^2(\mathbb{R})$. Any physical observable is associated with a self-adjoint operator in $L^2(\mathbb{R})$ and the possible results of the measure of

---

[1]See Petrini *et al.* (2017) and Appendix A for an introduction to Hilbert spaces, linear operators and the spectral theorem.

an observable are given by the proper and generalised eigenvalues of the corresponding operator. The spectral theorem gives the probability distribution for the results of such measure. In quantum mechanics the time evolution of the system is given by the Schrödinger equation

$$i\hbar\frac{d\psi}{dt} = -\frac{\hbar^2}{2m}\Delta\psi + V(\mathbf{x}, t)\psi = H\psi,$$

where H is the self-adjoint Hamiltonian operator. When the Hamiltonian does not depend on time, the Schrödinger equation is related to an eigenvalue equation for H. In this context the difference between discrete and continuous spectrum corresponds to the fact that some physical quantities, like the energy or the spin of a particle, can assume discrete (quantised) or continuous values. In some cases, it is possible to solve the Schrödinger equation exactly. In some other cases, we have to resort to approximation methods, like the series expansions discussed in Chapter 7 or the WKB method we discuss in Chapter 9.

8

Compact Operators and Integral Equations

In this chapter we study a special class of linear operators in a Hilbert space, the compact operators, that have very simple properties. In particular, up to subtleties, they have a purely discrete spectrum. Compact operators are the natural generalisation of matrices to infinite dimensions. They are particularly interesting since they appear in the theory of integral equations. The relation between compact operators and integral equations has been developed by Fredholm. In this book we will focus on equations of the form

$$\int_a^b k(x,y)\phi(y)\mathrm{d}y - \lambda\phi(x) = f(x), \tag{8.1}$$

which are called Fredholm equations of second type.

8.1. Compact Operators

In this section we discuss the general properties of compact operators in Hilbert spaces. We will not give many proofs, which can be found in Reed and Simon (1981), Rudin (1991) and Kolmogorov and Fomin (1999), and we will rather focus on examples.

An operator A from a Hilbert space $\mathcal{H}$ to itself is *compact* if it maps bounded sets into sets whose closure is compact. Recall that a set X in a metric space is compact if and only if any sequence in X contains a subsequence that converges in X. We can then rephrase the definition of compact operator as follows. An operator A is compact if any sequence $\{A\mathbf{x}_n\}$, where $\{\mathbf{x}_n\}$ is a bounded sequence in $\mathcal{H}$, contains a convergent subsequence. We can easily see that a compact operator is also bounded.

Proof. We prove it by contradiction. If A were not bounded, it would exist a sequence $\{\mathbf{x}_n\}$ such that $||A\mathbf{x}_n||/||\mathbf{x}_n|| \to \infty$ for $n \to \infty$. But then the sequence $||A\mathbf{e}_n||$, where $\mathbf{e}_n = \mathbf{x}_n/||\mathbf{x}_n||$, would also diverge and therefore $\{A\mathbf{e}_n\}$ does not contain any convergent subsequence, even if $\{\mathbf{e}_n\}$ is bounded. □

Compact operators are elements of $\mathcal{B}(\mathcal{H})$, the complete normed space consisting of all bounded operators on $\mathcal{H}$ (see Appendix A). One can show that, if a sequence of compact operators A_n converges to an operator A in the norm of $\mathcal{B}(\mathcal{H})$, namely $\lim_{n\to\infty} ||A_n - A|| = 0$, the limit A is also compact (see Exercise 8.1).

Standard examples of compact operators are finite rank operators and integral operators.

Example 8.1 (Finite rank operators). An operator A has *finite rank* if its image is a finite-dimensional subspace of $\mathcal{H}$. Any bounded finite rank operator A is compact. To prove it, we need to show that, if $S \subset \mathcal{H}$ is bounded, the closure of $A(S)$, $\overline{A(S)}$, is compact. The image $A(S)$ is bounded, because A is bounded. Then $\overline{A(S)} \subset \text{im}(A)$ is a closed bounded subspace of a finite-dimensional vector space. In a finite-dimensional vector space, a set is compact if and only if it is closed and bounded (Heine–Borel theorem).[1] Therefore $\overline{A(S)}$ is compact. More generally, one can show that an operator A is compact if and only if it is the limit of bounded operators of finite rank.

Example 8.2 (Integral operators). Consider the Hilbert space $L^2(\Omega)$, where Ω is a subset of $\mathbb{R}^n$ and the integral operator K

$$Kf(\mathbf{x}) = \int_\Omega k(\mathbf{x},\mathbf{y})f(\mathbf{y})d\mathbf{y}, \tag{8.2}$$

---

[1] Notice that the Heine–Borel theorem fails in infinite dimension (see Exercise 8.2). And indeed not all bounded operators are compact.

where the *integral kernel* $k(\mathbf{x}, \mathbf{y})$ is a square-integrable function of $\mathbf{x}$ and $\mathbf{y}$, namely $k(\mathbf{x}, \mathbf{y}) \in L^2(\Omega \times \Omega)$. The operator K maps $L^2(\Omega)$ into itself and it is bounded. Indeed, using Schwartz's inequality (A.8), we find

$$||Kf||^2 = \int_\Omega \left| \int_\Omega k(\mathbf{x}, \mathbf{y}) f(\mathbf{y}) d\mathbf{y} \right|^2 d\mathbf{x} \tag{8.3}$$

$$\leq \int_\Omega \left(\int_\Omega |k(\mathbf{x}, \mathbf{y})|^2 d\mathbf{y} \int_\Omega |f(\mathbf{u})|^2 d\mathbf{u} \right) d\mathbf{x} \leq ||f||^2 \int_{\Omega \times \Omega} |k(\mathbf{x}, \mathbf{y})|^2 d\mathbf{x} d\mathbf{y},$$

where $|| \cdot ||$ denotes the norm in $L^2(\Omega)$. We see that $||Kf||/||f||$ is bounded from above by $\int_{\Omega \times \Omega} |k(\mathbf{x}, \mathbf{y})|^2 d\mathbf{x} d\mathbf{y}$, which is finite since $k(\mathbf{x}, \mathbf{y}) \in L^2(\Omega \times \Omega)$. We now show that K is also compact. One can easily see that, if $\{\mathbf{u}_i(\mathbf{x})\}_{i \in \mathbb{N}}$ is a Hilbert basis for $L^2(\Omega)$, then $\{\mathbf{u}_i(\mathbf{x})\mathbf{u}_j(\mathbf{y})\}_{i,j \in \mathbb{N}}$ is a Hilbert basis for $L^2(\Omega \times \Omega)$. We can then expand $k(\mathbf{x}, \mathbf{y})$ as

$$k(\mathbf{x}, \mathbf{y}) = \sum_{i,j=0}^\infty k_{ij} \mathbf{u}_i(\mathbf{x}) \mathbf{u}_j(\mathbf{y}), \tag{8.4}$$

with $\int_{\Omega \times \Omega} |k(\mathbf{x}, \mathbf{y})|^2 d\mathbf{x} d\mathbf{y} = \sum_{i,j=0}^\infty |k_{ij}|^2 < \infty$. Consider now the sequence of operators

$$K_n f(\mathbf{x}) = \int_\Omega k_n(\mathbf{x}, \mathbf{y}) f(\mathbf{y}) d\mathbf{y}, \tag{8.5}$$

where $k_n(\mathbf{x}, \mathbf{y}) = \sum_{i,j=0}^n k_{ij} \mathbf{u}_i(\mathbf{x}) \mathbf{u}_j(\mathbf{y})$. Since $k_n(\mathbf{x}, \mathbf{y})$ is square-integrable, K_n are bounded. K_n are also of finite rank because the image of any function f

$$K_n f(\mathbf{x}) = \sum_{i=0}^n k_{ij} \left(\sum_{j=0}^n \int_\Omega \mathbf{u}_j(\mathbf{y}) f(\mathbf{y}) d\mathbf{y} \right) \mathbf{u}_i(\mathbf{x}), \tag{8.6}$$

lies in the finite-dimensional space spanned by $\mathbf{u}_0(\mathbf{x}), \ldots, \mathbf{u}_n(\mathbf{x})$. Thus K_n are bounded operators of finite rank and they converge to K. Indeed

$$||K_n - K||^2 \leq \int_{\Omega \times \Omega} |k_n(\mathbf{x}, \mathbf{y}) - k(\mathbf{x}, \mathbf{y})|^2 d\mathbf{x} d\mathbf{y} = \sum_{i,j=n+1}^\infty |k_{ij}|^2, \tag{8.7}$$

and the last expression tends to zero for large n since the series $\sum_{i,j=0}^{\infty} |k_{ij}|^2$ converges. Being the limit of compact operators, K is compact.

Integral operators behave like matrices in many respects. For example, it is immediate to see that the composition of two integral operators K and L with kernels $k(\mathbf{x}, \mathbf{y})$ and $l(\mathbf{x}, \mathbf{y})$ is the integral operator $M = KL$ with kernel

$$m(\mathbf{x}, \mathbf{y}) = \int_{\Omega} k(\mathbf{x}, \mathbf{z}) l(\mathbf{z}, \mathbf{y}) \mathrm{d}\mathbf{z}. \tag{8.8}$$

Moreover, given the integral operator K with kernel $k(\mathbf{x}, \mathbf{y})$, its adjoint $K^\dagger$ is the operator with kernel $\overline{k(\mathbf{y}, \mathbf{x})}$, as one easily verifies from

$$(f, Kg) = \int_{\Omega} \overline{f(\mathbf{x})} \left(\int_{\Omega} k(\mathbf{x}, \mathbf{y}) g(\mathbf{y}) \mathrm{d}\mathbf{y} \right) \mathrm{d}\mathbf{x}$$

$$= \int_{\Omega} \int_{\Omega} \overline{\left(\overline{k(\mathbf{x}, \mathbf{y})} f(\mathbf{x}) \mathrm{d}\mathbf{x} \right)} g(\mathbf{y}) \mathrm{d}\mathbf{y} = (K^\dagger f, g).$$

The kernel $k(\mathbf{x}, \mathbf{y})$ can be seen as a continuous version of a matrix where the indices are replaced with the continuous variables $\mathbf{x}$ and $\mathbf{y}$. The relation (8.8) is then analogous to the row by column multiplication of matrices. Similarly, the fact that $\overline{k(\mathbf{y}, \mathbf{x})}$ is the kernel of $K^\dagger$ is the analogue of the adjoint of a matrix being its conjugate transpose.

The importance of compact operators follows from the fact that their spectrum is very simple. This is summarised in the following properties. Consider a compact operator A in a separable Hilbert space.

I. The spectrum $\sigma(A)$ is a discrete set without accumulation points, with the possible exception of the point $\lambda = 0$. Every non-vanishing $\lambda \in \sigma(A)$ is an eigenvalue of finite multiplicity.

II. *Hilbert–Schmidt theorem.* If A is self-adjoint, there exists a complete orthonormal system consisting of eigenvectors, $A\mathbf{u}_n = \lambda_n \mathbf{u}_n$. Moreover, the set of distinct eigenvalues is either finite or $\lim_{n \to \infty} \lambda_n = 0$.

It is often said that a self-adjoint compact operator has a discrete spectrum. Notice, however, that for some operators the continuous spectrum is not empty, but it just consists of the single point $\lambda = 0$. This may happen when $\lambda = 0$ is the accumulation point of an infinite number of distinct non-zero eigenvalues λ_n. Since the spectrum $\sigma(A)$ of an operator must be a closed subset of $\mathbb{C}$, $\lambda = 0$ must then belong to the spectrum: if it is not an eigenvalue, it is necessarily a point in the continuous spectrum.

The spectral theorem for self-adjoint operators (A.10) becomes very simple for compact operators. All elements in $\mathcal{H}$ can be expanded in the basis of proper orthonormal eigenvectors $\{\mathbf{u}_n\}$,[2]

$$\mathbf{v} = \sum_n c_n \mathbf{u}_n, \tag{8.9}$$

and, in this basis, A is diagonal[3]

$$A\mathbf{v} = \sum_n \lambda_n c_n \mathbf{u}_n. \tag{8.10}$$

We can also easily define functions of self-adjoint compact operators. For any function $f(\lambda)$ that is bounded near $\lambda = 0$,[4] we can define the operator $f(A)$ as

$$f(A)\mathbf{v} = \sum_n f(\lambda_n) c_n \mathbf{u}_n, \tag{8.11}$$

when $\mathbf{v}$ is expressed as (8.9). Given the above properties, one can show that (see Exercise 8.3)

   (i) $f(A)$ is bounded;
  (ii) $f(\lambda_n)$ is an eigenvalue of $f(A)$;
 (iii) if $f(\lambda_n)$ are real, $f(A)$ is self-adjoint;
 (iv) if $f(\lambda_n) \to 0$ for $n \to \infty$, $f(A)$ is compact.

Example 8.3 (Operators with compact resolvent). An operator L is said to have *compact resolvent* if the resolvent operator $(L - \lambda \mathbb{I})^{-1}$ is compact for all $\lambda \in \mathbb{C}$ for which it exists. Since compact operators are also

---

[2]In the case of a non-separable Hilbert space, the set of eigenvectors of A is a basis only for the image of A. In this case, any vector $\mathbf{v} \in \mathcal{H}$ can be written as $\mathbf{v} = \sum_n c_n \mathbf{u}_n + \mathbf{w}$, where $\mathbf{w}$ is an element of the kernel of A. The formula (8.10) still holds since $A\mathbf{w} = 0$.

[3]Since the continuous spectrum consists at most of a point, there is no continuous part in the decomposition of a vector on the basis of eigenvectors. The integral over a point is zero.

[4]This is required to guarantee the convergence of the right-hand side of (8.11).

bounded such λ's cannot belong to the continuous spectrum. Therefore self-adjoint operators with compact resolvent have purely discrete spectrum. This result can be used to study the spectrum of Sturm–Liouville operators. Consider, for instance, the problem (5.19) associated with the operator L. When λ is not an eigenvalue, the operator $L - \lambda\mathbb{I}$ can be inverted and its inverse is an integral operator $(L - \lambda\mathbb{I})^{-1} f(x) = \int G(x, y; \lambda) f(y) dy$, where $G(x, y; \lambda)$ is the Green function for the boundary problem. If $G(x, y; \lambda)$ is square-integrable then the resolvent operator is compact and the operator L has purely discrete spectrum. This holds, for instance, for regular Sturm–Liouville problems (see Exercise 8.5).

8.1.1. *Fredholm alternative*

Compact operators satisfy the *Fredholm alternative theorem*. This states that, for a compact self-adjoint operator A, the equation

$$A\mathbf{v} - \lambda\mathbf{v} = \mathbf{w}, \quad \lambda \neq 0, \tag{8.12}$$

has a unique solution for any $\mathbf{w}$ if and only if the corresponding homogeneous equation has no non-trivial solutions. We can see it by expanding both the unknown $\mathbf{v}$ and the source $\mathbf{w}$ in the basis of eigenvectors $\{\mathbf{u}_n\}$ of A, $A\mathbf{u}_n = \lambda_n \mathbf{u}_n$,

$$\mathbf{v} = \sum_{n=1}^{\infty} (\mathbf{u}_n, \mathbf{v})\mathbf{u}_n, \quad \mathbf{w} = \sum_{n=1}^{\infty} (\mathbf{u}_n, \mathbf{w})\mathbf{u}_n. \tag{8.13}$$

Substituting in (8.12) we find

$$\sum_{n=1}^{\infty} (\mathbf{u}_n, \mathbf{v})(\lambda_n - \lambda)\mathbf{u}_n = \sum_{n=1}^{\infty} (\mathbf{u}_n, \mathbf{w})\mathbf{u}_n, \tag{8.14}$$

and we can solve the equation by equating the coefficients of $\mathbf{u}_n$ on the two sides,

$$(\lambda_n - \lambda)(\mathbf{u}_n, \mathbf{v}) = (\mathbf{u}_n, \mathbf{w}). \tag{8.15}$$

Thus, if $\lambda \neq \lambda_n$ for all $n \in \mathbb{N}$, there exists a solution for any $\mathbf{w}$ and it is unique. Indeed, (8.15) can be solved as follows

$$\mathbf{v} = \sum_{n=1}^{\infty} \frac{(\mathbf{u}_n, \mathbf{w})}{\lambda_n - \lambda} \mathbf{u}_n. \tag{8.16}$$

This expression is well defined since the λ_n are either a finite set or a sequence with $\lambda_n \to 0$,

$$\sum_{n=1}^{\infty} \left| \frac{(\mathbf{u}_n, \mathbf{w})}{\lambda_n - \lambda} \right|^2 \leq const \sum_{n=1}^{\infty} |(\mathbf{u}_n, \mathbf{w})|^2 = const \, ||\mathbf{w}||^2, \tag{8.17}$$

where we used Parseval's identity (A.9). If, on the contrary, $\lambda = \lambda_k$ for some k, we can find solutions only for those $\mathbf{w}$ for which $(\mathbf{u}_k, \mathbf{w}) = 0$. In this case, the solution is not unique. We can always add to a solution $\mathbf{v}$ any element of the eigenspace of λ_k, $\ker(A - \lambda_k \mathbb{I})$.

To understand better why the previous theorem is called Fredholm alternative, we can state it as follows: either the inhomogeneous equation $A\mathbf{v} - \lambda\mathbf{v} = \mathbf{w}$ has a unique solution for all $\mathbf{w}$, or the homogeneous equation $A\mathbf{v} - \lambda\mathbf{v} = 0$ has non-trivial solutions.

The Fredholm alternative theorem can be reformulated in a more abstract way that allows to extend it to generic compact operators, not necessarily self-adjoint. We know that, for a bounded operator (see Exercise 8.6),

$$\text{im}(A - \lambda\mathbb{I})^{\perp} = \ker(A^{\dagger} - \bar{\lambda}\mathbb{I}). \tag{8.18}$$

One can show that, if A is compact and $\lambda \neq 0$, $\text{im}(A - \lambda\mathbb{I})$ is closed, so that we have the orthogonal decomposition (see Petrini *et al.*, 2017, Section 5.6)

$$\mathcal{H} = \text{im}(A - \lambda\mathbb{I}) \oplus \ker(A^{\dagger} - \bar{\lambda}\mathbb{I}). \tag{8.19}$$

The equation $A\mathbf{v} - \lambda\mathbf{v} = \mathbf{w}$ with $\lambda \neq 0$ is equivalent to $\mathbf{w} \in \text{im}(A - \lambda\mathbb{I})$, and (8.19) tells us that it has a solution if and only if $\mathbf{w}$ is orthogonal to $\ker(A^{\dagger} - \bar{\lambda}\mathbb{I})$. The solution is unique up to the addition of an element of $\ker(A - \lambda\mathbb{I})$. If $\ker(A - \lambda\mathbb{I}) = \ker(A^{\dagger} - \bar{\lambda}\mathbb{I}) = \{0\}$, the solution exists and is unique for all $\mathbf{w}$. For self-adjoint operators, for which $A^{\dagger} = A$ and the eigenvalues are real, this is equivalent to what we have found before.

8.2. Fredholm Equation of Second Type

In this section we apply the theory of compact operators to the study of *Fredholm equations of second type*. These are integral equations of the form[5]

$$\int_a^b k(x,y)\phi(y)\mathrm{d}y - \lambda\phi(x) = f(x), \quad \lambda \neq 0, \qquad (8.20)$$

where $k(x,y) \in L^2([a,b] \times [a,b])$. It follows from Example 8.2 that the operator K

$$K\phi(x) = \int_a^b k(x,y)\phi(y)\mathrm{d}y, \qquad (8.21)$$

and its adjoint $K^\dagger$, which is obtained by replacing $k(x,y)$ with $\overline{k(y,x)}$, are bounded compact operators of norm smaller than

$$\kappa = \int_a^b \mathrm{d}x \int_a^b \mathrm{d}y\, |k(x,y)|^2. \qquad (8.22)$$

The general theory of compact operators tells us that the eigenvalues of K and $K^\dagger$ are two discrete sets λ_n and μ_m, with a possible accumulation point in zero. Every non-zero λ_n and μ_m has a finite number of corresponding eigenvectors $\phi_n(x)$ and $\psi_m(x)$. We can also apply the Fredholm alternative theorem to obtain information about the solutions of (8.20): a solution of (8.20) exists if and only if $f(x)$ is orthogonal to $\ker(K^\dagger - \bar{\lambda}\mathbb{I})$.

We now discuss some simple methods to solve equation (8.20).

8.2.1. *Method of iterations*

Equation (8.20) can be solved by successive iterations. Starting with

$$\phi(x) = -\frac{f(x)}{\lambda} + \frac{1}{\lambda}\int_a^b k(x,y)\phi(y)\mathrm{d}y, \qquad (8.23)$$

and substituting again the expression of $\phi(x)$ in the right-hand side we find

$$\phi(x) = -\frac{f(x)}{\lambda} - \frac{1}{\lambda^2}\int_a^b k(x,y)f(y)\mathrm{d}y$$

$$+ \frac{1}{\lambda^2}\int_a^b k(x,y)\left(\int_a^b k(y,z)\phi(z)\mathrm{d}z\right)\mathrm{d}y. \qquad (8.24)$$

---

[5]The equation is traditionally written as $\phi(x) - \mu\int_a^b \mathrm{d}y k(x,y)\phi(y) = F(x)$, with the obvious redefinition $\lambda = 1/\mu$, $F = -f/\lambda$. The same equation without the term $\phi(x)$ on the left-hand side it called *Fredholm equation of first type*. Analogous equations where the extremum b is replaced by the variable x are called *Volterra equations of second and first type*. See, for example, Tricomi (2012).

Notice that each term is again an integral operator acting on f or ϕ. Using (8.8) we can write (8.24) as $\phi = -f/\lambda - Kf/\lambda^2 + K^2\phi/\lambda^2$. We can now substitute again (8.23) into (8.24) and so on. At the pth step of the iteration we obtain

$$\phi = -\frac{f}{\lambda} - \sum_{n=1}^{p} \frac{K^n f}{\lambda^{n+1}} + \frac{K^{p+1}\phi}{\lambda^{p+1}}. \tag{8.25}$$

For large λ the last term will become more and more negligible when p grows, as K is bounded. Then, passing to the limit $p \to \infty$ we obtain

$$\phi = -\frac{f}{\lambda} - \sum_{n=1}^{\infty} \frac{K^n f}{\lambda^{n+1}}. \tag{8.26}$$

The same result can be recovered alternatively by inverting the operator $K - \lambda\mathbb{I}$. Indeed

$$\phi = (K - \lambda\mathbb{I})^{-1} f = -\frac{1}{\lambda}(\mathbb{I} - K/\lambda)^{-1} f = -\sum_{n=0}^{\infty} \frac{K^n f}{\lambda^{n+1}}, \tag{8.27}$$

where we have expanded the resolvent operator $(K - \lambda\mathbb{I})^{-1}$ as a geometric series. The series above is defined in $\mathcal{B}(\mathcal{H})$ and converges for $||K/\lambda|| < 1$. This is certainly true if $\lambda > \kappa$, where κ was given in (8.22). We see that for $\lambda > \kappa$ the solution exists and it is unique.[6]

Example 8.4. Let us find the solution of the equation

$$\int_0^1 dy(x - y)\phi(y) - \lambda\phi(x) = 1. \tag{8.28}$$

The kernel satisfies $k(x, y) = -\overline{k(y, x)}$ so the operator is anti-hermitian. To apply (8.26) we need to compute K^n. This is easily done using (8.8). Let us denote with $k^{(n)}(x, y)$ the kernel of K^n. We have

$$k^{(2)}(x, y) = \int_0^1 (x - z)(z - y)dz = \frac{x + y}{2} - xy - \frac{1}{3},$$

$$k^{(3)}(x, y) = \int_0^1 (x - z)k^{(2)}(z, y)dz = \frac{-x + y}{12} = -\frac{k(x, y)}{12}, \tag{8.29}$$

---

[6] Notice that we can also derive this result using Fredholm alternative theorem and the properties of bounded operators (see Petrini *et al.*, 2017, Section 10.1). Since the spectrum $\sigma(A)$ of a bounded operator A is contained in the disk $\{|z| \leq ||A||\}$, the eigenvalues of K and $K^\dagger$ are smaller or equal to κ. Therefore, any $\lambda > \kappa$ is not an eigenvalue of $K^\dagger$ nor of K. According to Fredholm alternative theorem, (8.20) has a unique solution for $\lambda > \kappa$.

and $k^{(2n)} = (-1/12)^{n-1}k^{(2)}$, $k^{(2n+1)} = (-1/12)^n k$ with $n \geq 1$. The solution (8.26) then can be written as

$$\phi(x) = -\frac{1}{\lambda} - \int_0^1 \left(\sum_{n=1}^{\infty} \frac{k^{(n)}(x,y)}{\lambda^{n+1}} \right) dy$$

$$= -\frac{1}{\lambda} - \int_0^1 \sum_{p=0}^{\infty} \frac{(-1)^p}{12^p \lambda^{2p+3}} (\lambda k(x,y) + k^{(2)}(x,y)) dy,$$

and, by explicit evaluation,

$$\phi(x) = \frac{6(1 - 2\lambda - 2x)}{1 + 12\lambda^2}. \tag{8.30}$$

The solution is singular for $\lambda = \pm 1/(2i\sqrt{3})$. This is to be expected since these λ are the eigenvalues of $K = -K^{\dagger}$ (see Exercise 8.7). They are purely imaginary numbers because K is anti-hermitian.

8.2.2. Degenerate kernels

Consider again the operator $K\phi(x) = \int_a^b k(x,y)\phi(y)dy$. We say that its kernel is *degenerate* if it can be written as the sum of products of functions of x and y

$$k(x,y) = \sum_{i=1}^{n} P_i(x)Q_i(y). \tag{8.31}$$

This form implies that K is of finite rank. Indeed, the operator K maps $L^2[a,b]$ into the subset spanned by the finite linear combination of the functions $P_i(x)$. For degenerate kernels the Fredholm equation, (8.20) becomes

$$\sum_{i=1}^{n} P_i(x)\zeta_i - \lambda\phi(x) = f(x), \tag{8.32}$$

where $\zeta_i = \int_a^b Q_i(y)\phi(y)dy$. Multiplying by $Q_j(x)$ and integrating over x, we reduce (8.32) to a system of linear algebraic equations for ζ_i

$$\sum_{i=1}^{n} (a_{ji} - \lambda\delta_{ji})\zeta_i = b_j, \tag{8.33}$$

where

$$a_{ij} = \int_a^b Q_i(x)P_j(x)\mathrm{d}x, \quad b_j = \int_a^b f(x)Q_j(x)\mathrm{d}x. \tag{8.34}$$

Once we have solved (8.33), the solution of the integral equation is given by (8.32)

$$\phi(x) = \frac{1}{\lambda}\left[\sum_{i=1}^n P_i(x)\zeta_i - f(x)\right]. \tag{8.35}$$

Notice that the problem can be reduced to a finite linear system because a degenerate kernel corresponds to an integral operator K of finite rank.

The set of linear equations (8.33) with $b_i = 0$ can also be used to find the eigenvalues of K. For $b_i = 0$, the system (8.33) becomes homogeneous

$$\sum_{i=1}^n (a_{ji} - \lambda\delta_{ji})\zeta_i = 0 \tag{8.36}$$

and its generic solution is $\zeta_i = 0$. Only for particular values of λ, those for which $\det(a - \lambda\mathbb{I}) = 0$, we can find non-trivial solutions ζ_i. These are the eigenvalues of a and are also the eigenvalues of K. The corresponding solutions ζ_i determine the eigenvectors $\phi_\lambda(x)$ of K through (8.35) after we set $f(x) = 0$

$$\phi_\lambda(x) = \frac{1}{\lambda}\sum_{i=1}^n P_i(x)\zeta_i. \tag{8.37}$$

Example 8.5. Consider the integral operator K in (8.21) with kernel $k(x, y) = xy(1 - xy)$. We want to solve the equation

$$\int_{-1}^1 \mathrm{d}y\, k(x, y)\phi(y) - \lambda\phi(x) = 5x^3 + x^2 - 3x. \tag{8.38}$$

As the kernel is real and symmetric, the operator K is self-adjoint. The kernel is also degenerate with $Q_1(x) = P_1(x) = x$ and $Q_2(x) = -P_2(x) = x^2$. It is easy to see that

$$b_i = \begin{pmatrix} 0 \\ \frac{2}{5} \end{pmatrix}, \quad a_{ij} = \begin{pmatrix} \frac{2}{3} & 0 \\ 0 & -\frac{2}{5} \end{pmatrix}. \tag{8.39}$$

The eigenvalues of the matrix a_{ij} correspond to the eigenvalues of K: $\lambda_1 = 2/3$ and $\lambda_2 = -2/5$. The corresponding eigenvectors are $(1,0)$ and $(0,1)$. Using (8.37) we find $\phi_{\lambda_1}(x) = x$ and $\phi_{\lambda_2}(x) = x^2$, where we rescaled the ϕ_{λ_i} by convenience. The inhomogeneous equations (8.33) give

$$\left(\frac{2}{3} - \lambda\right) \zeta_1 = 0, \quad \left(-\frac{2}{5} - \lambda\right) \zeta_2 = \frac{2}{5}. \tag{8.40}$$

If λ is different from the eigenvalues $2/3$ and $-2/5$, the system is solved by the vector $(\zeta_1, \zeta_2) = (0, -1/(1 + 5\lambda/2))$. The solution (8.35) is

$$\phi(x) = -\frac{5x^3 + x^2 - 3x}{\lambda} + \frac{2x^2}{(2 + 5\lambda)\lambda}. \tag{8.41}$$

If $\lambda = \lambda_1 = 2/3$, the Fredholm alternative theorem tells us that we can solve the equation since $f(x)$ is orthogonal to the eigenspace of λ_1, which is spanned by $\phi_{\lambda_1}(x) = x$. And indeed, the linear system (8.40) has solution $(\zeta_1, \zeta_2) = (C, -3/8)$, where C is an arbitrary constant, corresponding to

$$\phi(x) = -\frac{3(5x^3 + x^2 - 3x)}{2} + \frac{3Cx}{2} + \frac{9x^2}{16}. \tag{8.42}$$

The solution is not unique. The constant C parameterises the freedom to add the eigenvector $\phi_{\lambda_1}(x)$ associated with λ_1, again in agreement with the alternative theorem. Finally, if $\lambda = \lambda_2 = -2/5$, the alternative theorem tells us that there are no solutions since $f(x)$ is not orthogonal to the eigenspace of λ_2, which is spanned by $\phi_{\lambda_2}(x) = x^2$. And indeed, the linear system (8.40) has no solutions for $\lambda = \lambda_2 = -2/5$.

8.3. Exercises

* Exercise 8.1. Show that the limit of a sequence of compact operators is compact.

Exercise 8.2. Show that the unit sphere in a separable Hilbert space $\mathcal{H}$ is closed and bounded but not compact.

Exercise 8.3. Prove the properties (i)–(iv) of the operator $f(A)$ given in (8.11).

* **Exercise 8.4.** Consider the Hilbert space l^2 of sequences of complex numbers $\mathbf{z} = \{z_n\}$, with $n \in \mathbb{N}$, such that $\sum_{n=1}^{\infty} |z_n|^2 < \infty$. Let A be the operator in l^2 defined by

$$(A\mathbf{x})_{2n} = ix_{2n-1}/n, \quad (A\mathbf{x})_{2n-1} = -ix_{2n}/n,$$

where $(A\mathbf{x})_k$ is the k-component of the sequence $A\mathbf{x}$ and $\mathbf{x} = \{x_1, x_2, \ldots\}$.
(a) Show that A is bounded, compact and self-adjoint, and compute $\|A\|$.
(b) Find eigenvalues and eigenvectors of A.
(c) Find the spectrum of A.

* **Exercise 8.5.** Consider the regular Sturm–Liouville problem associated with the operator (5.13) with boundary conditions (5.15).
(a) Prove that the spectrum is purely discrete.
(b) Prove that the eigenvalues satisfy $\lim_{n \to \infty} \lambda_n = \infty$.

Exercise 8.6. Prove that, for a bounded operator A, $\mathrm{im}(A - \lambda\mathbb{I})^{\perp} = \ker(A^{\dagger} - \bar{\lambda}\mathbb{I})$.

Exercise 8.7. Find the eigenvalues of the operator K in Example 8.4 and solve again the integral equation with the method of degenerate kernels.

Exercise 8.8. Solve the Fredholm equation

$$\phi(x) - \mu \int_1^2 \left[\frac{1}{x} + \frac{1}{y}\right] \phi(y) dy = x$$

for $\phi(x) \in L^2[1, 2]$ and $\mu \in \mathbb{C}$.

Exercise 8.9. Discuss the spectrum of the Fredholm operator with integral kernel $k(x, y) = 6x^2 - 12xy$ in $L^2([0, 1] \times [0, 1])$. Solve, for any value of μ, the Fredholm equation

$$\phi(x) - \mu \int_0^1 k(x, y)\phi(y) dy = 1.$$

Exercise 8.10. Show that the Cauchy problem $y'' + xy = 1$ with $y(0) = y'(0) = 0$ is equivalent to the integral Volterra equation

$$y(x) = \frac{x^2}{2} + \int_0^x (t - x)ty(t) dt.$$

Exercise 8.11. Solve the Volterra equation

$$f(x) = \phi(x) - \mu \int_0^x e^{x-t}\phi(t)dt,$$

where $f(x)$ is a given function, using the method of iterations discussed in Section 8.2.1.

9

Hilbert Spaces and Quantum Mechanics

The theory of Hilbert spaces is central to the mathematical foundation of quantum physics. In quantum mechanics, the physical state of a system is given by a vector in a Hilbert space and the possible results of measures of physical quantities correspond to the eigenvalues of self-adjoint operators. The mathematical basis for the probabilistic interpretation of quantum mechanics is provided by the spectral theorem for self-adjoint operators. In this chapter we analyse all these concepts starting with the Schrödinger equation, a cornerstone of mathematical physics. This chapter is certainly better appreciated with some familiarity with the basic principles of quantum mechanics, but it can be read independently as a collection of applications of the theory of Hilbert spaces and self-adjoint operators. For a complete introduction to quantum mechanics, see Cohen-Tannoudji *et al.* (1992) and Griffiths (2016).

9.1. The Schrödinger Equation

In quantum mechanics the physical information about a particle is encoded in a *wave function* that gives the probability of finding the particle at a certain point in space. For a particle moving in $\mathbb{R}^3$, the wave function is a complex-valued function of the position $\mathbf{x} \in \mathbb{R}^3$ and the time t, and it is traditionally denoted as $\psi(\mathbf{x}, t)$. Its modulus squared $|\psi(\mathbf{x}, t)|^2$ is the probability density for the particle to be at the point $\mathbf{x}$ at the time t. More precisely,

$$\int_{\Omega} |\psi(\mathbf{x}, t)|^2 \mathrm{d}\mathbf{x} \tag{9.1}$$

is the probability density for the particle to be in the region of space $\Omega \subset \mathbb{R}^3$ at time t. Since the probability for the particle to be in the whole space must be one,

$$\int_{\mathbb{R}^3} |\psi(\mathbf{x}, t)|^2 \mathrm{d}\mathbf{x} = 1, \tag{9.2}$$

ψ must be a square-integrable function of $\mathbf{x}$. Then, for any given t, $\psi(\mathbf{x}, t)$ is an element of the Hilbert space $L^2(\mathbb{R}^3)$. Notice that the time t appears just as a parameter. We will simply write $\psi(t)$ when we want to emphasise that the wave function is a vector in a Hilbert space. Every $\psi(t)$ satisfying (9.2) determines a possible *state* of the particle and, as we will see, contains all information about the probability of the results of a measure of the position, or any other physical quantity, like energy or momentum.[1]

The time evolution of the wave function of a particle of mass m moving in a potential $V(\mathbf{x}, t)$ is governed by the *Schrödinger equation*

$$i\hbar \frac{d\psi}{dt} = -\frac{\hbar^2}{2m} \Delta \psi + V(\mathbf{x}, t)\psi, \tag{9.3}$$

where Δ is the Laplacian operator in $\mathbb{R}^3$ and $\hbar$ is the Planck constant. For simplicity, we set $\hbar = 1$ in the following. Defining the *Hamiltonian operator*

$$H(t) = -\frac{1}{2m} \Delta + V(\mathbf{x}, t) \tag{9.4}$$

we can write the Schrödinger equation as

$$i \frac{d\psi}{dt} = H(t)\psi. \tag{9.5}$$

It is important to notice that $H(t)$ is a self-adjoint operator. It is easy to see that $H(t)$ is symmetric in any domain that contains functions vanishing

---

[1] Notice that $\tilde{\psi}(x, t) = e^{i\alpha}\psi(x, t)$ with $\alpha \in \mathbb{R}$ defines the same probability distribution as $\psi(x, t)$.

at infinity with their derivatives. Indeed, by integrating by parts, we find

$$(f, H(t)g) = \int_{\mathbb{R}^3} \mathrm{d}\mathbf{x} \overline{f(\mathbf{x})} \left(-\frac{1}{2m} \Delta g(\mathbf{x}) + V(\mathbf{x}, t)g(\mathbf{x}) \right)$$

$$= \int_{\mathbb{R}^3} \mathrm{d}\mathbf{x} \overline{\left(-\frac{1}{2m} \Delta f(\mathbf{x}) + V(\mathbf{x}, t)f(\mathbf{x}) \right)} g(\mathbf{x}) = (H(t)f, g),$$

since the boundary terms vanish. When $V(\mathbf{x}, t)$ is sufficiently regular, one can find a domain $\mathcal{D}(H) \in L^2(\mathbb{R}^3)$ where $H(t)$ is self-adjoint.

In order to make contact with classical physics, we can write the Hamiltonian operator $H(t)$ in terms of *momentum operators* P_i and *position operators* Q_i, where $i = 1, 2, 3$. They are defined as follows. Q_i is the multiplication operator in $L^2(\mathbb{R}^3)$,

$$Q_i f(\mathbf{x}) = x_i f(\mathbf{x}), \tag{9.6}$$

with domain $\mathcal{D}(Q_i) = \{f(\mathbf{x}) \in L^2(\mathbb{R}^3) \,|\, x_i f(\mathbf{x}) \in L^2(\mathbb{R}^3)\}$. Q_i is self-adjoint. Indeed[2]

$$(f, Q_i g) = \int_{\mathbb{R}^3} \overline{f(\mathbf{x})} (x_i g(\mathbf{x})) \mathrm{d}\mathbf{x} = \int_{\mathbb{R}^3} \overline{(x_i f(\mathbf{x}))} g(\mathbf{x}) \mathrm{d}\mathbf{x} = (Q_i f, g).$$

P_i is instead the derivative operator[3]

$$P_i f(\mathbf{x}) = -i \frac{\partial}{\partial x_i} f(\mathbf{x}), \tag{9.7}$$

with domain $\mathcal{D}(P_i) = \{f(\mathbf{x}) \in L^2(\mathbb{R}^3) \,|\, f(\mathbf{x})_{\mathrm{ac}}, \partial_{x_i} f(\mathbf{x}) \in L^2(\mathbb{R}^3)\}$. P_i is also self-adjoint since, integrating by parts, we find

$$(f, P_i g) = \int_{\mathbb{R}^3} \overline{f(\mathbf{x})} (-i\partial_{x_i} g(\mathbf{x})) \mathrm{d}\mathbf{x} = \int_{\mathbb{R}^3} \overline{(-i\partial_{x_i} f(\mathbf{x}))} g(\mathbf{x}) \mathrm{d}\mathbf{x} = (P_i f, g).$$

The boundary terms vanish as all functions $f, g \in \mathcal{D}(P_i)$ and their derivatives vanish at infinity. The Hamiltonian operator can be formally written

---

[2] Strictly speaking, here we only prove that Q_i and P_i are symmetric. However one can show that they are also self-adjoint in the domains $\mathcal{D}(P_i)$ and $\mathcal{D}(Q_i)$ (see Petrini *et al.*, 2017, Section 10.2).

[3] Reintroducing the Planck constant we have $P_i = -i\hbar \partial/\partial x_i$.

in terms of P_i and Q_i as

$$H(t) = -\frac{1}{2m} \sum_{i=1}^{3} \frac{\partial^2}{\partial x_i^2} + V(\mathbf{x}, t) = \sum_{i=1}^{3} \frac{P_i^2}{2m} + V(\mathbf{Q}, t) = \frac{\mathbf{P}^2}{2m} + V(\mathbf{Q}, t).$$

(9.8)

Notice that $H(t)$ looks like the Hamiltonian of a particle of mass m moving in the potential $V(\mathbf{x}, t)$ in classical mechanics. Indeed, in quantum mechanics, the operators $\mathbf{P}$, $\mathbf{Q}$ and H correspond to the momentum, position and energy of a particle.

Notice also that the operators P_i and Q_i do not commute. Their commutator[4] is given by $[Q_i, P_j] = i\delta_{ij}\mathbb{I}$.[5] Indeed, we have

$$\left(Q_i P_j - P_j Q_i\right) f(\mathbf{x}) = x_i \left(-i\frac{\partial f(\mathbf{x})}{\partial x_j}\right) + i\frac{\partial}{\partial x_j}\left(x_i f(\mathbf{x})\right) = i\delta_{ij} f(\mathbf{x}).$$ (9.9)

On the other hand, P_i commutes with P_j and Q_i commutes with Q_j for any i and j.

The general rule of quantum mechanics is that the quantisation of a classical system is obtained by formally replacing the phase space variables $(\mathbf{q}, \mathbf{p})$ in the classical Hamiltonian function $H_{\text{class}}(\mathbf{p}, \mathbf{q}, t)$ with the operators $(\mathbf{P}, \mathbf{Q})$ and the classical Poisson brackets[6] with the commutators of the operators P_i and Q_i defined above. Since operators do not commute, some caution is needed in passing from classical variables to operators. For the Hamiltonian (9.8) there is no ambiguity. Indeed, each term only contains commuting operators and is separately self-adjoint. Using $(AB)^\dagger = B^\dagger A^\dagger$, we find

$$(P_i^2)^\dagger = (P_i P_i)^\dagger = P_i^\dagger P_i^\dagger = P_i P_i = P_i^2$$ (9.10)

since P_i is self-adjoint. Analogously, $V(\mathbf{Q}, t)$ is self-adjoint if the Q_i's are. For example, restricting for simplicity to polynomial potentials in one

---

[4]The *commutator* of two operators A and B is defined as $[A, B] = AB - BA$. The commutator is zero if and only if the operators commute.

[5]This identity is only valid in the intersection of the domains of P_i and Q_i.

[6]The commutation relations $[Q_i, P_j] = i\delta_{ij}\mathbb{I}$, $[Q_i, Q_j] = [P_i, P_j] = 0$ are the quantum generalisation of the classical Poisson brackets of the variables p_i and q_i, $\{q_i, p_j\} = \delta_{ij}$, $\{q_i, q_j\} = \{p_i, p_j\} = 0$.

dimension, $V(Q,t) = \sum_n c_n(t)Q^n$ with real $c_n(t)$, we find

$$V(Q,t)^\dagger = \left(\sum_n c_n(t)Q^n\right)^\dagger = \sum_n c_n(t)(Q^\dagger)^n = V(Q,t). \qquad (9.11)$$

However, for more complicated Hamiltonians we can have problems. Suppose, for example, that there is a term $p_i q_i$ in the classical Hamiltonian. As P_i and Q_i do not commute, the three different operators

$$P_i Q_i, \qquad Q_i P_i \qquad \frac{P_i Q_i + Q_i P_i}{2} \qquad (9.12)$$

all correspond to the same classical term $p_i q_i$. The correct operator is selected requiring that $H(t)$ is self-adjoint. In the above example, only $(P_i Q_i + Q_i P_i)/2$ is self-adjoint.[7] In more general situations, there might exist different operators $H(t)$ that reduce to the same classical Hamiltonian: they correspond to physically inequivalent quantisations of the same classical system. Experiments will tell us which one is realised in nature.

In quantum physics there are degrees of freedom that have no classical counterpart, like, for example, the spin of a particle. In these cases the previous discussion must be generalised by replacing $L^2(\mathbb{R}^3)$ with a more general Hilbert space. The main principle of quantum mechanics is that the possible states of a physical system are given by vectors of unit norm in a Hilbert space $\mathcal{H}$,[8] and the time evolution is governed by a self-adjoint operator $H(t)$ through the Schrödinger equation (9.5).

Example 9.1. The *spin* of a particle is an intrinsic angular momentum that has no classical counterpart. The degrees of freedom related to the spin are described using a finite-dimensional Hilbert space. For a particle of spin $1/2$, like the electron, we use $\mathcal{H} = \mathbb{C}^2$. We define three spin operators S_i acting on $\mathcal{H} = \mathbb{C}^2$ and describing the three components of an (intrinsic)

---

[7]$(P_i Q_i)^\dagger = Q_i^\dagger P_i^\dagger = Q_i P_i \neq P_i Q_i.$

[8]More precisely, physical states can be identified with *rays* in the Hilbert space, namely equivalence classes of vectors under the identification $\tilde\psi = \lambda\psi$ with $\lambda \in \mathbb{C}$ and $\lambda \neq 0$. This is because each equivalence class contains a unit norm vector. Since unit norm vectors differing by multiplication by a complex number of modulus one give the same probabilities, they contain the same physical information.

angular momentum vector **S**. Consider the Pauli matrices

$$\sigma_1 = \begin{pmatrix} 0 & 1 \\ 1 & 0 \end{pmatrix}, \qquad \sigma_2 = \begin{pmatrix} 0 & -i \\ i & 0 \end{pmatrix}, \qquad \sigma_3 = \begin{pmatrix} 1 & 0 \\ 0 & -1 \end{pmatrix}. \qquad (9.13)$$

These are self-adjoint operators, $\sigma_i^\dagger = \sigma_i$ with $i = 1, 2, 3$ and satisfy $\sigma_i^2 = \mathbb{I}$ and $\sigma_i \sigma_j = i \sum_{k=1}^{3} \epsilon_{ijk} \sigma_k$ for $i \neq j$, where $\mathbb{I}$ is the two-dimensional identity matrix and ϵ_{ijk} is a totally antisymmetric tensor with $\epsilon_{123} = 1$. The spin operators are defined as[9] $S_i = \frac{1}{2}\sigma_i$ and satisfy $[S_i, S_j] = i \sum_{j,k=1}^{3} \epsilon_{ijk} S_k$. The Hamiltonian for a spin $1/2$ particle in an external magnetic field **B** is given by

$$H = -\mathbf{B} \cdot \mathbf{S} = -\frac{1}{2}(B_1 \sigma_1 + B_2 \sigma_2 + B_3 \sigma_3), \qquad (9.14)$$

and it generalises the standard potential energy $-\mathbf{B} \cdot \mathbf{L}$ of a particle of angular momentum **L** in a magnetic field. The wave function describing a particle of spin $1/2$ is a two-dimensional complex vector, $\psi = (a, b)$, of unit norm, $|a|^2 + |b|^2 = 1$, and the Schrödinger equation (9.5) is a system of two ordinary linear differential equations for a and b.

9.2. Quantum Mechanics and Probability

As already discussed, in quantum mechanics the state of a system is given by a unit norm vector ψ in a Hilbert space $\mathcal{H}$. The physical quantities (observables) that we measure are given instead by self-adjoint operators. For instance, position, momentum and energy are associated with the operators Q, P and H, respectively. It is a postulate of quantum mechanics that any self-adjoint operator corresponds to a physically observable quantity.

Quantum mechanics is an intrinsically probabilistic theory: we can only predict the probability of the result of an experiment. Given the state ψ we can ask what are the possible results of a measure of the observable associated with the self-adjoint operator A and what are the probabilities of the results of such a measure. The answer to the two questions is in the following assertions, which can be viewed as postulates of quantum mechanics.

---

[9]Restoring $\hbar$ this relation becomes $S_i = \frac{\hbar}{2}\sigma_i$.

- The possible results of a measure of the observable associated with a self-adjoint operator A are the elements of the spectrum of A. These are the proper (discrete) eigenvalues a_n or generalised (continuous) eigenvalues a of A

$$Au_n = a_n u_n, \quad Au_a = au_a, \qquad (9.15)$$

where $u_n \in \mathcal{H}$ and u_a are suitable distributions in a distributional space containing $\mathcal{H}$.

- The probability that a measure of the observable associated with A gives the results a_n or a is given by

$$P_\psi(a_n) = |(u_n, \psi)|^2, \quad P_\psi(a) = |(u_a, \psi)|^2. \qquad (9.16)$$

The consistency of these postulates relies on the spectral theorem for self-adjoint operators.[10] First, the eigenvalues a_n or a are real, as appropriate for physical quantities. Moreover, any vector state ψ can be expanded in the generalised basis of eigenvectors of A

$$\psi = \sum_n c_n u_n + \int da \, c(a) u_a, \qquad (9.17)$$

where $c_n = (u_n, \psi)$ and $c(a) = (u_a, \psi)$. The probability of a measure of A is encoded in the square of the modulus of the Fourier coefficients of ψ with respect to the basis of generalised eigenvectors of A

$$P_\psi(a_n) = |c_n|^2, \quad P_\psi(a) = |c(a)|^2. \qquad (9.18)$$

Notice that if the state ψ coincides with an eigenvector u_n (or u_a) of A the results of a measure of A are uniquely determined: we obtain a_n (or a) with probability one.

The spectral theorem guarantees that the expansion (9.17) exists for all ψ and that it defines a probability distribution correctly normalised to one.

---

[10]Notice that the spectral theorem and, therefore, some of the following results do not hold for operators that are only symmetric. Quantum mechanics requires A to be self-adjoint. There are also various subtleties in writing (9.17) and (9.20). We refer to Gelfand and Vilenkin (1964), Petrini *et al.* (2017) and Appendix A for more precise statements.

Indeed by the generalised Parseval identity we find

$$\sum_n P_\psi(a_n) + \int da P_\psi(a) = \sum_n |c_n|^2 + \int da |c(a)|^2 = ||\psi||^2 = 1. \quad (9.19)$$

Here we are assuming that the eigenvectors are normalised as

$$(u_n, u_m) = \delta_{nm}, \quad (u_n, u_a) = 0, \quad (u_a, u_{a'}) = \delta(a - a'). \quad (9.20)$$

In the previous discussion we have implicitly assumed that the eigenvalues of A are non-degenerate. If the eigenvalues are degenerate, $Au_{n\beta} = a_n u_{n\beta}$ and $Au_{a\beta} = a u_{a\beta}$, (9.16) should be replaced by

$$P_\psi(a_n) = \sum_\beta |(u_{n\beta}, \psi)|^2, \quad P_\psi(a) = \sum_\beta |(u_{a\beta}, \psi)|^2. \quad (9.21)$$

Given the probability distribution P_ψ, we can define the *expectation value* of the observable associated with A on the state ψ as

$$\langle A \rangle = \sum_n a_n P_\psi(a_n) \quad (9.22)$$

and the *standard deviation* ΔA as

$$(\Delta A)^2 = \langle (A - \langle A \rangle)^2 \rangle. \quad (9.23)$$

These notions are familiar from probability theory. Notice that the average value of A can be written as

$$\langle A \rangle = (\psi, A\psi). \quad (9.24)$$

Proof. Assume for simplicity that the spectrum is discrete and non-degenerate. Expand $\psi = \sum_n c_n u_n$, where $Au_n = a_n u_n$ and $|c_n|^2 = P_\psi(a_n)$ by (9.18). Then

$$(\psi, A\psi) = \sum_{n,m} \bar{c}_m c_n (u_m, Au_n) = \sum_{n,m} \bar{c}_m c_n a_n (u_m, u_n) = \sum_n |c_n|^2 a_n = \langle A \rangle. \quad \square$$

Example 9.2. Consider a particle moving on the real axis. The appropriate Hilbert space is $L^2(\mathbb{R})$ and the state of the particle at a specific time t is specified by a square-integrable function $\psi(x, t) \in L^2(\mathbb{R})$ with unit norm $||\psi||^2 = \int_\mathbb{R} |\psi(x, t)|^2 dx = 1$. The self-adjoint operator corresponding to the position is the multiplication operator $Qf(x) = xf(x)$, with domain $\mathcal{D}(Q) = \{f(x) \in L^2(\mathbb{R}) \mid xf(x) \in L^2(\mathbb{R})\}$. The eigenvector equation $Qu(x) = \lambda u(x)$ becomes

$$xu(x) = \lambda u(x), \qquad (9.25)$$

and cannot be solved for any square-integrable function $u(x)$. It has instead the distributional solution $u(x) = c\delta(x - \lambda)$ for all $\lambda \in \mathbb{R}$. The spectrum of Q is then $\sigma(Q) = \mathbb{R}$ and the generalised eigenfunctions are $u_\lambda(x) = \delta(x-\lambda)$. We have normalised[11] the eigenvectors as in (9.20)

$$(u_\lambda, u_{\lambda'}) = \int_{\mathbb{R}} \delta(x - \lambda)\delta(x - \lambda')dx = \delta(\lambda - \lambda'). \qquad (9.26)$$

As expected, the possible values of a measure of the position are all the points in $\mathbb{R}$. The probability that a measure of the position of a particle prepared in the state ψ gives the value x_0, namely the probability that the particle is in x_0, is

$$P_\psi(Q = x_0) = |(u_{x_0}, \psi)|^2 = \left| \int_{\mathbb{R}} \delta(x - x_0)\psi(x,t)dx \right|^2 = |\psi(x_0,t)|^2 \quad (9.27)$$

thus reproducing our starting point (9.1). Notice that, as expected from the spectral theorem, the u_λ are a generalised basis in $L^2(\mathbb{R})$. Indeed, any $f \in L^2(\mathbb{R})$ can be expanded as

$$f(x) = \int_{\mathbb{R}} (u_\lambda, f)u_\lambda d\lambda = \int_{\mathbb{R}} f(\lambda)\delta(x - \lambda)d\lambda, \qquad (9.28)$$

which is just a tautology.

Example 9.3. Consider again a particle on the real axis. The self-adjoint operator corresponding to the momentum is the derivative operator $Pf(x) = -i\frac{d}{dx}f(x)$, with domain $\mathcal{D}(P) = \{f(x) \in L^2(\mathbb{R}) \mid f(x)_{\mathrm{ac}}, f'(x) \in L^2(\mathbb{R})\}$. The eigenvector equation $Pu(x) = \lambda u(x)$ is

$$-iu'(x) = \lambda u(x), \qquad (9.29)$$

and has solutions $u_\lambda(x) = c_\lambda e^{i\lambda x}$. The u_λ are not square-integrable but are tempered distributions for all $\lambda \in \mathbb{R}$. Thus the spectrum is $\sigma(P) = \mathbb{R}$ and

---

[11]Notice that this is a formal statement. The proper meaning of the normalisation conditions (9.20) is discussed in Petrini *et al.* (2017), Section 10.3, and Appendix A.

the generalised eigenfunctions are $u_\lambda(x) = \frac{1}{\sqrt{2\pi}}e^{i\lambda x}$. We have normalised the eigenvectors as in (9.20)

$$(u_\lambda, u_{\lambda'}) = \frac{1}{2\pi}\int_{\mathbb{R}} e^{ix(\lambda'-\lambda)}\mathrm{d}x = \delta(\lambda - \lambda'), \tag{9.30}$$

where we used (A.13). As expected, the possible values of a measure of the momentum are all the real numbers. The probability that a measure of the momentum of a particle prepared in the state ψ gives the value p is

$$P_\psi(P = p) = |(u_p, \psi)|^2 = \left|\int_{\mathbb{R}} \frac{1}{\sqrt{2\pi}}e^{-ipx}\psi(x, t)\mathrm{d}x\right|^2 = |\hat\psi(p, t)|^2, \tag{9.31}$$

where $\hat\psi$ is the Fourier transform of ψ with respect to the variable x. As expected from the spectral theorem, the u_λ are a generalised basis in $L^2(\mathbb{R})$. Indeed any $f \in L^2(\mathbb{R})$ can be expanded as

$$f(x) = \int_{\mathbb{R}} (u_p, f)u_p\mathrm{d}p = \frac{1}{\sqrt{2\pi}}\int_{\mathbb{R}} \hat f(p)e^{ipx}\mathrm{d}p. \tag{9.32}$$

This is just the Fourier integral of f.

From the previous two examples, we see that the wave function $\psi(x, t)$ determines the probability of a measure of the position while its Fourier transform determines the probability of a measure of the momentum; all information about any observable is contained in ψ.

Example 9.4. Consider the particle of spin $1/2$ described in Example 9.1. The possible results of a measure of a component S_i of the spin are given by the eigenvalues of $\sigma_i/2$. Since $\sigma_i^2 = \mathbb{I}$, the eigenvalues of σ_i are ± 1 and correspondingly those of S_i are $\pm 1/2$. Consider, for example, the component of the spin along the z-axis, S_3. Its eigenvectors are $e_+^{(z)} = (1, 0)$ and $e_-^{(z)} = (0, 1)$, corresponding to the eigenvalues $\pm 1/2$, respectively. If a system is in the state $\psi = e_+^{(z)}$ ($\psi = e_-^{(z)}$), a measure of S_3 will give $1/2$ ($-1/2$) with probability one. On the other hand, a measure of S_3 in a general state of unit norm

$$\psi = (a, b) = a(1, 0) + b(0, 1) = ae_+^{(z)} + be_-^{(z)} \tag{9.33}$$

with $|a|^2 + |b|^2 = 1$, can give either of the two values $1/2$ and $-1/2$. The probability of obtaining $\pm 1/2$ is given by (9.16)

$$P_\psi(1/2) = |(e_+^{(z)}, \psi)|^2 = |a|^2, \quad P_\psi(-1/2) = |(e_-^{(z)}, \psi)|^2 = |b|^2. \quad (9.34)$$

The two probabilities correctly sum to one. The other components of the spin are treated similarly. The eigenvectors of S_1 corresponding to the eigenvalues $\pm 1/2$ are $e_\pm^{(x)} = (1, \pm 1)/\sqrt{2}$ and those of S_2 are $e_\pm^{(y)} = (1, \pm i)/\sqrt{2}$. Consider again a system in the state $\psi = e_+^{(z)}$. A measure of S_3 will give $1/2$ with probability one, while a measure of S_1 will give two possible values $\pm 1/2$ with probability $|(e_\pm^{(x)}, \psi)|^2 = 1/2$ and, similarly, a measure of S_2 will give two possible values $\pm 1/2$ with probability $|(e_\pm^{(y)}, \psi)|^2 = 1/2$. We see that, if one component of the spin is completely determined, the other two are undetermined. This is due to the fact that S_i do not commute and it is a particular instance of the Heisenberg uncertainty principle that we discuss in Section 9.4.

9.3. Spectrum of the Hamiltonian Operator

An important role in quantum mechanics is played by the Hamiltonian operator $H(t)$. Consider a *conservative system* where H is independent of time. As in classical physics, in a conservative system energy is conserved. The possible energies of the system are given by the spectrum of the Hamiltonian operator, $\sigma(H)$, which can be found by solving

$$Hu = Eu. \quad (9.35)$$

This is called the *time-independent Schrödinger equation*. The spectral theory for self-adjoint operators tells us that $\sigma(H)$ is the disjoint union of a discrete spectrum and a continuous spectrum. The distinction between discrete and continuous spectrum reflects what we observe in nature. It is well known that the energy levels of atoms and molecules are discrete. More generally, the energy is discrete when the particle forms a bound state with other particles or is sitting in a potential well. On the other hand, when the particle is free to move in space, its energy can assume a continuum of values.

For a particle moving in one dimension in a potential $V(x)$, the eigenvalue equation (9.35) is just a second-order ordinary differential equation

$$-\frac{1}{2m}u''(x) + V(x)u(x) = Eu(x). \tag{9.36}$$

The solutions that belong to $L^2(\mathbb{R})$ give the discrete spectrum and those that belong to the space of tempered distributions, $\mathcal{S}'(\mathbb{R})$, the continuous spectrum. There is a simple qualitative way of determining the properties of the spectrum, in particular whether it is discrete or continuous, by looking at the shape of the function $V(x)$. Examples of potentials with different kinds of spectrum are given in Figure 9.1. Since the kinetic energy $\frac{p^2}{2m} = E - V(x)$ must be positive, the classical motion takes place in the region of space where the energy is greater than the potential, $E \geq V(x)$. One finds that

- there is a discrete spectrum consisting of a finite, or countable, number of eigenvalues E_n corresponding to energy values for which the classical motion takes place in a bounded region of space. All these eigenvalues satisfy $E_n > \min V(x)$;
- there is a continuous spectrum consisting of all E for which the classical motion takes place in an unbounded region of space.

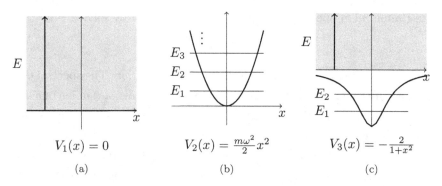

$$V_1(x) = 0 \qquad\qquad V_2(x) = \frac{m\omega^2}{2}x^2 \qquad\qquad V_3(x) = -\frac{2}{1+x^2}$$

(a) \qquad\qquad\qquad (b) \qquad\qquad\qquad (c)

Fig. 9.1. (a) The free particle with purely continuum spectrum $\sigma(H) = [0, \infty)$ (see Example 9.5). (b) The harmonic oscillator with purely discrete spectrum $E_n = \omega(n+1/2)$ (see Example 9.7). (c) A potential with a finite number N of negative eigenvalues $E_1 \leq E_2 \cdots \leq E_N < 0$ and a continuous spectrum $E \in [0, \infty)$.

Proof. Consider, for simplicity, a Hamiltonian where $V(x)$ is smooth and approaches a constant at infinity. Without loss of generality, we can assume that this constant is zero as for the potential $V_3(x)$ in Figure 9.1.(c). Since $V(x)$ is smooth, the existence and uniqueness theorem for second-order linear differential equations guarantees that (9.36) has two linearly independent continuous solutions for every E. The spectrum is given by those E for which there exists a solution that belongs to $L^2(\mathbb{R})$ (discrete spectrum) or to $\mathcal{S}'(\mathbb{R})$ (continuous spectrum). Since the solutions are continuous, they are square-integrable on any finite interval and we need only to understand how they behave at infinity. To this purpose, we can look at the asymptotic form of (9.36). For large $|x|$, $V(x) \sim 0$ and the equation reduces to $u'' = -2mEu$, which has two linearly independent solutions $u \sim \exp(\pm\sqrt{-2mE}x)$. For $E < 0$, corresponding to the regions where the classical motion is bounded, the asymptotic solutions are real exponentials at both infinities: one goes to zero and is $L^2(\mathbb{R})$, the other blows up and is not $L^2(\mathbb{R})$ nor $\mathcal{S}'(\mathbb{R})$. We can choose a relation between the two arbitrary constants appearing in the general solution such that at $x \sim -\infty$ this is $L^2(\mathbb{R})$, $u \sim Ae^{\sqrt{-2mE}x}$. At $x \sim +\infty$, the solution will be given again by a linear combination of exponentials with coefficients proportional to A and depending on the energy, $u \sim A(c_1(E)e^{-\sqrt{-2mE}x} + c_2(E)e^{\sqrt{-2mE}x})$. In order for it to be square-integrable also at $x \sim \infty$ we need to set $c_2(E) = 0$. The details of the function $c_2(E)$ depend on the particular $V(x)$, but, in general, this is an equation for E which selects a finite or countable number of values E_n: $c_2(E_n) = 0$. These values correspond to the discrete spectrum. Notice that, by multiplying (9.36) by $\bar{u}$, integrating over $\mathbb{R}$ and integrating by parts we find

$$E \int_{\mathbb{R}} |u|^2 = \int_{\mathbb{R}} \bar{u}\left(-\frac{1}{2m}u'' + Vu\right) = \int_{\mathbb{R}} \left(\frac{1}{2m}|u'|^2 + V|u|^2\right) \geq \min V(x) \int_{\mathbb{R}} |u|^2$$

which tells us that $E \geq \min V(x)$. Therefore, there are no solutions for values of the energy for which the classical motion is forbidden. In classical physics the minimum value of the energy is $E = \min V(x)$ and corresponds to a particle at rest at the bottom of the potential well. In quantum mechanics, instead, $E = \min V(x)$ is not a possible value of the energy since it requires $u' = 0$ for all x and therefore u constant, which is not square integrable. Consider now $E > 0$, corresponding to the regions where the classical motion is unbounded. The solutions of (9.36) have an oscillatory behaviour at infinity, $u \sim \exp(\pm i\sqrt{2mE}x)$, typical of a wave. These solutions are not $L^2(\mathbb{R})$ but are $\mathcal{S}'(\mathbb{R})$. For every $E > 0$ we thus find two solutions that are tempered distributions. The continuous spectrum is therefore $[0, \infty)$. The proof can be generalised to other forms of potential $V(x)$.   □

Example 9.5 (Free particle). Consider a free particle of mass m in one dimension. The potential is zero and the two independent solutions of (9.36) are $u_{E_\pm} = e^{\pm i\sqrt{2mE}x}$. For every $E \in [0, \infty)$, $u_{E_\pm}$ are tempered distributions, while for all other values of E they are not in $L^2(\mathbb{R})$ nor in $\mathcal{S}'(\mathbb{R})$. Therefore, the spectrum of the Hamiltonian is $[0, \infty)$. The possible values of the energy of a free particle are all positive numbers E. The generalised eigenvectors $u_{E_\pm}$ are also eigenvectors of the momentum operator P with eigenvalues $p = \pm\sqrt{2mE}$. This is not surprising since $H = p^2/2m$. Each generalised eigenvalue E is doubly degenerate. The two independent solutions with momentum $p = \pm\sqrt{2mE}$ correspond to particles moving in the positive or negative direction of the real axis.

Example 9.6 (Particle in an infinite well). Consider a particle of mass m that is constrained to live on the segment $[0, a]$ by two impenetrable barriers at $x = 0$ and $x = a$ but is otherwise free. This corresponds to the potential $V(x) = 0$ for $0 < x < a$ and $V(x) = \infty$ otherwise, as in Figure 9.2. The potential is singular and we need to understand how to treat the differential equation. What we will do is to consider the eigenvalue problem in $L^2[0, a]$ and mimic the effect of the infinite walls of the well by appropriate boundary conditions. The natural ones are reflection conditions where u vanishes at the boundary. We then consider the operator H in the domain

$$\mathcal{D}(H) = \{f_{ac}, f', f'' \in L^2(\mathbb{R}) | f(0) = 0, f(a) = 0\}. \qquad (9.37)$$

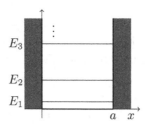

Fig. 9.2.   The infinite well has a purely discrete spectrum of energies.

One can easily check that H is self-adjoint (this is a particular Sturm–Liouville operator of the form discussed in Section 5.1.1). The classical motion is allowed for $E > 0$ and restricted to the interval $[0, a]$. We expect a discrete spectrum $E_1 \leq E_2 \leq \cdots$ with $E_1 > 0$. The eigenvalue equation $u'' = -2mEu$ is solved in terms of exponentials

$$u = A_1 e^{ikx} + A_2 e^{-ikx}, \qquad k = \sqrt{2mE}. \tag{9.38}$$

The condition $u(0) = 0$ requires $A_1 = -A_2$ and $u(a) = 0$ further requires $\sin ka = 0$, which has solutions $k = n\pi/a$ for $n = 1, 2, 3, \ldots$. The spectrum

$$\sigma(H) = \left\{ \frac{n^2 \pi^2}{2ma^2}, \quad n = 1, 2, 3, \ldots \right\} \tag{9.39}$$

is discrete. As prescribed by the spectral theorem, the set of eigenvectors

$$u_n(x) = \sqrt{\frac{2}{a}} \sin \frac{n\pi x}{a}, \tag{9.40}$$

is an orthonormal system $(u_m, u_n) = \delta_{mn}$, and is complete (the Fourier basis on the half-period). Any function in $L^2(\mathbb{R})$ can be expanded as $u(x) = \sum_n c_n u_n(x)$.

Example 9.7 (Harmonic oscillator). Consider the harmonic potential $V(x) = m\omega^2 x^2/2$ (see Figure 9.1(b)). The classical motion is allowed and bounded for any $E \geq 0$. We expect a discrete spectrum $E_1 \leq E_2 \leq \cdots$, with $E_1 > 0$. The stationary Schrödinger equation is

$$-\frac{1}{2m} u'' + \frac{m\omega^2}{2} x^2 u = Eu. \tag{9.41}$$

In terms of the new independent variable $\zeta = \sqrt{m\omega}x$ and rescaled energy $\epsilon = E/\omega$ it becomes

$$-u''(\zeta) + \zeta^2 u(\zeta) = 2\epsilon u(\zeta). \tag{9.42}$$

By a further change of variable[12] $u = ye^{-\zeta^2/2}$ we obtain the equation

$$y'' - 2\zeta y' + (2\epsilon - 1)y = 0, \tag{9.43}$$

---

[12]This change of variable is suggested by the asymptotic behaviour of the solutions. For large ζ, we can neglect the term ϵu compared to $\zeta^2 u$ and find the approximate solutions $e^{\pm \zeta^2/2}$.

with the condition that $u = ye^{-\zeta^2/2} \in L^2(\mathbb{R})$ or, equivalently, $y \in L^2_\omega(\mathbb{R})$ with $\omega(\zeta) = e^{-\zeta^2}$. This is just the Sturm–Liouville problem associated with the Hermite equation discussed in Example 5.4 with x replaced by ζ and λ by $(2\epsilon - 1)$. We thus find square-integrable solutions if and only if

$$2\epsilon = 2n + 1, \tag{9.44}$$

where $n \geq 0$ is an integer. The corresponding eigenvectors are the Hermite polynomials $H_n(\zeta)$ and therefore

$$u_n(x) = c_n H_n(\sqrt{m\omega}x)e^{-m\omega x^2/2}, \tag{9.45}$$

where the constants c_n are determined by the normalisation condition $||u_n|| = 1$. The spectrum of the Hamiltonian is then $E_n = \omega(n+1/2)$.[13] We see that the spectrum is purely discrete and the eigenvectors form a basis for the Hilbert space. Notice that all eigenvalues E_n are strictly positive, in agreement with the general statements. The lowest energy $E_0 = \omega/2$ is called the zero point energy.

Example 9.8 (The delta function potential). Consider a potential well $V(x)$ that goes to zero at infinity, like for example Figure 9.1(c). For a generic $V(x)$ the Schrödinger equation cannot be analytically solved. We consider then the limiting case of a delta function potential $V(x) = -\frac{\mu}{2m}\delta(x)$, with $\mu > 0$, which corresponds to an infinitely narrow and deep well. For $x < 0$ and $x > 0$ the potential is identically zero since the delta function is supported in $x = 0$. So we divide the real axis in two regions, I and II, corresponding to $x < 0$ and $x > 0$, respectively, and we solve the free Schrödinger equation in each of them. The solutions in region I and II are just those of the free particle

$$u_I(x) = A_1 e^{+\sqrt{-2mE}x} + A_2 e^{-\sqrt{-2mE}x}, \quad x < 0,$$
$$u_{II}(x) = A_3 e^{+\sqrt{-2mE}x} + A_4 e^{-\sqrt{-2mE}x}, \quad x > 0. \tag{9.46}$$

The effect of the delta function potential can be summarised in a matching condition that states that u must be continuous in $x = 0$ and u' must have

---

[13] Reinstating $\hbar$, we recover the well-known formula $E_n = \hbar\omega(n + 1/2)$.

a jump of $-\mu u(0)$. Indeed, integrating the equation

$$-\frac{1}{2m}u'' - \frac{\mu}{2m}\delta(x)u = Eu \qquad (9.47)$$

on a small interval $[-\epsilon, \epsilon]$ around $x = 0$, we find

$$-\frac{1}{2m}\left[u'(\epsilon) - u'(-\epsilon) + \mu u(0)\right] = \int_{-\epsilon}^{\epsilon} Eu. \qquad (9.48)$$

If we assume that u is continuous, the right-hand side vanishes for $\epsilon \to 0$, and we obtain

$$\lim_{\epsilon \to 0}\left(u'(\epsilon) - u'(-\epsilon)\right) = -\mu u(0). \qquad (9.49)$$

This equation implies that the first derivative of u has a jump at $x = 0$. Consider now the dependence of (9.46) on the values of the energy E. For $E < 0$, the solutions are real exponentials. The exponentials with coefficients A_2 and A_3 blow up at infinity and we need to set $A_2 = A_3 = 0$. The continuity of u in $x = 0$ requires $A_1 = A_4$ and the jump (9.49) in the first derivative give

$$u'_{II}(0) - u'_{I}(0) = -\sqrt{-2mE}A_4 - \sqrt{-2mE}A_1 = -\mu A_1 \qquad (9.50)$$

and therefore $2\sqrt{-2mE}A_1 = \mu A_1$. We are forced to set $E = -\mu^2/8m$, since, setting $A_1 = 0$, the solution would vanish. We thus find one discrete eigenvalue $E_0 = -\mu^2/8m$ of the Hamiltonian. The wavefunction can be written as

$$u_0(x) = \sqrt{\frac{\mu}{2}}e^{-\mu|x|/2}, \qquad (9.51)$$

where the coefficient has been determined by imposing $||u_0|| = 1$. For $E \geq 0$, the solutions (9.46) are not square-integrable at infinity but are tempered distributions. Continuity of u in $x = 0$ and the jump (9.49) in its first derivative give two algebraic equations for the four coefficients A_i, which determine, for example, A_3 and A_4 as functions of A_1 and A_2. Thus for all $E \geq 0$ we find two linearly independent distributional solutions, $u_{E_\pm}(x)$, that oscillate at infinity. The spectrum is then

$$\sigma(H) = \{E_0 = -\mu^2/8m\} \cup [0, \infty). \qquad (9.52)$$

The spectral theorem also implies that all functions in $L^2(\mathbb{R})$ can be expanded as

$$f = c_0 u_0 + \int_0^{+\infty} dE \left(c_+(E)u_{E_+} + c_-(E)u_{E_-}\right) \qquad (9.53)$$

with $c_0 = (u_0, f)$ and $c_\pm(E) = (u_{E_\pm}, f)$.

Example 9.9 (Hydrogen atom). Consider now the time-independent Schrödinger equation for a particle in a Coulomb potential $V(\mathbf{x}) = -e^2/r$, where $r = |\mathbf{x}|$. This describes, for example, an electron moving in the potential of a proton in a hydrogen atom.[14] The potential is singular in $\mathbf{x} = 0$ but one can show that the Hamiltonian operator in (9.8), $H = -\frac{1}{2m}\Delta - \frac{e^2}{r}$, is self-adjoint in the same domain $\mathcal{D}(\Delta)$ of the free particle Hamiltonian.[15] Because of the symmetry of the problem, we use spherical coordinates. The time-independent Schrödinger equation becomes

$$-\frac{1}{2m}\left(\frac{1}{r}\frac{\partial^2}{\partial r^2}(ru) + \frac{1}{r^2}\hat{\Delta}u\right) - \frac{e^2}{r}u = Eu, \qquad (9.54)$$

where $\hat{\Delta}$ is the Laplace operator on the unit sphere S^2 defined in (5.59). We can solve (9.54) by separation of variables as in Section 5.3. Setting $u(r, \theta, \phi) = r^{-1}U(r)Y(\theta, \phi)$, we find

$$-\frac{1}{2m}U''(r) + \left(-\frac{e^2}{r} + \frac{\lambda}{2mr^2}\right)U(r) = EU(r), \qquad (9.55)$$

$$\hat{\Delta}Y(\theta, \phi) + \lambda Y(\theta, \phi) = 0. \qquad (9.56)$$

The solution u must be square-integrable

$$\int |u(\mathbf{x})|^2 d\mathbf{x} = \int_0^{+\infty} |U(r)|^2 dr \int_0^{2\pi} d\phi \int_0^\pi d\theta \sin\theta |Y(\theta, \phi)|^2 \leq \infty, \qquad (9.57)$$

where the measure in spherical coordinates is $d\mathbf{x} = r^2 \sin\theta dr d\theta d\phi$. We can satisfy (9.57) by requiring that $U(r) \in L^2[0, \infty)$ and $Y \in L^2(S^2)$. As discussed in Section 5.3, $L^2(S^2)$ solutions of (9.56) exist only for $\lambda = l(l+1)$

---

[14]More precisely, we obtain this equation in the center of mass system. $\pm e$ are the charges of the proton and the electron, respectively, and m the reduced mass.
[15]The *Kato-Rellich theorem* states that the operator $H = -\Delta + V$ with $V = f + g$, where $f \in L^2(\mathbb{R}^3)$ and g is bounded, is self-adjoint in the same domain as the operator $H_0 = -\Delta$.

and are given by the spherical harmonics $Y_{lm}(\theta, \phi)$ with $l = 0, 1, \dots$ and $m = -l, -l+1, \dots, l$. We are left with equation (9.55) for $U(r)$ to be solved with the condition $U(r) \in L^2[0, \infty)$.[16] This is equivalent to solving a one-dimensional time-independent Schrödinger equation on the half-line $[0, \infty)$ with an effective potential

$$V_{\text{eff}}(r) = -\frac{e^2}{r} + \frac{l(l+1)}{2mr^2}. \tag{9.58}$$

For large r the effective potential (9.58) goes to zero, while for $r \to 0$ it goes to plus infinity if $l \geq 1$ and to minus infinity if $l = 0$. For $E < 0$ the region of classical motion is bounded for any l, and therefore we expect a set of negative discrete eigenvalues E_n. For $E \geq 0$ the classical motion is unbounded and we expect a continuous spectrum $E \in [0, \infty)$. We focus here on the discrete spectrum, which corresponds to the discrete energy levels of the hydrogen atom. Setting $\rho = me^2 r$ and $\mu = \sqrt{-2E/me^4} > 0$ we can write (9.55) as

$$U''(\rho) - \left(\frac{l(l+1)}{\rho^2} - \frac{2}{\rho} + \mu^2\right) U(\rho) = 0. \tag{9.59}$$

For large ρ the equation can be approximated by $U'' = \mu^2 U$, which has the square-integrable solutions $U(\rho) \sim e^{-\mu\rho}$. Setting $U(\rho) = e^{-\mu\rho} y(\rho)$ the equation takes the form (7.7) with a Fuchsian singularity in $\rho = 0$

$$y''(\rho) - 2\mu y'(\rho) - \left(\frac{l(l+1)}{\rho^2} - \frac{2}{\rho}\right) y(\rho) = 0. \tag{9.60}$$

We solve it by power series around $\rho = 0$. The indicial equation (7.19), $\alpha(\alpha - 1) - l(l+1) = 0$, has solutions $\alpha_1 = l + 1$ and $\alpha_2 = -l$. Only the solution corresponding to α_1 is square-integrable near $\rho = 0$.[17] Substituting $y(\rho) = \rho^{l+1} \sum_{p=0}^{\infty} c_p \rho^p$ into (9.60) we find the recursion relation

$$\frac{c_{p+1}}{c_p} = 2\frac{\mu(p+l+1) - 1}{(p+1)(p+2l+2)}, \tag{9.61}$$

---

[16] One needs to impose also the condition $U(0) = 0$ in order to avoid spurious solutions, as we will see below.

[17] For $l = 0$ also the solution corresponding to α_2, being bounded, is normalisable near $\rho = 0$. However, for such solution, the three-dimensional wave function would behave near $r = 0$ as $u = Y_{00} U(r)/r \sim const/r$, and this is not a solution of the Schrödinger equation (9.35) since $\Delta\frac{1}{r} = -4\pi\delta(\mathbf{x})$. These spurious solutions can be eliminated by requiring $U(0) = 0$.

which allows to determine all the coefficients c_p for a given c_0. In order to see if the original function $U(\rho)$ is square-integrable on $[0, \infty)$ we need to know how $y(\rho)$ grows at infinity. To estimate its behaviour for large ρ we need to estimate the behaviour of the coefficients c_p for large p. From (9.61) we see that, for large p, $c_{p+1}/c_p \sim 2\mu/(p+1)$, which implies $c_p \sim (2\mu)^p/p!$. Then

$$y(\rho) \sim \rho^{l+1} \sum_{p=0}^{\infty} \frac{(2\mu)^p}{p!} \rho^p = \rho^{l+1} e^{2\mu\rho}. \tag{9.62}$$

Therefore, for a generic μ, $U(\rho) = e^{-\mu\rho} y(\rho) \sim \rho^{l+1} e^{\mu\rho}$ is not square-integrable on $[0, \infty)$. However, for specific values of μ the power series $y(\rho)$ reduces to a polynomial. Indeed, from (9.61), we see that if $\mu = 1/(k+l+1)$ for some integer $k \geq 0$, $c_p = 0$ for all $p \geq k + 1$. Then $U(\rho)$ is square-integrable on $[0, \infty)$ only for $\mu = 1/n$, where $n = k+l+1 = 1, 2, 3, \ldots$. For such values of μ, the recursion relation (9.61) is solved by

$$c_p = (-1)^p \frac{2^p}{n^p p!} \frac{\Gamma(n-l)\Gamma(2l+2)}{\Gamma(n-l-p)\Gamma(2l+2+p)}$$

for $0 \leq p \leq n - l - 1$ and $c_p = 0$ for $p > n - l - 1$. One can check that the result can be expressed in terms of the *associated Laguerre polynomials*

$$L_q^r(\rho) = \rho^{-r} \frac{e^\rho}{q!} \frac{d^q}{d\rho^q} \left(e^{-\rho} \rho^{q+r} \right), \tag{9.63}$$

which reduce to the ordinary Laguerre polynomials for $m = 0$. Combining all information, we can write the solution as[18]

$$U_{nl}(\rho) = c_{nl} \rho^{l+1} e^{-\rho/n} L_{n-l-1}^{2l+1} \left(\frac{2\rho}{n} \right), \tag{9.64}$$

where c_{nl} is a normalisation constant. The polynomial L_{n-l-1}^{2l+1} is of degree $k = n - l - 1$ and the solution exists only if $l \leq n - 1$. In summary, we find

---

[18]It is interesting to notice that we could have obtained the same result in a different way. By further setting $y(\rho) = \rho^{l+1} h(2\mu\rho)$ into (9.60) we obtain a confluent hypergeometric equation (7.48) for h, with $c = 2l + 2$ and $a = l + 1 - 1/\mu$. Regularity at $\rho = 0$ selects the solution $\Phi(l + 1 - 1/\mu, 2l + 2, \rho)$ which reduces to a polynomial when $\mu = 1/n$ with $n \geq l + 1$ integer. One can check that $L_q^r(x) = \frac{(q+r)!}{q!r!} \Phi(-q, r + 1; x)$.

that the discrete energy levels of the hydrogen atom are $E_n = -\frac{me^4}{2n^2}$ and the corresponding eigenfunctions

$$u_{nlm}(r, \theta, \phi) = \tilde{c}_{nlm} r^l e^{-\frac{me^2 r}{n}} L_{n-l-1}^{2l+1} \left(\frac{2me^2 r}{n} \right) Y_{lm}(\theta, \phi). \tag{9.65}$$

Notice that the energy levels of the hydrogen atom are degenerate. The eigenvalues E_n only depend on the quantum number n, while the eigenvectors depend on n, l and m. Since $l \leq n - 1$ and $m = -l, \ldots, l$, the total degeneration of the level E_n is $\sum_{l=0}^{n-1}(2l + 1) = n^2$. One can show that the solutions are also eigenvectors of the angular momentum operators $\mathbf{L}^2$ and L_3 with eigenvalues $l(l + 1)$ and m, respectively (see Exercise 9.7).

Example 9.10. Consider a particle of spin $1/2$ in a magnetic field $\mathbf{B} = B_0 \mathbf{n}$ with intensity B_0 and direction $\mathbf{n} = (\sin \theta \cos \phi, \sin \theta \sin \phi, \cos \theta)$. The Hamiltonian is given by

$$H = -\mathbf{B} \cdot \mathbf{S} = -\frac{B_0}{2}(\sin \theta \cos \phi \sigma_1 + \sin \theta \sin \phi \sigma_2 + \cos \theta \sigma_3)$$

$$= -\frac{B_0}{2} \begin{pmatrix} \cos \theta & \sin \theta e^{-i\phi} \\ \sin \theta e^{i\phi} & -\cos \theta \end{pmatrix}. \tag{9.66}$$

The eigenvalues of H are $\mp B_0/2$ and the corresponding eigenvectors are given by

$$e_+^{(\mathbf{n})} = \begin{pmatrix} \cos \frac{\theta}{2} e^{-i\phi/2} \\ \sin \frac{\theta}{2} e^{i\phi/2} \end{pmatrix}, \quad e_-^{(\mathbf{n})} = \begin{pmatrix} -\sin \frac{\theta}{2} e^{-i\phi/2} \\ \cos \frac{\theta}{2} e^{i\phi/2} \end{pmatrix}. \tag{9.67}$$

These are also the eigenvectors of the projection $\mathbf{S} \cdot \mathbf{n}$ of the spin in the direction $\mathbf{n}$. The eigenstates of the Hamiltonian correspond to a spin parallel or anti-parallel to the magnetic field.

9.4. Heisenberg Uncertainty Principle

The Heisenberg uncertainty principle is one of the fundamental principles of quantum mechanics. It can be stated as follows. It is impossible to simultaneously measure with infinite precision the values of two observables whose operators do not commute. The *Heisenberg inequality* says that, in any state ψ and for any pairs of self-adjoint operators A and B,

$$(\Delta A)^2 \, (\Delta B)^2 \geq \frac{1}{4} \langle i[A, B] \rangle^2, \qquad (9.68)$$

where ΔA and ΔB are the standard deviations defined in (9.23) and are a measure of the uncertainty in the determination of a variable. Notice that the equation is valid also when the left-hand side diverges. If the two variables commute, the inequality provides no information and the uncertainties can both vanish. Vice versa, if the variables do not commute there is an intrinsic minimum value of the uncertainties when we try to simultaneously measure them. The Heisenberg principle is usually applied to position and momentum. Since $[Q, P] = i\mathbb{I}$, the uncertainty principle can be written as

$$\Delta P \, \Delta Q \geq \frac{1}{2}, \qquad (9.69)$$

or, by reintroducing the Planck constant ($P = -i\hbar d/dx$, $Q = x$), in the more familiar form

$$\Delta P \, \Delta Q \geq \frac{\hbar}{2}. \qquad (9.70)$$

The Heisenberg principle implies that if we know with absolute precision the position (momentum) of a particle, its momentum (position) is totally undetermined. The last statement means that the result of a measure of the momentum (position) can give any possible value with the same probability.

Proof. We compute averages using (9.24). Since $[A - \langle A \rangle \mathbb{I}, B - \langle B \rangle \mathbb{I}] = [A, B]$ there is no loss of generality in setting $\langle A \rangle = \langle B \rangle = 0$, so that $(\Delta A)^2 = \langle A^2 \rangle$ and $(\Delta B)^2 = \langle B^2 \rangle$. Let us now consider the operator $R = A + i\lambda B$, with $\lambda \in \mathbb{R}$. We have

$$0 \leq (R\psi, R\psi) = \lambda^2 (\Delta B)^2 + (\Delta A)^2 + \lambda \langle i[A, B] \rangle. \qquad (9.71)$$

Since $i[A, B]$ is self-adjoint,[19] $\langle i[A, B] \rangle$ is real. The right-hand side of (9.71) is then a real positive second-order polynomial in the variable λ. Therefore its discriminant can only be either zero or negative, $\langle i[A, B] \rangle^2 - 4(\Delta A)^2(\Delta B)^2 \leq 0$, and we find (9.68). □

---

[19] $(i[A, B])^\dagger = -i(AB - BA)^\dagger = -i(B^\dagger A^\dagger - A^\dagger B^\dagger) = -i(BA - AB) = i[A, B]$.

The Heisenberg principle can be reformulated as a statement in Fourier analysis. As we saw in Examples 9.2 and 9.3, the probability distributions for the position and momentum of a particle in one dimension at a fixed time are given by the modulus square of the wave function $|\psi(x)|^2$ and of its Fourier transform $|\hat{\psi}(p)|^2$, respectively. For simplicity, we omitted the time dependence. The average values of position and momentum are given by (9.24)

$$\langle Q \rangle = \int_{\mathbb{R}} x|\psi(x)|^2 \mathrm{d}x, \quad \langle P \rangle = \int_{\mathbb{R}} p|\hat{\psi}(p)|^2 \mathrm{d}p,$$

and the standard deviations by (9.23)

$$(\Delta Q)^2 = \int_{\mathbb{R}} (x - \langle Q \rangle)^2 |\psi(x)|^2 \mathrm{d}x, \quad (\Delta P)^2 = \int_{\mathbb{R}} (p - \langle P \rangle)^2 |\hat{\psi}(p)|^2 \mathrm{d}p.$$

The standard deviations give a measure of how large the distributions are. The Heisenberg principle (9.68)

$$\Delta Q \Delta P \geq 1/2 \tag{9.72}$$

implies that a non-vanishing function and its Fourier transform cannot be both simultaneously sharply localised around a point. The more one distribution is peaked around its average value, the more the other should be spread (see Exercise 9.8). The limiting case of this statement is the fact that the Fourier transform of a delta function is a constant and vice versa.

9.5. Compatible Observables

Two observables are called *compatible* if the corresponding self-adjoint operators commute. The uncertainty principle (9.68) gives no constraints on the product of their standard deviations and they can be simultaneously measured. To formulate better this concept, we need a general result about self-adjoint operators.

Let A and B be two commuting self-adjoint operators, $[A, B] = 0$, in a Hilbert space $\mathcal{H}$. They can be simultaneously diagonalised, in the sense that there exists an orthonormal basis $u_{nm\beta}$ in $\mathcal{H}$ such that

$$A u_{nm\beta} = a_n u_{nm\beta}, \quad B u_{nm\beta} = b_m u_{nm\beta}. \tag{9.73}$$

The previous equation is straightforwardly generalised to the case where the spectrum has a continuous part. The index β allows for a degeneration of the eigenspaces.

Proof. Assume for simplicity that the operators have a discrete spectrum. Being self-adjoint, A can be diagonalised. This means that $\mathcal{H}$ can be decomposed in an orthogonal sum of eigenspaces $\mathcal{H} = \bigoplus_n \mathcal{H}_n$, and A acts as the multiplication by a_n on $\mathcal{H}_n$. The action of B leaves $\mathcal{H}_n$ invariant. Indeed, if $u \in \mathcal{H}_n$ we have $Au = a_n u$ and $A(Bu) = B(Au) = B(a_n u) = a_n(Bu)$. Then $Bu \in \mathcal{H}_n$. Moreover, the restriction of B to $\mathcal{H}_n$ is a self-adjoint operator. We can then find a basis $\{u_{nm\beta}\}$ in $\mathcal{H}_n$ where the restriction of B is diagonal, $Bu_{nm\beta} = b_m u_{nm\beta}$, with β a possible index of degeneration. Since the restriction of A to $\mathcal{H}_n$ is the multiplication by a_n, we have $Au_{nm\beta} = a_n u_{nm\beta}$. Repeating this operation for all $\mathcal{H}_n$ we find a basis in $\mathcal{H}$ satisfying (9.73). $\square$

The theorem can be used to define the probability of a simultaneous measure of the observables associated with commuting operators A and B. Given the state ψ, the probability that a simultaneous measure of the observables associated with A and B gives a_n and b_m, respectively, is

$$P_\psi(a_n, b_m) = \sum_\beta |(u_{nm\beta}, \psi)|^2, \qquad (9.74)$$

where $u_{nm\beta}$ is a basis of common eigenvectors of A and B.

9.6.   Time Evolution for Conservative Systems

As already mentioned, the time evolution in quantum mechanics is governed by the Schrödinger equation (9.5). When the Hamiltonian is time-independent, $H(t) = H$, there is a very simple way to solve (9.5). The solution is

$$\psi(t) = U(t)\psi(0), \qquad (9.75)$$

where $\psi(0)$ is the state at time $t = 0$ and

$$U(t) = e^{-iHt} \qquad (9.76)$$

is the time-evolution operator. $U(t)$ is unitary if H is self-adjoint (see Exercise 9.9). Notice that the Schrödinger equation is first order and deterministic in time: if we know the state at time $t = 0$, the initial condition, the state at time t is uniquely determined.

If the state $\psi(0) = u$ is an eigenvector of the Hamiltonian with eigenvalue E, $Hu = Eu$, then the state at time t is simply given by

$$\psi(t) = e^{-iEt}u. \tag{9.77}$$

Obviously, $H\psi(t) = E\psi(t)$. We see that if a system is in the energy level E it will remain in the same energy level forever. As in classical mechanics, in a conservative system energy is conserved. A state of the form (9.77) is called *stationary state*. This is because, for a particle in a stationary state the probability of a measure of its position in space, $|\psi(\mathbf{x}, t)|^2 = |e^{-iEt}u(\mathbf{x})|^2 = |u(\mathbf{x})|^2$, is independent of time.

The time evolution of a generic state $\psi(0)$ is more complicated. Any $\psi(0)$ can be expanded in the basis of eigenvectors of the Hamiltonian H,

$$\psi(0) = \sum_n c_n u_n + \int dE c(E) u_E, \tag{9.78}$$

where $Hu_n = E_n u_n$ and $Hu_E = Eu_E$. The state at time t is given by (9.75)

$$\psi(t) = e^{-itH}\psi(0) = \sum_n c_n e^{-iE_n t} u_n + \int dE c(E) e^{-iEt} u_E, \tag{9.79}$$

where we used the linearity of the operator e^{-itH} and

$$e^{-itH}u_n = e^{-itE_n}u_n, \quad e^{-itH}u_E = e^{-itE}u_E. \tag{9.80}$$

We see that the energy basis is well adapted to compute the time evolution of a state. For any time-independent Hamiltonian and in any state ψ the probability of a measure of the energy at time t is then given by

$$P(E_n) = |c_n e^{-iE_n t}|^2 = |c_n|^2, \qquad P(E) = |c_E e^{-iEt}|^2 = |c_E|^2, \tag{9.81}$$

and is independent of time. This is the incarnation of the conservation of energy in quantum mechanics.

The probability of the measure of a generic observable A obviously depends on time. There are few observables, however, whose probability is time-independent. These are the *constants of motion*. They can be defined as the observables corresponding to self-adjoint operators A that are time-independent and commute with the Hamiltonian, $[A, H] = 0$. Indeed, as we saw in Section 9.5, if $[A, H] = 0$, we can find a common basis of eigenvectors for H and A

$$H u_{nm\beta} = E_n u_{nm\beta}, \qquad A u_{nm\beta} = a_m u_{nm\beta}, \qquad (9.82)$$

where β is a degeneration index and, for simplicity of notations, we assumed discrete spectrum. The state at time $t = 0$ can be decomposed in the common basis $\psi(0) = \sum_{nm\beta} c_{nm\beta} u_{nm\beta}$. Its time evolution is given as before by

$$\psi(t) = \sum_{nm\beta} c_{nm\beta} u_{nm\beta} e^{-iE_n t}. \qquad (9.83)$$

The probability that a measure of A gives the value a_m is the given by (9.21)

$$P(a_m) = \sum_{n\beta} |(u_{nm\beta}, \psi(t))|^2 = \sum_{n\beta} |c_{nm\beta}|^2, \qquad (9.84)$$

and it is clearly independent of time.

Example 9.11. The state $\psi(x, 0) = u_p(x) = \frac{1}{\sqrt{2\pi}} e^{ipx}$ describes a particle in one dimension with momentum p. If the particle is free, $\psi(x, 0)$ is also an eigenvector of the Hamiltonian $H = \frac{p^2}{2m}$ with eigenvalue $E = \frac{p^2}{2m}$. The time evolution is then given by

$$\psi(x, t) = e^{-iHt} \psi(x, 0) = \frac{1}{\sqrt{2\pi}} e^{ipx - iEt}, \qquad (9.85)$$

which is just a monochromatic wave propagating along the real axis. The wave number k and the frequency ω of the wave are related to the momentum and energy of the particle by *de Broglie relations*[20]

$$k = p, \qquad \omega = E. \qquad (9.86)$$

These relations are the basis of the wave/particle duality of quantum physics. Notice that the probability of finding a particle described by the monochromatic wave (9.85) at the position x is constant and given by

---

[20]Reintroducing $\hbar$, we have $p = \hbar k$ and $E = \hbar \omega$.

$|\psi(x,t)|^2 = \frac{1}{2\pi}$. This is a manifestation of the Heisenberg principle: since the particle has determined momentum p, the probability for the position is completely undetermined. A generic wave function $\psi(x,0)$ is a superposition of waves u_p,

$$\psi(x,0) = \frac{1}{\sqrt{2\pi}} \int \hat{\psi}(p,0)e^{ipx}dp, \tag{9.87}$$

where $\hat{\psi}(p,0)$ is the Fourier transform of $\psi(x,0)$. Its time evolution is given by (9.76)

$$\psi(x,t) = \frac{1}{\sqrt{2\pi}} \int \hat{\psi}(p,0)e^{ipx-i\frac{p^2}{2m}t}dp. \tag{9.88}$$

Equation (9.88) represents a wave-packet.

Example 9.12. Consider the Hamiltonian $H_0 = E\mathbb{I}$ describing two degenerate energy levels of a system. A basis of eigenvectors is given by $e_1 = (1,0)$ and $e_2 = (0,1)$. Suppose that the system is in the stationary state $\psi(0) = e_1$ at time $t = 0$ and turn on an off-diagonal perturbation so that the full Hamiltonian is $H = \begin{pmatrix} E & A \\ A & E \end{pmatrix}$. Since $(1,0)$ is not an eigenvector of the Hamiltonian H, the state is no longer stationary. The eigenvalues of H are $E_\pm = E \pm A$ with corresponding eigenvectors $e_\pm = (1,\pm1)/\sqrt{2}$. We can write the initial state as $\psi(0) = e_1 = (e_+ + e_-)/\sqrt{2}$. The state at time t is given by (9.76) and reads $\psi(t) = (e^{-iE_+t}e_+ + e^{-iE_-t}e_-)/\sqrt{2} = e^{-iEt}(\cos At, -i\sin At)$. We see that there is a non-zero probability that, due to the perturbation A, the system originally in the state e_1 jumps to e_2: $|(\psi(t), e_2)|^2 = \sin^2(At)$. The system actually oscillates between the two states.

9.7. Dirac Notation

The *Dirac notation* is a widely used notation in quantum mechanics introduced by Dirac and based on Riesz's theorem. Recall that Riesz's theorem states that a Hilbert space $\mathcal{H}$ and its dual $\mathcal{H}^*$[21] are "canonically isomorphic" (see Petrini *et al.*, 2017, Section 10.2.1). More precisely, the map between

---

[21]The dual space $\mathcal{H}^*$ is the vector space of continuous linear functionals $\alpha_{\mathbf{x}}$, $\alpha_{\mathbf{x}}(\mathbf{y}) = (\mathbf{x}, \mathbf{y})$ where $\mathbf{x}, \mathbf{y} \in \mathcal{H}$.

$\mathcal{H}$ and $\mathcal{H}^*$ that sends the vector $\mathbf{x} \in \mathcal{H}$ to the continuous linear functional $\alpha_{\mathbf{x}} \in \mathcal{H}^*$

$$\alpha_{\mathbf{x}}(\mathbf{y}) = (\mathbf{x}, \mathbf{y}), \qquad (9.89)$$

is anti-linear and bijective. In Dirac's notation, the vector $\mathbf{x} \in \mathcal{H}$ is denoted as $|\mathbf{x}\rangle$ and the dual vector $\alpha_{\mathbf{x}}$ with $\langle \mathbf{x}|$. Given two vectors $\mathbf{x}, \mathbf{y} \in \mathcal{H}$, the action of the functional $\langle \mathbf{x}|$ on the vector $|\mathbf{y}\rangle$ is denoted with $\langle \mathbf{x}|\mathbf{y}\rangle$ and, according to (9.89), is given by the scalar product

$$\langle \mathbf{x}|\mathbf{y}\rangle = (\mathbf{x}, \mathbf{y}). \qquad (9.90)$$

Dirac's notation is thus a sort of deconstruction of the scalar product. Dirac also called $\langle \mathbf{y}|$ a *bra* and $|\mathbf{x}\rangle$ a *ket*, since their juxtaposition is a bracket.

In Dirac's notation it is common to denote vectors in $\mathcal{H}$ just with letters, numbers or symbols, like $|A\rangle, |n\rangle, |\uparrow\rangle$. For examples, the eigenvector equation, $A\mathbf{u}_n = a_n \mathbf{u}_n$, is usually abbreviated as $A|n\rangle = a_n|n\rangle$. Similarly the eigenvectors of the hydrogen atom (9.65) are commonly written as $|nlm\rangle$ using only their quantum numbers.

Dirac's notation is useful for formal computations. For example, given a Hilbert basis labelled by the integers n, $|n\rangle$ with $\langle n|m\rangle = \delta_{nm}$, the expression

$$A = \sum_{nm} a_{nm} |n\rangle\langle m|, \qquad (9.91)$$

defines a linear operator. Its action on a Hilbert basis $|p\rangle$ is given by

$$A|p\rangle = \sum_{nm} a_{nm} |n\rangle\langle m|p\rangle = \sum_{nm} a_{nm} |n\rangle \delta_{pm} = \sum_{n} a_{np} |n\rangle, \qquad (9.92)$$

and it is extended by linearity to all vectors $|\mathbf{x}\rangle = \sum_p c_p |p\rangle$,

$$A|\mathbf{x}\rangle = \sum_{p} c_p \sum_{nm} a_{nm} |n\rangle\langle m|p\rangle = \sum_{np} c_p a_{np} |n\rangle. \qquad (9.93)$$

The coefficients a_{nm} are just the matrix elements of the operator A, $a_{nm} = \langle n|A|m\rangle$. A very interesting and useful identity is the following:

$$\sum |n\rangle\langle n| = \mathbb{I}, \qquad (9.94)$$

where $|n\rangle$ is a basis and $\mathbb{I}$ is the identity operator. This follows from (9.91) since the matrix element of the identity $\mathbb{I}$ are $\langle n|\mathbb{I}|m\rangle = \langle n|m\rangle = \delta_{nm}$. Equation (9.94) is usually referred to in quantum mechanics textbooks as the completeness condition for the orthonormal system $|n\rangle$. A standard trick is to insert expression (9.94) into an equation in order to simplify it or to do formal manipulations. For example, given a vector $|\mathbf{x}\rangle$, we have

$$\langle \mathbf{x}|\mathbf{x}\rangle = \langle \mathbf{x}|\mathbb{I}|\mathbf{x}\rangle = \sum_n \langle \mathbf{x}|n\rangle\langle n|\mathbf{x}\rangle = \sum_n |\langle \mathbf{x}|n\rangle|^2, \tag{9.95}$$

which is just Parseval's identity.

As already mentioned, the map $|\mathbf{x}\rangle \to \langle \mathbf{x}|$ between vectors and dual vectors is anti-linear. Under this map

$$|\mathbf{w}\rangle = \alpha|\mathbf{x}\rangle + \beta|\mathbf{y}\rangle \to \langle \mathbf{w}| = \bar{\alpha}\langle \mathbf{x}| + \bar{\beta}\langle \mathbf{y}|, \tag{9.96}$$

$$|\mathbf{w}\rangle = A|\mathbf{x}\rangle \to \langle \mathbf{w}| = \langle \mathbf{x}|A^\dagger. \tag{9.97}$$

The last equation follows by writing the identity $(A\mathbf{x}, \mathbf{y}) = (\mathbf{x}, A^\dagger\mathbf{y})$ in Dirac's notation. In particular, the norm square $\langle \mathbf{w}|\mathbf{w}\rangle$ of the vector $|\mathbf{w}\rangle = A|\mathbf{x}\rangle$ is given by

$$||A|\mathbf{x}\rangle||^2 = \langle \mathbf{x}|A^\dagger A|\mathbf{x}\rangle. \tag{9.98}$$

9.8. WKB Method

The WKB method[22] is a procedure to find approximate solutions to linear differential equations. It applies to equations where the highest derivative is multiplied by a small parameter. As its main application is quantum mechanics, we will illustrate how it works for the time-independent Schrödinger equation. In this case the small parameter is $\hbar$ and, therefore, the WKB limit is associated with a semi-classical approximation.

Let us consider the one-dimensional time-independent Schrödinger equation

$$-\frac{\hbar^2}{2m}\frac{d^2\psi}{dx^2} + V(x)\psi = E\psi \tag{9.99}$$

---

[22]The WKB method owes the name to G. Wentzel, H. A. Kramers, and L. Brillouin, who applied it to quantum mechanics.

for a particle in a potential $V(x)$, where we have reinstated the Planck constant $\hbar$. Using the ansatz

$$\psi(x) = e^{\frac{i}{\hbar}S(x)}, \tag{9.100}$$

with S not necessarily real, we obtain the equation

$$\frac{1}{2m}\left(\frac{dS}{dx}\right)^2 - \frac{i\hbar}{2m}\frac{d^2S}{dx^2} = E - V(x). \tag{9.101}$$

The idea is then to study the limit of small $\hbar$ expanding S in power series,[23]

$$S = S_0 + \hbar S_1 + \hbar^2 S_2 + \cdots, \tag{9.102}$$

and solving the equation recursively order by order in $\hbar$. For $\hbar = 0$, the equation becomes

$$\frac{1}{2m}\left(\frac{dS_0}{dx}\right)^2 = E - V(x). \tag{9.103}$$

In classical mechanics (9.103) is known as the stationary Hamilton–Jacobi equation, and determines the action of a particle of conserved energy E. The Hamilton–Jacobi equation (9.103) is solved by

$$S_0(x) = \pm \int dx' \sqrt{2m[E - V(x')]} = \pm \int dx'\, p(x'),$$

where $p(x)$ indicates the "classical" momentum. The quantity S_0 can be identified with the classical action of the particle. The classical motion is allowed only for $E - V(x) > 0$. We can then identify two regions

- *classically allowed region:* $E - V(x) > 0$, $p(x)$ is real and

$$S_0 = \pm \int dx'\, p(x').$$

- *classically forbidden region:* $E - V(x) < 0$, $p(x)$ is imaginary and

$$S_0 = \pm i \int dx'\, |p(x')|.$$

---

[23]It should be noticed that the series is typically only an asymptotic one.

The WKB approximation holds when the term with $\hbar$ in equation (9.101) is very small with respect to the rest. Assuming that the potential is slowly varying compared to the variation of S, the condition is

$$\left|\frac{S''}{\hbar}\right| \ll \left|\frac{S'}{\hbar}\right|^2, \tag{9.104}$$

or, by introducing the *de Broglie wavelength* $\lambda(x) = 2\pi\hbar/p(x)$ (see Example 9.11),

$$\left|\frac{d\lambda}{dx}\right| \ll 1. \tag{9.105}$$

The physical meaning of the WKB approximation is then that the wavelength of the particle must vary slowly over distances of the order of itself.

We can now go to higher order in $\hbar$ in the expansion of (9.101). At order $\hbar$ we find

$$S_0' S_1' - \frac{i}{2} S_0'' = 0,$$

which can be solved as

$$S_1 = \frac{i}{2} \log(S_0').$$

Up to first order in $\hbar$, the WKB solutions can thus be written as

$$\psi_c(x) = \frac{A}{\sqrt{p(x)}} e^{\frac{i}{\hbar} \int^x p(x')dx'} + \frac{B}{\sqrt{p(x)}} e^{-\frac{i}{\hbar} \int^x p(x')dx'} \tag{9.106}$$

in the classically allowed region and

$$\psi_f(x) = \frac{C}{\sqrt{|p(x)|}} e^{\frac{1}{\hbar} \int^x |p(x')|dx'} + \frac{D}{\sqrt{|p(x)|}} e^{-\frac{1}{\hbar} \int^x |p(x')|dx'} \tag{9.107}$$

in classically forbidden region. While in classical physics the motion is confined to the region $E \geq V(x)$, in quantum mechanics there is a non-zero probability of finding the particle also in the forbidden region if the potential energy barrier is not infinite.[24] It is also interesting to observe that $|\psi|^2 \sim \frac{C}{p(x)}$ in the classical region, as expected classically: the particle spend less time where it moves more rapidly.

---

[24] In physical applications this probability goes to zero exponentially with the size and height of the potential barrier.

The WKB approximation breaks down in the regions close to the classical *turning points*, namely the roots of the equation $V(x) = E$, since the potential is rapidly varying and the conditions (9.104) and (9.105) are not satisfied. As the classical momentum vanishes at the turning points, we also see that the wave functions (9.106) and (9.107) diverge there. On the other hand, the Schrödinger equation is regular also at the turning points. Then the idea behind the WKB approximation is to solve exactly the Schrödinger equation in a neighbourhood of the turning points, choosing a suitable approximation of the potential, and to patch the solution found this way to the solutions (9.106) and (9.107). To see how this works we consider the case where the potential $V(x)$ is replaced by its linear approximation.

Let x^* be a turning point, $V(x^*) = E$. In its neighbourhood, assuming $V'(x^*) \neq 0$, the potential can be approximated by

$$\frac{2m}{\hbar^2}[E - V(x)] = \gamma(x^* - x), \tag{9.108}$$

where $\gamma = \frac{2mV'(x^*)}{\hbar^2}$ is a constant. The classically forbidden region is on the right of the turning point if $\gamma > 0$, or on the left if $\gamma < 0$ (see Figure 9.3). Changing variable to $x - x^* = \gamma^{-1/3}\xi$, the Schrödinger equation linearised around the turning point x^* takes the Airy's form (see Example 3.10)

$$\psi''(\xi) - \xi\psi(\xi) = 0. \tag{9.109}$$

Then the exact solution around the turning point can be written in terms of the Airy and Bairy functions as

$$\psi_e(x) = C_1 Ai(\xi) + C_2 Bi(\xi). \tag{9.110}$$

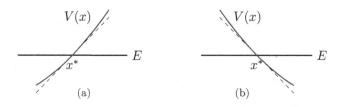

Fig. 9.3.   Turning points for $\gamma > 0$ (a) and $\gamma < 0$ (b).

This exact solution should be patched with (9.106) and (9.107) for small $\hbar$. Since $\gamma \sim \hbar^{-2}$, the limit $\hbar \to 0$ corresponds to $\xi \to \infty$ and we can use for ψ_e the asymptotic expansions of Ai and Bi given in Example 3.10. By patching the asymptotic expansions of ψ_e to the WKB solutions ψ_c and ψ_f in the neighbourhood of x^*, we find the *WKB connection formulae*. For $V'(x^*) > 0$, the solutions in the classical and the forbidden regions are related as follows:

$$\psi_c(x) = \frac{2D}{\sqrt{p(x)}} \cos\left[\frac{1}{\hbar}\int_x^{x^*} p(x')\mathrm{d}x' - \frac{\pi}{4}\right]$$

$$- \frac{C}{\sqrt{p(x)}} \sin\left[\frac{1}{\hbar}\int_x^{x^*} p(x')\mathrm{d}x' - \frac{\pi}{4}\right], \tag{9.111}$$

$$\psi_f(x) = \frac{D}{\sqrt{|p(x)|}} e^{-\frac{1}{\hbar}\int_{x^*}^x |p(x')|\mathrm{d}x'} + \frac{C}{\sqrt{|p(x)|}} e^{\frac{1}{\hbar}\int_{x^*}^x |p(x')|\mathrm{d}x'}. \tag{9.112}$$

For $V'(x^*) < 0$, we have instead

$$\psi_c(x) = \frac{2C}{\sqrt{p(x)}} \cos\left[\frac{1}{\hbar}\int_{x^*}^x p(x')\mathrm{d}x' - \frac{\pi}{4}\right]$$

$$- \frac{D}{\sqrt{p(x)}} \sin\left[\frac{1}{\hbar}\int_{x^*}^x p(x')\mathrm{d}x' - \frac{\pi}{4}\right], \tag{9.113}$$

$$\psi_f(x) = \frac{C}{\sqrt{|p(x)|}} e^{-\frac{1}{\hbar}\int_x^{x^*} |p(x')|\mathrm{d}x'} + \frac{D}{\sqrt{|p(x)|}} e^{\frac{1}{\hbar}\int_x^{x^*} |p(x')|\mathrm{d}x'}. \tag{9.114}$$

Proof. Consider the case $V'(x^*) > 0$. The classically allowed and forbidden regions are $x < x^*$ and $x > x^*$, respectively. The limit $\hbar \to 0$ corresponds to $\xi \to +\infty$ and $\xi \to -\infty$ for $x > x^*$ and $x < x^*$, respectively. The asymptotic behaviour of (9.110) in the classically forbidden region can be read from (3.54) and (3.58),

$$\psi_e(x) \sim \frac{C_1}{2\sqrt{\pi}} \frac{1}{\xi^{1/4}} e^{-\frac{2}{3}\xi^{3/2}} + \frac{C_2}{\sqrt{\pi}} \frac{1}{\xi^{1/4}} e^{+\frac{2}{3}\xi^{3/2}}. \tag{9.115}$$

We need to compare it with the WKB solution (9.107) for ψ_f. Since $|p(x)| = \hbar\sqrt{\gamma}\sqrt{x - x^*} = \hbar\gamma^{1/3}\sqrt{\xi}$, (9.107) becomes

$$\psi_f(x) \sim \frac{C}{\sqrt{\hbar}\gamma^{1/6}\xi^{1/4}} e^{\frac{2}{3}\xi^{3/2}} + \frac{D}{\sqrt{\hbar}\gamma^{1/6}\xi^{1/4}} e^{-\frac{2}{3}\xi^{3/2}}. \tag{9.116}$$

Comparing (9.115) and (9.116) we find $C_1 = 2D\sqrt{\frac{\pi}{\hbar}}\gamma^{-1/6}$ and $C_2 = C\sqrt{\frac{\pi}{\hbar}}\gamma^{-1/6}$. The same reasoning can be applied to the classical region, where now we take $\xi \to -\infty$. Using (3.56) and (3.59) we find

$$\psi_e(x) \sim \frac{C_1}{\sqrt{\pi}|\xi|^{1/4}} \cos\left(\frac{2}{3}|\xi|^{3/2} - \frac{\pi}{4}\right) - \frac{C_2}{\sqrt{\pi}|\xi|^{1/4}} \sin\left(\frac{2}{3}|\xi|^{3/2} - \frac{\pi}{4}\right). \quad (9.117)$$

Since $p(x) = \hbar\sqrt{\gamma}\sqrt{x^* - x} = \hbar\gamma^{1/3}\sqrt{|\xi|}$, the WKB solution (9.106) gives

$$\psi_c(x) \sim \frac{A}{\sqrt{\hbar}\gamma^{1/6}|\xi|^{1/4}} e^{-\frac{2}{3}i|\xi|^{3/2}} + \frac{B}{\sqrt{\hbar}\gamma^{1/6}|\xi|^{1/4}} e^{\frac{2}{3}i|\xi|^{3/2}}. \quad (9.118)$$

Comparing (9.117) and (9.118) and using the relation between C_1, C_2 and C, D, we find

$$A = \frac{2D - iC}{2}e^{i\frac{\pi}{4}}, \qquad B = \frac{2D + iC}{2}e^{-i\frac{\pi}{4}}. \quad (9.119)$$

Substituting in (9.106) and rearranging terms we find the connection formulae (9.111) and (9.112). The same kind of reasoning, or a simple symmetry argument, leads to the connection formulas for $V'(x^*) < 0$, where the forbidden region precedes the classical one. $\qquad \square$

The connection formulae must be used with some care. In the forbidden region one of the two terms is exponentially small compared to the other and, in general, it is not correct to keep both terms. Indeed, we expanded the result to first order in $\hbar$. Any powers of $\hbar$ that we have neglected in the expansion of the exponentially large term is certainly bigger than the exponentially small term. As an example of wrong use of the connection formulae consider the following situation. Suppose that the solution in the classical region is given by (9.111) with $C = 0$. Using the connection formula (9.112) we would conclude that, in the forbidden region $x > x^*$, the solution is exponentially decreasing with x. However, we cannot necessarily conclude that the coefficient of the exponentially large term is zero. Our result only shows that it is at least of order $\hbar$ and a more refined analysis is necessary. For this reason sometimes the WKB connection formula are written with a *direction*. It is certainly true that, in going from $x > x^*$ to $x < x^*$,

$$\frac{D}{\sqrt{|p(x)|}}e^{-\frac{1}{\hbar}\int_{x^*}^{x}|p(x')|dx'} \to \frac{2D}{\sqrt{p(x)}}\cos\left[\frac{1}{\hbar}\int_{x}^{x^*}p(x')dx' - \frac{\pi}{4}\right], \quad (9.120)$$

while the opposite direction is not necessarily true. A similar problem occurs in going from the forbidden region to the classical one. Since, in the forbidden region, one term is exponentially small compared to the other, it would be reasonable to neglect it. However, if C and D in (9.112) are of the same order in $\hbar$, we need to keep both terms in order to obtain the right expression in (9.111), since the two trigonometric functions are of comparable magnitude. In other words, in order to use correctly the connection formulae in this case we need to know with sufficient precision both coefficients C and D. We can nevertheless reliably use the WKB method in a number of interesting situations. For instance, the WKB method can be used to find the approximate energy levels without knowing the exact solution of the Schrödinger equation or to study scattering problems.

Example 9.13. Consider a particle in the potential well in Figure 9.4(a). We know that the spectrum is discrete. The energy levels are given by those E for which there exists a square-integrable solution. Let x_1 and x_2 be the turning points corresponding to the value E of the energy. In the classically forbidden regions $x < x_1$ and $x > x_2$, which we denote as regions I and III, the solution is given by (9.107). A square integrable solution must vanish for $x \to \pm\infty$, and therefore, in each region, we keep only the exponentials that decrease for large $|x|$

$$\psi_I(x) = \frac{A}{\sqrt{p(x)}} \exp\left(-\frac{1}{\hbar} \int_x^{x_1} |p(x')| \mathrm{d}x'\right), \tag{9.121}$$

$$\psi_{III}(x) = \frac{A'}{\sqrt{p(x)}} \exp\left(-\frac{1}{\hbar} \int_{x_2}^x |p(x')| \mathrm{d}x'\right). \tag{9.122}$$

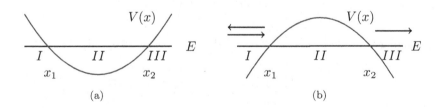

Fig. 9.4.   Potential well (a) and potential barrier (b).

Starting from region I and using the connection formulae (9.113) and (9.114), we find the solution in the classical region $x_1 < x < x_2$, which we denote as region II,

$$\psi_{II}(x) = \frac{2A}{\sqrt{p(x)}} \cos \left(\frac{1}{\hbar} \int_{x_1}^x p(x') \mathrm{d}x' - \frac{\pi}{4} \right). \tag{9.123}$$

Similarly, starting from region III and using the connection formulae (9.111) and (9.112), we find an alternative form for the solution in region II,

$$\psi_{II}(x) = \frac{2A'}{\sqrt{p(x)}} \cos \left(\frac{1}{\hbar} \int_x^{x_2} p(x') \, \mathrm{d}x' - \frac{\pi}{4} \right)$$

$$= \frac{2A'}{\sqrt{p(x)}} \cos \left(\frac{1}{\hbar} \int_{x_2}^x p(x') \mathrm{d}x' + \frac{\pi}{4} \right). \tag{9.124}$$

The two expressions (9.123) and (9.124) must be equal. Clearly, this is only possible if the arguments of the cosines coincide modulo π and we find the condition[25]

$$\int_{x_1}^{x_2} p(x') \mathrm{d}x' = \left(n + \frac{1}{2} \right) \pi \hbar. \tag{9.125}$$

Since $p(x) > 0$ in the classical region, $n = 0, 1, 2, \ldots$. This is the *Bohr–Sommerfeld quantisation rule* of the so-called *Old Quantum Theory*, and implicitly determines the eigenvalues E_n. As an example, consider the harmonic oscillator potential

$$V(x) = \frac{1}{2} m\omega^2 x^2.$$

Here the turning points $-x_1 = x_2 = \sqrt{\frac{2E}{m\omega^2}}$ are symmetric with respect to the origin and we find

$$\int_{x_1}^{x_2} p(x') \mathrm{d}x' = 2\sqrt{2m} \int_0^{x_2} \mathrm{d}x' \sqrt{E - \frac{1}{2} m\omega^2 x'^2}.$$

Changing variable to $y = \sqrt{\frac{m\omega^2}{2E}} x'$ and using (9.125), we find

$$\frac{4E}{\omega} \int_0^1 \mathrm{d}y \sqrt{1 - y^2} = \frac{\pi E}{\omega} = \left(n + \frac{1}{2} \right) \pi \hbar, \tag{9.126}$$

---

[25] The constants are related by $A' = (-1)^n A$.

and the eigenvalues

$$E_n = \left(n + \frac{1}{2}\right)\hbar\omega.$$

Notice that these are the exact energy levels of the harmonic oscillator (see Example 9.7), even though the wave functions are not. For a generic potential the WKB method only gives approximate energy levels. In general, one can show that, the larger is the number n, the better is the approximation (correspondence principle).[26]

Example 9.14 (Scattering from a potential barrier). Consider a potential of the form in Figure 9.4(b). The classically allowed regions are now $x < x_1$ and $x > x_2$, which are denoted by I and III, respectively. We want to study the motion of a particle moving from left to right with energy E smaller than the maximum of V. Classically the particle would hit the barrier and be bounced back. Quantum mechanically there is a non-zero probability that the particle is transmitted beyond the barrier. This is called *tunnel effect.* We look for solutions of the form

$$\psi_I(x) = \frac{A_1}{\sqrt{p(x)}} e^{\frac{i}{\hbar}\int^x p(x')\mathrm{d}x'} + \frac{A_2}{\sqrt{p(x)}} e^{-\frac{i}{\hbar}\int^x p(x')\mathrm{d}x'} \qquad (9.127)$$

in the classical region $x < x_1$, and

$$\psi_{III}(x) = \frac{A'}{\sqrt{p(x)}} e^{\frac{i}{\hbar}\int^x p(x')\mathrm{d}x'} \qquad (9.128)$$

in the classical region $x > x_2$. In the region $x < x_1$ both an incident and a reflected wave are present, while in the region $x > x_2$ there is only a transmitted wave. We can use the connection formulae to determine the solution in region II. At the right of the point $x = x_2$ the solution is of the form (9.113) with $D = -2iC = -iA'e^{i\pi/4}$. Using (9.114) we find the

---

[26] In this example, this is due to the fact that the eigenvalue E_n and therefore also $S_0 = \int \mathrm{d}x \sqrt{2m(E_n - V(x))}$ grow with n. The WKB approximation requires $S_0 \gg \hbar S_1$. This can either be obtained by taking $\hbar$ small, or, at fixed $\hbar$, E_n large. So the larger is E_n the better is the WKB approximation. The argument is similar for other types of potentials.

solution in the forbidden region,

$$\psi_{II} = \frac{A'e^{i\frac{\pi}{4}}}{2\sqrt{|p(x)|}}[e^{-\frac{1}{\hbar}\int_x^{x_2}|p(x')|dx'} - 2ie^{+\frac{1}{\hbar}\int_x^{x_2}|p(x')|dx'}]$$

$$= \frac{A'e^{i\frac{\pi}{4}}}{2\sqrt{|p(x)|}}[Be^{\frac{1}{\hbar}\int_{x_1}^x|p(x')|dx'} - 2iB^{-1}e^{-\frac{1}{\hbar}\int_{x_1}^x|p(x')|dx'}], \quad (9.129)$$

where we introduced the quantity

$$B = e^{-\frac{1}{\hbar}\int_{x_1}^{x_2}|p(x')|dx'}, \quad (9.130)$$

called the *barrier penetration factor*. Using now the connection formulae (9.111) and (9.112) around the point x_1, we see that the solution is of the form (9.127) with[27]

$$A_1 = -iA'\left(\frac{1}{B} - \frac{B}{4}\right), \qquad A_2 = -iA'\left(\frac{1}{B} + \frac{B}{4}\right). \quad (9.131)$$

The ratio $|A_2/A_1|^2$ is proportional to the probability that the particle is reflected, while $|A'/A_1|^2$ is proportional to the probability that the particle is transmitted. These quantities are determined in terms of the barrier penetration factor B. In the semiclassical approximation, the quantity B is exponentially small and it is further exponentially suppressed the larger and taller the barrier is. Indeed, for a barrier of size L with constant potential V_0, we find $B = \exp(-\sqrt{2m(V_0 - E)}L/\hbar)$, which goes to zero exponentially with V_0 and L. We see that the reflection probability from a large and tall barrier $|A_2/A_1|^2$ is of order one, while the transmission probability $|A'/A_1|^2 \sim B^2$ is exponentially small.

9.9. Exercises

Exercise 9.1. Consider the spin operators S_i of Example 9.1. Show that the projection $\mathbf{S} \cdot \mathbf{n}$ of the spin in an arbitrary direction $\mathbf{n} = (\sin\theta\cos\phi, \sin\theta\sin\phi, \cos\theta)$ has eigenvalues $\pm 1/2$.

---

[27]Notice that the term $1/B$ is exponentially large when $\hbar \to 0$, while B is exponentially small. In this computation we have neglected corrections in power of $\hbar$ to the term $1/B$ that come from higher-order terms in the WKB expansion and these are certainly bigger than the term B. Such terms should be included but, as it is easy to see, they do not change the conclusion that the transmission probability is suppressed compared to the reflection probability.

Exercise 9.2. Properties of one-dimensional motion. Show that for the one-dimensional equation (9.36) on $\mathbb{R}$

(a) the proper eigenvalues are non-degenerate;

(b) the proper eigenvectors can be chosen to be real;

(c) if $V(x) = V(-x)$ the proper eigenvectors are either even or odd functions.

Exercise 9.3. Find the discrete energy levels of a one-dimensional well of finite depth: $V(x) = 0$ for $|x| > a$ and $V(x) = -V_0$ for $-a < x < a$, with $V_0 > 0$.

* Exercise 9.4. Algebraic approach to the harmonic oscillator. Consider the harmonic oscillator of Example 9.7 and define the *raising* and *lowering operators*

$$ a = \sqrt{\frac{m\omega}{2}}Q + i\sqrt{\frac{1}{2m\omega}}P, \qquad a^\dagger = \sqrt{\frac{m\omega}{2}}Q - i\sqrt{\frac{1}{2m\omega}}P, $$

and the *number operator* $N = a^\dagger a$. Call $|n\rangle$ the eigenstates of N, $N|n\rangle = n|n\rangle$, and assume that they are non-degenerate[28] and normalised, $\langle n'|n\rangle = \delta_{nn'}$. Show that

(a) $[a, a^\dagger] = \mathbb{I}$ and $H = \omega(N + \frac{1}{2}\mathbb{I})$;

(b) $[N, a] = -a$ and $[N, a^\dagger] = a^\dagger$;

(c) $n \geq 0$, and $n = 0$ if and only if $a|0\rangle = 0$ [Hint: compute $||a|n\rangle||^2$ using part (b)];

(d) $a^\dagger|n\rangle$ and $a|n\rangle$ are eigenvectors of N with eigenvalues $n+1$ and $n-1$, respectively. In particular, $a^\dagger|n\rangle = \sqrt{n+1}|n+1\rangle$ and $a|n\rangle = \sqrt{n}|n-1\rangle$;

(e) $n \in \mathbb{N}$ and $|n\rangle = \frac{(a^\dagger)^n}{\sqrt{n!}}|0\rangle$ [Hint: apply a to $|n\rangle$ repeatedly. Use part (c)];

(f) the spectrum of H is $E_n = \omega(n + 1/2)$.

Exercise 9.5. Find the discrete energy levels of the one-dimensional potential $V(x) = -\frac{\mu}{2m}\delta(x+l) - \frac{\mu}{2m}\delta(x-l)$ with $\mu, l > 0$.

Exercise 9.6. Consider the angular momentum operator $\mathbf{L} = \mathbf{Q} \times \mathbf{P}$. In components we have $L_i = \sum_{j,k=1}^3 \epsilon_{ijk}Q_jP_k$ where ϵ_{ijk} is a totally antisymmetric tensor with $\epsilon_{123} = 1$. Define also $\mathbf{L}^2 = L_1^2 + L_2^2 + L_3^2$. Show that

(a) L_i and $\mathbf{L}^2$ are self-adjoint;

(b) $[L_i, L_j] = i\sum_{k=1}^3 \epsilon_{ijk}L_k$;

(c) $[L_i, \mathbf{L}^2] = 0$;

(d) $[L_i, Q_j] = i\sum_{k=1}^3 \epsilon_{ijk}Q_k$ and $[L_i, P_j] = i\sum_{k=1}^3 \epsilon_{ijk}P_k$.

---

[28] One can show that the eigenstates are actually non-degenerate.

* Exercise 9.7. Using spherical coordinates show that the spherical harmonics are simultaneous eigenvectors of the angular momentum operators L_3 and $\mathbf{L}^2$ defined in Exercise 9.6

$$L_3 Y_{lm} = m Y_{lm}, \qquad \mathbf{L}^2 Y_{lm} = l(l+1) Y_{lm}.$$

Exercise 9.8. Verify the Heisenberg principle (9.72) by computing the Fourier transforms of the wave functions

(a) $\psi(x) = \dfrac{\sqrt{a}}{\pi^{1/4}} e^{-a^2 x^2/2}$,   (b) $\psi(x) = \sqrt{\dfrac{2a^3}{\pi}} \dfrac{1}{x^2 + a^2}$,   (c) $\psi(x) = \dfrac{1}{\sqrt{2\pi}} e^{i p_0 x}$.

Show that the inequality (9.72) is saturated only for a Gaussian wave function. [Hint: use $R\psi = 0$ in (9.71).]

Exercise 9.9. Show that (9.76) is unitary.

Exercise 9.10. Particle of spin 1/2 in a magnetic field. In the notations of Example 9.1, consider the Hilbert space $\mathcal{H} = \mathbb{C}^2$ and the Hamiltonian $H = B\sigma_3$. Consider at $t = 0$ the state $\psi(0) = (1,1)/\sqrt{2}$ with spin S_1 equal to 1/2. Determine the state $\psi(t)$ at time $t > 0$ and find the probabilities for a measure of the three components of the spin.

* Exercise 9.11. General theory of angular momentum. Consider three self-adjoint operators J_i satisfying $[J_i, J_j] = i \sum_{k=1}^{3} \epsilon_{ijk} J_k$ and define $J_\pm = J_1 \pm i J_2$. Denote by $|jm\rangle$ the common eigenvectors of $\mathbf{J}^2 = J_1^2 + J_2^2 + J_3^2$ and J_3 and call $j(j+1)$ and m their eigenvalues

$$\mathbf{J}^2 |jm\rangle = j(j+1)|jm\rangle, \qquad J_3|jm\rangle = m|jm\rangle.$$

Assume for simplicity that they are non-degenerate[29] and normalised, $\langle jm|j'm'\rangle = \delta_{jj'}\delta_{mm'}$. Show that
(a) $[\mathbf{J}^2, J_i] = 0$;
(b) $[J_3, J_\pm] = \pm J_\pm$ and $\mathbf{J}^2 = J_+ J_- + J_3^2 - J_3 = J_- J_+ + J_3^2 + J_3$;
(c) $-j \le m \le j$ [Hint: compute $||J_\pm|jm\rangle||^2$ using part (b)];
(d) $J_\pm|jm\rangle$ are eigenvectors of J_3 with eigenvalues $m \pm 1$, respectively. In particular, show that $J_\pm|jm\rangle = \sqrt{(j \mp m)(j \pm m + 1)}|j, m \pm 1\rangle$;
(e) $j = 0, 1/2, 1, 3/2, \dots$ and $m = -j, -j+1, \dots, j$. [Hint: start with $|jm\rangle$ and apply $J_\pm$ repeatedly. Use part (c).]

---

[29] One can show that the eigenstates are actually non-degenerate.

Exercise 9.12. Consider the WKB expansion $y(x) = \exp\left(\frac{1}{\hbar}\sum_{n=0}^{\infty}\hbar^n S_n(x)\right)$ for the equation $\hbar^2 y''(x) = v(x)y(x)$.

(a) Find the differential equations satisfied by the functions $S_n(x)$.

(b) Solve them for $v(x) = x$ and $\hbar = 1$, and recover the first few terms of the asymptotic expansion of the solutions of the Airy equation for large positive x.

PART IV
Appendices

Appendix A

Review of Basic Concepts

In this appendix we give a brief review of the basic concepts and formulae about holomorphic functions, Hilbert spaces and distributions that are used in the text. This also serves to establish our notations. See Petrini *et al.* (2017) and Rudin (1987, 1991) for details and proofs.

Holomorphic functions. A function of complex variable $f(z)$ is holomorphic in a open set Ω if it is differentiable in complex sense in Ω: $f'(z_0) = \lim_{z \to z_0}(f(z) - f(z_0))/(z - z_0)$ exists for all $z_0 \in \Omega$. This implies, and it is basically equivalent to, the Cauchy–Riemann conditions $\partial_x u = \partial_y v, \partial_x v = -\partial_x u$ for the real and imaginary parts of $f(z) = u(x, y) + iv(x, y)$, with $z = x + iy$. Equivalently, $\partial_{\bar{z}} f = 0$ where $\partial_{\bar{z}} = (\partial_x + i\partial_y)/2$. We can also write $f'(z) = \partial_x f = -i\partial_y f$. A holomorphic function is infinitely differentiable and can be expanded in Taylor series

$$f(z) = \sum_{n=0}^{\infty} a_n (z - z_0)^n \qquad (A.1)$$

with $a_n = f^{(n)}(z_0)/n!$ in a neighbourhood of any point $z_0 \in \Omega$. The radius of convergence of the Taylor series is given by

$$R = \left(\lim_{n \to \infty} \sup_{k \ge n} |a_k|^{1/k} \right)^{-1} = \lim_{n \to \infty} |a_n/a_{n+1}|, \qquad (A.2)$$

where the second expression is valid only when the limit exists. A holomorphic function $f(z)$ can be also expanded in a Laurent series

$$f(z) = \sum_{n=-\infty}^{\infty} a_n(z - z_0)^n$$

in any annulus $r < |z - z_0| < R$ contained in its domain of holomorphicity. The coefficients can be computed as $a_n = \frac{1}{2\pi i} \oint_\gamma \frac{f(z)}{(z-z_0)^{n+1}} dz$ for any simple[1] and positively oriented curve surrounding once z_0 and contained in the annulus. The Laurent series defined in a punctured disk $D = \{0 < |z - z_0| < R\}$ allows to classify isolated singularities of a holomorphic function: if $a_n = 0$ for all $n < 0$, z_0 is a removable singularity and $f(z)$ can be extended to a holomorphic function defined in z_0; if $a_n = 0$ for $n < -k$ with $k > 0$, z_0 is a pole of order k and we can write $f(z) = g(z)/(z - z_0)^n$ with $g(z)$ holomorphic in z_0 and $g(z_0) \neq 0$; if $a_n \neq 0$ for infinitely many negative n, z_0 is an essential singularity, the limit $\lim_{z \to z_0} f(z)$ does not exists and it depends on the path that we use to approach z_0. The coefficient a_{-1} of the Laurent expansion of $f(z)$ near an isolated singularity z_0 is called residue of f in z_0 and denoted $\mathrm{Res}[f, z_0]$. An important property of holomorphic functions is that their line integrals do not depend on the path of integration. More precisely, if $f(z)$ is holomorphic in Ω, we can continuously deform the contour $\gamma \subset \Omega$ without changing $\int_\gamma f(z)dz$, provided γ remains in the domain of holomorphicity of f (if γ is an open path the endpoints must be kept fixed). Moreover, for any closed curve γ the integral $\int_\gamma f(z)dz$ is zero if Ω is simply connected or, more generally, if γ can be continuously contracted to a point without leaving Ω (Cauchy's theorem). If γ is a closed, simple and positively oriented curve that surrounds only a finite number of isolated singularities $z_1, \ldots, z_n$ of the holomorphic function $f(z)$, the residue theorem tells us that

$$\int_\gamma f(z)dz = 2\pi i \sum_{i=1}^n \mathrm{Res}[f, z_i]. \tag{A.3}$$

If f is holomorphic in z_0 and γ is a closed, simple curve surrounding z_0 and no singularities, we have the Cauchy's integral formula for f and its derivatives

$$f(z_0) = \frac{1}{2\pi i} \int_\gamma \frac{f(z)}{z - z_0} dz, \quad f^{(n)}(z_0) = \frac{n!}{2\pi i} \int_\gamma \frac{f(z)}{(z - z_0)^{n+1}} dz. \tag{A.4}$$

---

[1] A simple curve is a curve that has no self-intersections.

A useful trick to evaluate integrals of $f(z)$ on the real axis is to compute $\lim_{R\to\infty} \int_{\Gamma_R} f(z)\mathrm{d}z$ using the residue theorem, where $\Gamma_R = [-R, R] + \gamma_R$ with γ_R a semi-circle of radius R in the upper or lower half-plane such that $\lim_{R\to\infty} \int_{\gamma_R} f(z)\mathrm{d}z = 0$. Jordan's lemma ensures that $\lim_{R\to\infty} \int_{\gamma_R} f(z)e^{i\alpha z}\mathrm{d}z = 0$, if $\lim_{z\to\infty} f(z) = 0$ and γ_R is chosen in the upper half-plane if $\alpha > 0$ and lower half-plane if $\alpha < 0$, and it can be used to compute with the same trick integrals of the type $\int_{-\infty}^{+\infty} f(x)e^{i\alpha x}\mathrm{d}x$.

Distributions. A distribution is a continuous linear functional from a space of test functions $\mathcal{F}$ to $\mathbb{C}$. The space $\mathcal{F}$ can be either the space of bump functions $\mathcal{D}(\mathbb{R})$ ($C^\infty(\mathbb{R})$ functions with compact support) or the space of rapidly decreasing functions $\mathcal{S}(\mathbb{R})$ ($C^\infty(\mathbb{R})$ functions that vanish at infinity with all derivatives more rapidly than any power) endowed with suitable topologies. The linear space of all distributions is denoted with $\mathcal{F}'$. The action of a distribution T on a test function ϕ is denoted as $T(\phi)$ or $\langle T, \phi \rangle$. When $\mathcal{F} = \mathcal{S}(\mathbb{R})$ the distribution is called tempered. A distribution of the form $T(\phi) = \int_{\mathbb{R}} f(x)\phi(x)\mathrm{d}x$, where f is a locally integrable function f, is called regular and can be identified with the function f itself. With this identification, we have the inclusion $\mathcal{D}(\mathbb{R}) \subset \mathcal{S}(\mathbb{R}) \subset \mathcal{S}'(\mathbb{R}) \subset \mathcal{D}'(\mathbb{R})$ where each space is dense in those containing it. A commonly used tempered distributions is the Dirac delta function δ, defined as $\langle \delta, \phi \rangle = \phi(0)$. We often write $\langle \delta, \phi \rangle = \int \delta(x)\phi(x)\mathrm{d}x$. Notice however that this expression is only formal since the delta cannot be written as the integral of any locally integrable function. Other examples of tempered distributions are the principal part $P\frac{1}{x}$ and the distribution $\frac{1}{x\pm i0}$

$$\langle P\frac{1}{x}, \phi \rangle = \lim_{\epsilon\to 0} \left(\int_{-\infty}^{-\epsilon} \frac{\phi(x)}{x}\mathrm{d}x + \int_{\epsilon}^{\infty} \frac{\phi(x)}{x}\mathrm{d}x \right), \qquad (A.5)$$

$$\langle \frac{1}{x\pm i0}, \phi \rangle = \lim_{\epsilon\to 0} \int_{\mathbb{R}} \frac{\phi(x)}{x\pm i\epsilon}\mathrm{d}x. \qquad (A.6)$$

A useful identity is

$$\frac{1}{x\pm i0} = P\frac{1}{x} \mp \pi i\delta(x). \qquad (A.7)$$

We can define operations on distributions by moving the operation on the test functions. The multiplication by a $C^\infty(\mathbb{R})$ function g is defined by $\langle gT, \phi \rangle = \langle T, g\phi \rangle$. The complex conjugate of a distribution is defined as $\langle \bar{T}, \phi \rangle = \overline{\langle T, \bar{\phi} \rangle}$.

The derivative T' of a distribution is defined by $\langle T', \phi \rangle = -\langle T, \phi' \rangle$. Distributions are infinitely differentiable.

Normed spaces.   A norm on a vector space V is a map from V to $[0, +\infty)$ that satisfies: $||\mathbf{x}|| = 0$ if and only if $\mathbf{x} = 0$, $||\lambda \mathbf{x}|| = |\lambda|\,||\mathbf{x}||$ and $||\mathbf{x} + \mathbf{y}|| \leq ||\mathbf{x}|| + ||\mathbf{y}||$ for all $\lambda \in \mathbb{C}$ and $\mathbf{x}, \mathbf{y} \in V$. We say that the sequence $\{\mathbf{v}_n\}$ converges to $\mathbf{v}$ in the norm of V and we write $\mathbf{v}_n \to \mathbf{v}$ if $\lim_{n\to\infty} ||\mathbf{v}_n - \mathbf{v}|| = 0$. A sequence $\{\mathbf{v}_n\}$ is Cauchy if, for every $\epsilon > 0$, $||\mathbf{v}_n - \mathbf{v}_m|| < \epsilon$ for n and m sufficiently large. Every convergent sequence is Cauchy. V is complete if every Cauchy sequence converges. A complete normed space is also called a Banach space. An example of Banach space is $C[K]$, the space of complex continuous functions defined on a compact set $K \in \mathbb{R}^n$, endowed with the norm $||f|| = \max_K |f(x)|$. Another important example is $L_\omega^p(\Omega)$, the set complex functions defined on a subset Ω of $\mathbb{R}^n$ such that $\int_\Omega |f(x)|^p \omega(x) dx < \infty$, where $\omega(x)$ is a smooth measure density, endowed with the norm $||f||_p = (\int_\Omega |f(x)|^p \omega(x) dx)^{1/p}$.

Hilbert spaces.   A scalar product on a vector space V is a map from $V \times V$ to $\mathbb{C}$ that is linear in the second entry, hermitian, non-degenerate and positive definite. In other words, for all $\mathbf{v}, \mathbf{w}, \mathbf{u} \in V$ and $\alpha, \beta \in \mathbb{C}$ we have: (i) $(\mathbf{v}, \alpha \mathbf{w} + \beta \mathbf{u}) = \alpha(\mathbf{v}, \mathbf{w}) + \beta(\mathbf{v}, \mathbf{u})$; (ii) $(\mathbf{v}, \mathbf{w}) = \overline{(\mathbf{w}, \mathbf{v})}$; (iii) $(\mathbf{v}, \mathbf{v}) \geq 0$ and $(\mathbf{v}, \mathbf{v}) = 0$ implies $\mathbf{v} = 0$. A scalar product induces a norm through $||\mathbf{v}|| = \sqrt{(\mathbf{v}, \mathbf{v})}$. Schwarz's inequality states that

$$|(\mathbf{v}, \mathbf{w})| \leq ||\mathbf{v}||\,||\mathbf{w}||, \qquad \mathbf{v}, \mathbf{w} \in \mathcal{H}. \tag{A.8}$$

A Hilbert space $\mathcal{H}$ is a space with scalar product that is complete in such norm. An example of Hilbert space is $L_\omega^2(\Omega)$ endowed with the scalar product $(f, g) = \int_\Omega \overline{f(x)} g(x) \omega(x) dx$. A Hilbert basis, or complete orthonormal system, in $\mathcal{H}$ is a maximal set of orthonormal vectors: $(\mathbf{u}_\alpha, \mathbf{u}_\beta) = \delta_{\alpha\beta}$, where $(\mathbf{u}_\alpha, \mathbf{v}) = 0$ for all α implies $\mathbf{v} = 0$. A Hilbert space is called separable if there exists a complete orthonormal system, $\{\mathbf{u}_n\}$, $n = 1, 2, 3, \ldots$, that consists of a countable number of elements. In this book we only consider separable Hilbert spaces. Any vector $\mathbf{v} \in \mathcal{H}$ can be expanded in the basis as $\mathbf{v} = \sum_{n=1}^\infty \alpha_n \mathbf{u}_n$. The coefficients α_n are complex and are given by $\alpha_n = (\mathbf{u}_n, \mathbf{v})$. We also have Parseval's identity: if $\mathbf{v} = \sum_{n=1}^\infty \alpha_n \mathbf{u}_n$ and $\mathbf{u} = \sum_{n=1}^\infty \beta_n \mathbf{u}_n$ then

$$||\mathbf{v}||^2 = \sum_{n=1}^\infty |\alpha_n|^2, \quad (\mathbf{v}, \mathbf{u}) = \sum_{n=1}^\infty \bar{\alpha}_n \beta_n. \tag{A.9}$$

Examples of Hilbert basis are

$L^2[a,b]$: Fourier basis $\qquad \dfrac{1}{\sqrt{b-a}} e^{\frac{2\pi inx}{b-a}}, \quad n \in \mathbb{Z},$

$L^2[a,b]$: half-period Fourier basis $\qquad \sqrt{\dfrac{2}{b-a}} \sin \dfrac{\pi inx}{b-a}, \quad n = 1, 2, \ldots,$

$L^2[-1,1]$: Legendre basis $\qquad P_l(x) = \dfrac{1}{2^l l!} \dfrac{d^l}{dx^l}(x^2-1)^l, \quad l \in \mathbb{N},$

$L^2[0,\infty]$: Laguerre basis $\qquad L_n(x) = \dfrac{e^x}{n!} \dfrac{d^n}{dx^n}[e^{-x}x^n], \quad n \in \mathbb{N},$

$L^2(\mathbb{R})$: Hermite basis $\qquad H_n(x) = (-1)^n e^{x^2} \dfrac{d^n}{dx^n} e^{-x^2}, \quad n \in \mathbb{N},$

where the Legendre, Laguerre and Hermite polynomials P_l, L_n, L_n, are orthogonal but, for historical reasons, are not of unit norm. The orthogonal complement $S^\perp$ of a set S is defined as $S^\perp = \{\mathbf{v} \in \mathcal{H} | (\mathbf{v}, \mathbf{s}) = 0, \mathbf{s} \in \mathcal{H}\}$ and is a closed vector space. If S is a subspace of $\mathcal{H}$, then $\mathcal{H} = \bar{S} \oplus S^\perp$.

Linear operators. Given two normed spaces X and Y, an operator $A : X \to Y$ is a linear map from X to Y. The operator A is continuous if and only if it is bounded, which means that there exists a constant K such that $||A\mathbf{x}|| \leq K||\mathbf{x}||$ for any $\mathbf{x} \in X$. The smallest value of K for which this inequality holds is called the norm of A and is computed as $||A|| = \sup_{\mathbf{x} \in X} ||A\mathbf{x}||/||\mathbf{x}||$. We will mostly consider operators on a Hilbert space, $A : \mathcal{H} \to \mathcal{H}$. The set of bounded operators on a Hilbert space $\mathcal{H}$ is denoted as $\mathcal{B}(\mathcal{H})$ and is a complete normed vector space, when endowed with the norm $||A||$. The unbounded operators are usually defined only on a dense subset of $\mathcal{H}$ called the domain of A and denoted with $\mathcal{D}(A)$. Riesz's theorem states that any continuous linear functional $L : \mathcal{H} \to \mathbb{C}$ can be written as $L(\mathbf{v}) = (\mathbf{v}_L, \mathbf{v})$ for some $\mathbf{v}_L \in \mathcal{H}$. The map $L \to \mathbf{v}_L$ defines an anti-linear bijection of $\mathcal{H}$ with its dual $\mathcal{H}^*$ that allows to identify the two. The adjoint $A^\dagger$ of a densely defined linear operator A with domain $\mathcal{D}(A)$ is defined in the domain $\mathcal{D}(A^\dagger)$ consisting of all vectors $\mathbf{v}$ such that the linear functional $L_\mathbf{v}(\mathbf{w}) = (\mathbf{v}, A\mathbf{w})$ is continuous, and satisfies $(A^\dagger\mathbf{v}, \mathbf{w}) = (\mathbf{v}, A\mathbf{w})$. If A is bounded, $A^\dagger$ also is, $\mathcal{D}(A^\dagger) = \mathcal{H}$ and $(A^\dagger)^\dagger = A$. An operator A is symmetric if $(A\mathbf{v}, \mathbf{w}) = (\mathbf{v}, A\mathbf{w})$ for $\mathbf{v}, \mathbf{w} \in \mathcal{D}(A)$. The operator A is self-adjoint, or hermitian, if $A^\dagger = A$, which includes the fact that $\mathcal{D}(A) = \mathcal{D}(A^\dagger)$. The properties of operators depend on the domain. For example, the operator $P = -i\partial_x$ is symmetric but non-self-adjoint when defined in $\mathcal{S}(\mathbb{R}) \subset L^2(\mathbb{R})$; it becomes self-adjoint in the densely defined domain of square-integrable functions f that are also absolutely continuous and with $f' \in L^2(\mathbb{R})$. An operator U is unitary if $U^\dagger U = UU^\dagger = \mathbb{I}$; a unitary operator is bounded, $||U|| = 1$, and preserves the scalar product, $(U\mathbf{v}, U\mathbf{w}) = (\mathbf{v}, \mathbf{w})$. The spectrum $\sigma(A)$ of an operator A is the set of complex numbers λ such that the resolvent operator $(A - \lambda\mathbb{I})^{-1}$ either does not exist or is not an element of $\mathcal{B}(\mathcal{H})$; $\sigma(A)$ decomposes in the disjoint union of

- discrete spectrum: the set of eigenvalues, $A\mathbf{v} = \lambda\mathbf{v}$ for some $\mathbf{v} \neq 0$,
- continuous spectrum: $(A - \lambda\mathbb{I})^{-1}$ exists, is densely defined but not bounded,
- residual spectrum: $(A - \lambda\mathbb{I})^{-1}$ exists but is not densely defined.

The spectrum $\sigma(A)$ is always a closed subset of $\mathbb{C}$. When A is bounded, $\sigma(A)$ is contained in the disk of radius $||A||$. If λ belongs to the continuous spectrum, there exists a sequence of vectors $\mathbf{v}_n \in \mathcal{H}$ that approximates the eigenvalue equation arbitrarily well, $\lim_{n\to\infty} ||A\mathbf{v}_n - \lambda\mathbf{v}_n|| = 0$. The spectrum of a self-adjoint operator A is a subset of the real axis, the eigenvectors are mutually orthogonal and the residual spectrum is empty. The spectral theorem states that, for any self-adjoint operator A on $L^2(\mathbb{R})$ there exists, in addition to the (proper) eigenvectors $\mathbf{u}_{n\beta}$, $A\mathbf{u}_{n\beta} = \lambda_n\mathbf{u}_{n\beta}$, a family of tempered distributions $\mathbf{u}_{\lambda\beta}$ satisfying $A\mathbf{u}_{\lambda\beta} = \lambda\mathbf{u}_{\lambda\beta}$ for any λ in the continuous spectrum (generalised eigenvectors) such that any vector $\mathbf{u}$ in $L^2(\mathbb{R})$ can be written as (see Gelfand and Vilenkin, 1964)

$$\mathbf{u} = \sum_{n\beta} c_{n\beta}\mathbf{u}_{n\beta} + \sum_{\beta} \int c_\beta(\lambda)\mathbf{u}_{\lambda\beta}\mathrm{d}\lambda, \qquad (A.10)$$

where $c_{n\beta} = (\mathbf{u}_{n\beta}, \mathbf{u})$ and $c_\beta(\lambda) = (\mathbf{u}_{\lambda\beta}, \mathbf{u})$.[2] Here β denotes one or more discrete or continuous indices of multiplicity. The set $\{\mathbf{u}_{n\beta}, \mathbf{u}_{\lambda\beta}\}$ is called a generalised basis for $L^2(\mathbb{R})$. Equation (A.10) only makes sense for $\mathbf{u} \in \mathcal{S}(\mathbb{R})$ and it should be extended to $L^2(\mathbb{R})$ by continuity. Proper and generalised eigenvectors can be normalised in such a way that $(\mathbf{u}_{n\beta}, \mathbf{u}_{m\beta'}) = \delta_{nm}\delta_{\beta\beta'}$, $(\mathbf{u}_{\lambda\beta}, \mathbf{u}_{n\gamma}) = 0$, $(\mathbf{u}_{\lambda\beta}, \mathbf{u}_{\lambda',\beta'}) = \delta(\lambda - \lambda')\delta_{\beta\beta'}$. This is a useful but formal statement. It is equivalent to the fact that, if $\mathbf{u}$ and $\tilde{\mathbf{u}}$ have an expansion as in (A.10), their scalar product is given by $(\tilde{\mathbf{u}}, \mathbf{u}) = \sum_{n\beta} \overline{\tilde{c}_{n\beta}} c_{n\beta} + \sum_{\beta} \int \overline{\tilde{c}_\beta(\lambda)} c_\beta(\lambda)\mathrm{d}\lambda$ (generalised Parseval's identity). The above results about the spectrum and the spectral theorem hold also for unitary

---

[2]The scalar product between the distribution $\mathbf{u}_{\lambda\beta}$ and the vector $\mathbf{u}$ can be defined as $(\mathbf{u}_{\lambda\beta}, \mathbf{u}) = \langle\overline{\mathbf{u}_{\lambda\beta}}, \mathbf{u}\rangle$, where $\overline{\mathbf{u}_{\lambda\beta}}$ is the complex conjugate distribution. We also define $(\mathbf{u}, \mathbf{u}_{\lambda\beta}) = \overline{(\mathbf{u}_{\lambda\beta}, \mathbf{u})}$.

operators, with the only difference that the spectrum is a subset of the unit circle. They also hold for more general Hilbert spaces.

Fourier transform. We define the Fourier transform of a function $f \in \mathcal{S}(\mathbb{R})$ as

$$\hat{f}(p) = \frac{1}{\sqrt{2\pi}} \int_{\mathbb{R}} f(x) e^{-ipx} \mathrm{d}x. \tag{A.11}$$

The Fourier transform is a bijective map from $\mathcal{S}(\mathbb{R})$ to itself that preserves norm and scalar product, $||\hat{f}|| = ||f||$ and $(\hat{f}, \hat{g}) = (f, g)$. The Fourier transform of $f'(x)$ is $ip\hat{f}(p)$ and the Fourier transform of $xf(x)$ is $i\hat{f}'(p)$. The Fourier transform of the convolution $h(x) = (f * g)(x) = \int f(x-y)g(y)\mathrm{d}y$ is $\hat{h}(p) = \sqrt{2\pi}\hat{f}(p)\hat{g}(p)$. The Fourier transform can be inverted using

$$f(x) = \frac{1}{\sqrt{2\pi}} \int_{\mathbb{R}} \hat{f}(p) e^{ipx} \mathrm{d}p. \tag{A.12}$$

The Fourier transform is extended to $L^2(\mathbb{R})$ by continuity, where it becomes a unitary operator. The Fourier transform can also be extended to tempered distributions using $\langle \hat{T}, \phi \rangle = \langle T, \hat{\phi} \rangle$.

Notable formulae. The following Fourier transforms and integrals are widely used in the text

$$\int_{\mathbb{R}} e^{ipx} \mathrm{d}x = 2\pi\delta(p), \quad \text{in } \mathcal{S}'(\mathbb{R}) \tag{A.13}$$

$$\int_{\mathbb{R}} e^{-\frac{x^2}{2a^2} + ipx} \mathrm{d}x = \sqrt{2\pi} a e^{-\frac{a^2 p^2}{2}}, \quad a \in \mathbb{R}, \tag{A.14}$$

$$\int_0^{+\infty} e^{-zt} t^a \mathrm{d}t = \frac{\Gamma(a+1)}{z^{a+1}}, \quad \operatorname{Re} z > 0, a > -1, \tag{A.15}$$

$$\int_{\mathbb{R}} e^{-zy^2} y^{2n} \mathrm{d}y = \frac{\Gamma(n+1/2)}{z^{n+1/2}}, \quad \operatorname{Re} z > 0, n \in \mathbb{N}. \tag{A.16}$$

Recall that the Gamma function is defined as $\Gamma(w) = \int_0^{+\infty} e^{-t} t^{w-1} \mathrm{d}t$ for $\operatorname{Re} w > 0$ and it can be analytically continued to $\mathbb{C} - \{0, -1, -2, \ldots\}$. It has simple poles with residue $(-1)^n/n!$ in $w = -n$ with $n \in \mathbb{N}$. Useful identities are $\Gamma(z+1) = z\Gamma(z)$, $\Gamma(z)\Gamma(1-z) = \pi/\sin(\pi z)$ and $\Gamma(z)\Gamma(z+1/2) = 2^{1-2z}\sqrt{\pi}\Gamma(2z)$. In particular, $\Gamma(k+1) = k!$ and $\Gamma(k+1/2) = (2k)!\sqrt{\pi}/(4^k k!)$ for $k \in \mathbb{N}$. Another useful identity is the binomial expansion

$$(1-x)^{-\alpha} = \sum_{n=0}^{\infty} \frac{\Gamma(n+\alpha)}{\Gamma(\alpha)\Gamma(n+1)} x^n. \tag{A.17}$$

The digamma function is defined as $\psi(z) = \frac{d}{dz} \ln \Gamma(z) = \frac{\Gamma'(z)}{\Gamma(z)}$. It satisfies $\psi(z+1) = 1/z + \psi(z)$ and $\psi(n+1) = \sum_{k=1}^{n} \frac{1}{k} - \gamma$, where $\gamma = 0.577215$ is the Euler constant.

Appendix B

Solutions of the Exercises

Chapter 1

Exercise 1.1. Using (1.4) with $z_0 = w_0 = 0$ we have $b_0 = 0$ and $b_n = \frac{1}{n!}\frac{d^{n-1}}{dz^{n-1}}(\frac{z}{e^z-1})^n|_{z=0}$. We find $b_1 = 1$, $b_2 = -1/2$ and $b_3 = 1/3$. And indeed $z = \ln(w+1) = \sum_{n=1}^{\infty}(-1)^{n+1}w^n/n = w - w^2/2 + w^3/3 + O(w^4)$.

Exercise 1.2. Since any linear fractional transformation is a combination of dilations, translations and inversions, it is enough to check that these transformations map the set of lines and circles into itself. This is obvious for translations. Consider a dilation $w = \beta z$ with $0 \neq \beta \in \mathbb{C}$. Writing $\beta = |\beta|e^{i\phi}$ we see that $e^{i\phi}$ acts as a rotation in the plane that sends lines into lines and circles into circles. Without loss of generality we can then take β real. The dilation then acts as $x' = \beta x, y' = \beta y$ with $w = x' + iy'$ and $z = x + iy$. By a dilation, a line in the plane, $Ax + By + C = 0$ is mapped into $Ax' + By' + C\beta = 0$ which is again a line, and a circle $(x - x_0)^2 + (y - y_0)^2 = R^2$ is mapped into $(x' - x_0\beta)^2 + (y' - y_0\beta)^2 = R^2\beta^2$ which is again a circle. The inversion $w = 1/z$ acts as $x' = x/(x^2 + y^2), y' = -y/(x^2 + y^2)$. Thus, a line in the plane, $Ax + By + C = 0$ is mapped into $C(x' + A/2C)^2 + C(y' - B/2C)^2 = (A^2 + B^2)/4C$ which is a circle. The circle $(x - x_0)^2 + (y - y_0)^2 = R^2$ is mapped into $(x' - x_0(x'^2 + y'^2))^2 + (y' + y_0(x'^2 + y'^2))^2 = R^2(x'^2 + y'^2)^2$, which, after some manipulations, can be written again as a circle $(x' - x_0')^2 + (y' - y_0')^2 = R'^2$ of centre $x_0' = -x_0/(R^2 - x_0^2 - y_0^2), y_0' = y_0/(R^2 - x_0^2 - y_0^2)$ and radius $R' = R/(R^2 - x_0^2 - y_0^2)$.

Notice that, in general, the image of a line is a circle. This is because a line should be thought of as a circle passing through the point at infinity.

Exercise 1.3.   (a) Consider the cross ratio

$$B(F(z_1), F(z_2), F(z_3), F(z_4)) = \frac{(F(z_1) - F(z_2))(F(z_3) - F(z_4))}{(F(z_1) - F(z_4))(F(z_3) - F(z_2))}.$$

From a direct computation, we find $F(z_i) - F(z_j) = \frac{(ad-bc)(z_i-z_j)}{(cz_i+d)(cz_j+d)}$. The denominators in this expression cancel out when we plug it back in the cross ratio, and we immediately find $B(F(z_1), F(z_2), F(z_3), F(z_4)) = B(z_1, z_2, z_3, z_4)$.

(b) Consider the cross ratio (1.13) as a function of z. It maps the point z_1 into $B(z, z_1, z_2, z_3)|_{z=z_1} = 0$ and, similarly, the points z_2 and z_3 into 1 and ∞. Since $B(w, w_1, w_2, w_3)$ also maps the points w_1, w_2 and w_3 into $0, 1$ and ∞, the transformation defined implicitly by $B(w, w_1, w_2, w_3) = B(z, z_1, z_2, z_3)$ maps z_1, z_2 and z_3 into w_1, w_2 and w_3. It is easy to see that the transformation defined by $B(w, w_1, w_2, w_3) = B(z, z_1, z_2, z_3)$ is again a linear fractional transformation.

Exercise 1.4.   Consider two transformations of the form (1.11), $F_i(z) = \frac{a_i z + b_i}{c_i z + d_i}$. The composition is $(F_2 \circ F_1)(z) = \frac{a_3 z + b_3}{c_3 z + d_3}$ where $a_3 = a_2 a_1 + b_2 c_1$, $b_3 = a_2 b_1 + b_2 d_1$, $c_3 = c_2 a_1 + d_2 c_1$ and $d_3 = c_2 b_1 + d_2 d_1$ and corresponds to the product of the corresponding matrices (1.14),

$$\begin{pmatrix} a_3 & b_3 \\ c_3 & d_3 \end{pmatrix} = \begin{pmatrix} a_2 & b_2 \\ c_2 & d_2 \end{pmatrix} \begin{pmatrix} a_1 & b_1 \\ c_1 & d_1 \end{pmatrix}.$$

The inverse of (1.11), $F^{-1}(w) = (dw - b)/(-cw + a)$, is again of the form (1.11). The corresponding matrix $\begin{pmatrix} d & -b \\ -c & a \end{pmatrix}$ is the inverse of (1.14) since $ad - bc = 1$.

Exercise 1.5.   According to part (b) of Exercise 1.3, we can find the required linear fractional transformation by solving $B(w, i, 1, 2) = B(z, 1, 0, i)$. We find $w = \frac{(1+2i) - (2+3i)z}{(1+2i) - (2+i)z}$. Alternatively, we can determine $F(z)$ by imposing $F(1) = i, F(0) = 1, F(i) = 2$. These three equations determine the ratios $a/d, b/d, c/d$ of the constants in (1.11) and, therefore, uniquely fix $F(z)$.

Exercise 1.6.   The level sets of a holomorphic function $f(z) = u(x, y) + iv(x, y)$, with $z = x + iy$, are the families of curves $u(x, y) = const$ and $v(x, y) = const$. They are mutually orthogonal, except at the points where $f'(z) = 0$, as discussed in Example 1.5.

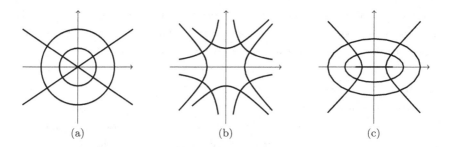

Fig. B.1. Level sets for Exercise 1.6.

(a) $\ln z = \ln|z| + i \arg z$ and therefore the level sets are circles, $|z| = const$, and lines passing through the origin, $\arg z = const$ (see Figure B.1(a)).

(b) $z^2 = (x^2 - y^2) + 2ixy$ and therefore the level sets are two families of mutually orthogonal hyperbolae, $x^2 - y^2 = const$ and $xy = const$ (see Figure B.1(b)).

(c) We can invert the function $w = \ln(z + \sqrt{z^2 - 1})$ and find $z = (e^w + e^{-w})/2 = \cosh w$. Define $w = u + iv$. The level set $u(x, y) = c$ is the set of complex numbers $z = x + iy = \cosh(c + iv) = (e^c e^{iv} + e^{-c} e^{-iv})/2$ parameterised by $v \in \mathbb{R}$. By equating real and imaginary parts, we find $x = a \cos v, y = b \sin v$ where $a = \cosh c \geq 1$ and $b = \sinh c$. This is a family of ellipses $x^2/a^2 + y^2/b^2 = 1$ with $a \geq 1$. The level set $v(x, y) = d$ is the set of complex numbers $z = x + iy = \cosh(u + id) = (e^{id} e^u + e^{-id} e^{-u})/2$ parameterised by $u \in \mathbb{R}$. By equating real and imaginary parts, we find $x = A \cosh u, y = B \sinh u$ where $A = \cos d$ and $B = \sin d$. This is a family of hyperbolae $x^2/A^2 - y^2/B^2 = 1$. The result is depicted in Figure B.1(c).

Exercise 1.7. The Riemann mapping theorem tells that the image of $\Omega = \{\operatorname{Re} z > \operatorname{Re} z_0, \operatorname{Im} z > 0\}$ should be a simply connected domain. To find it, it is enough to determine the image of its boundary. We know that a linear fractional transformation sends the set of lines and circles of the plane into itself. The real axis $z = x \in \mathbb{R}$ is mapped into the unit circle since $|f(z)| = |x - z_0|/|x - \bar{z}_0| = 1$. The vertical line parameterised by $z = \operatorname{Re} z_0 + iy$ with $y \in \mathbb{R}$ is mapped into the real axis, since $f(z) = (y - \operatorname{Im} z_0)/(y + \operatorname{Im} z_0) \in \mathbb{R}$. The shape of $f(\Omega)$ depends on the sign of $\operatorname{Im} z_0$. For $\operatorname{Im} z_0 > 0$, the point z_0 belongs to the boundary of the domain Ω and it is mapped into the origin of the plane $f(z)$. The semi-axis $z = \operatorname{Re} z_0 + iy$ with $y \geq 0$ is mapped into the segment $[-1, 1]$. Then $f(\Omega)$ can only be the upper-half or the lower-half unit disk. We can determine which is the case by looking at the orientation of the boundary: when we go along the semi-axis $z = \operatorname{Re} z_0 + iy$ with y increasing from zero to infinity, Ω is on the right. The same must be true for $f(\Omega)$ where we go along the segment $[-1, 1]$ from -1 to 1. We conclude that $f(\Omega)$ is the lower-half unit disk. For $\operatorname{Im} z_0 < 0$, the semi-axis $z = \operatorname{Re} z_0 + iy$ with $y \geq 0$ is mapped into the segments $(-\infty, -1]$ and $[1, \infty)$. Considering again the orientation, it is easy to see that $f(\Omega)$ must contain part of the upper half-plane. There are just two possibilities: the exterior of the unit disk in the upper half-plane, or the union of the upper half-plane with the lower half unit disk. Since, for $\operatorname{Im} z_0 < 0$, z_0 is not contained in Ω we conclude that $f(\Omega)$ is the exterior of the unit disk in the upper half-plane, $f(\Omega) = \{\operatorname{Im} z > 0, |z| > 1\}$. For $\operatorname{Im} z_0 = 0$, the function degenerates to $f(z) = 1$ and the image of Ω is a point.

Exercise 1.8.   As in the previous exercise, it is enough to determine the image of the boundary of the strip. The two boundaries of the strip, $z = \pm\frac{\pi}{2} + iy$ with $y \in \mathbb{R}$, are mapped into $f(z) = \pm 6\cosh y$, corresponding to the segments $(-\infty, -6]$ and $[6, +\infty)$. Notice that each segment is traversed twice. When y goes from minus infinity to zero, we go from $\pm\infty$ to ± 6 and then, when y goes from zero to infinity we go back from ± 6 to $\pm\infty$. As in Example 1.12, the two boundaries are folded on themselves in the mapping. We conclude that the interior of the strip is mapped to the complex plane with two cuts, $\mathbb{C} - \{(-\infty, -6] \cup [6, +\infty)\}$.

Exercise 1.9.   The horizontal flow of Figure 1.6 past a circle centred in the origin can be obtained by considering the inverse (1.31) of the Joukowski transformation. The streamlines are given by the inverse image of horizontal lines in the plane w, $z(t) = (w(t) \pm \sqrt{w(t)^2 - 4})/2$, where $w(t) = t + ic$ for fixed c. The streamlines for a flow moving at an angle of 45 degrees past a circle of centre $(-1 + i)/2$ and radius $\sqrt{5/2}$ is obtained by a rotation of 45 degrees and a translation combined with a dilation: $\tilde{z}(t) = \sqrt{5/2}e^{i\pi/4}z(t) + (-1 + i)/2$. The image of the circle under the Joukowski map is an airfoil as in Figure 1.7. The streamlines past the airfoil are obtained by applying the Joukowski map to $\tilde{z}(t)$: $\tilde{w}(t) = \tilde{z}(t) + 1/\tilde{z}(t)$. The result is plotted in Figure B.2.

Exercise 1.10.   Under the map $w = z^2$, the first quadrant is mapped into the upper half-plane. The problem is then reduced to finding a harmonic function $u(s, t)$ in the upper half-plane $w = s + it$ that has boundary value $u(s, 0) = \theta(s)$. The solution of the problem is given by the Poisson formula for the upper

Fig. B.2. Flow past the airfoil of Exercise 1.9.

half-plane (1.22) and reads

$$u(s,t) = \int_{-\infty}^{+\infty} \frac{t\theta(v)}{(s-v)^2 + t^2} \frac{dv}{\pi} = \int_0^{+\infty} \frac{t}{(s-v)^2 + t^2} \frac{dv}{\pi} = \frac{1}{\pi}\left(\frac{\pi}{2} + \arctan\frac{s}{t}\right).$$

Going back to the original variables, $s + it = w = z^2 = (x^2 - y^2) + 2ixy$, we find
$u(x,y) = \frac{1}{\pi}\left(\frac{\pi}{2} + \arctan\frac{x^2-y^2}{2xy}\right)$.

Chapter 2

Exercise 2.1. By applying the inversion formula (2.6) to the Laplace transform obtained in Example 2.2, we find $\frac{1}{2\pi i}\int_\gamma \frac{e^{sx}n!}{s^{n+1}}ds$ where γ is a vertical line with $\operatorname{Re}s = \alpha > 0$. We proceed as in Example 2.1. We compute the integral using the residue theorem. By Jordan's lemma, we close the contour with a semi-circle on the right of the line $\operatorname{Re}s = \alpha$ for $x < 0$ and on the left for $x > 0$. For $x < 0$ the integral is zero since there are no singularities inside the integration contour. For $x > 0$, we pick the residue in $s = 0$, obtaining $\frac{1}{n!}\frac{d^n}{ds^n}(n!e^{sx})|_{s=0} = x^n$. The final result is $\theta(x)x^n$, as expected.

Exercise 2.2. (a) By the definition of Laplace transform (2.2) we find
$\mathcal{L}[e^{ax}f](s) = \int_0^{+\infty} dx f(x)e^{ax}e^{-sx} = \mathcal{L}[f](s-a)$.
(b) By a change of variables we find $\mathcal{L}[f(ax)](s) = \int_0^{+\infty} dx f(ax)e^{-sx} = \int_0^{+\infty} \frac{dy}{a} f(y)e^{-sy/a} = \frac{1}{a}\mathcal{L}[f](\frac{s}{a})$.
(c) $\int_0^t f(\tau)d\tau$ can be written as the convolution $(1 * f)(t)$ (see (2.18)). Using the convolution theorem (2.17) and the fact that the Laplace transform of 1 is $1/s$ (see Example (2.2)), we find $\mathcal{L}[\int_0^t f(\tau)d\tau](s) = \frac{1}{s}\mathcal{L}[f]$.

(d) By using the definition of Laplace transform (2.2) and exchanging the order of integration, we find $\int_s^{+\infty} \mathcal{L}[f](u)\,du = \int_0^{+\infty} dx f(x) \int_s^{+\infty} e^{-xu}\,du = \int_0^{+\infty} dx \frac{f(x)}{x} e^{-sx} = \mathcal{L}[\frac{f(x)}{x}](s)$.

Exercise 2.3. (a) We find $\mathcal{L}[x^p](s) = \int_0^{+\infty} x^p e^{-sx} dx = s^{-p-1} \int_0^{+\infty} y^p e^{-y} dy = \Gamma(p+1)/s^{p+1}$ where we used (A.15). The convergence abscissa is $\alpha = 0$. Notice that the result is consistent with Example 2.2, since $\Gamma(n+1) = n!$ for n integer.

(b) This is the same function as in Example 2.1 but with a complex: the Laplace transform of e^{ax} is $1/(s-a)$.

(c) Using $\cos \omega x = (e^{i\omega x} + e^{-i\omega x})/2$, the linearity of the Laplace transform and point (b), we find $\mathcal{L}[\cos \omega x](s) = \frac{s}{s^2+\omega^2}$.

(d) From $\sin \omega x = (e^{i\omega x} - e^{-i\omega x})/(2i)$ we find $\mathcal{L}[\sin \omega x](s) = \frac{\omega}{s^2+\omega^2}$.

Exercise 2.4. The Laplace transform of $f(x)$ can be written as a sum of integrals over intervals of the length of a period, $F(s) = \int_0^{+\infty} f(x) e^{-sx} dx = \sum_{n=0}^\infty \int_{nL}^{(n+1)L} f(x) e^{-sx} dx$. By performing the change of variable $y = x - nL$ in each integral, and using $f(y + nL) = f(y)$, we find $F(s) = \sum_{n=0}^\infty e^{-nLs} \int_0^L f(y) e^{-sy} dy$. By summing the geometric series we obtain $F(s) = (\int_0^L f(y) e^{-sy} dy)(1 - e^{-sL})^{-1}$. Consider now the function of period $L = 1$ whose restriction to the interval $[0, 1)$ is $f(x) = x$. Our formula for the Laplace transform gives $F(s) = (\int_0^1 y e^{-sy} dy)(1 - e^{-s})^{-1} = (1 - e^{-s}(s+1))/(s^2(1-e^{-s}))$.

Exercise 2.5. The Laplace transform of $g(t)$ is

$$G(s) = \int_0^{+\infty} dt \frac{e^{-st}}{2\sqrt{\pi t^3}} \int_0^{+\infty} dy\, y\, f(y) e^{-\frac{y^2}{4t}} = \int_0^{+\infty} dy\, y f(y) \int_0^{+\infty} dt \frac{e^{-st-\frac{y^2}{4t}}}{2\sqrt{\pi t^3}}.$$

With the change of variables $\tau = y/\sqrt{4t}$ in the second integral we obtain

$$\frac{2}{\sqrt{\pi}} \int_0^{+\infty} dy f(y) \int_0^{+\infty} d\tau e^{-\tau^2 - \frac{sy^2}{4\tau^2}}$$

$$= \frac{2}{\sqrt{\pi}} \int_0^{+\infty} dy f(y) e^{-\sqrt{sy}} \int_0^{+\infty} d\tau e^{-(\tau - \frac{\sqrt{sy}}{2\tau})^2}.$$

The last integral can be evaluated with the change of variables $u = \tau - \sqrt{sy}/(2\tau)$. We obtain $\int_0^{+\infty} d\tau e^{-(\tau-\sqrt{sy}/2\tau)^2} = \frac{1}{2} \int_{-\infty}^{+\infty} du\, e^{-u^2}(1 + \frac{u}{\sqrt{u^2+2\sqrt{sy}}}) = \frac{\sqrt{\pi}}{2}$. The term proportional to u in the integral vanishes being odd. We finally obtain $G(s) = \int_0^{+\infty} dy f(y) e^{-\sqrt{sy}} = F(\sqrt{s})$, as requested.

(a) We saw in Example 2.3 that the Laplace transform of $f(t) = \delta(t - a)$ is $F(s) = e^{-as}$. It follows that the inverse Laplace transform of $F(\sqrt{s}) = e^{-a\sqrt{s}}$ is $(4\pi t^3)^{-1/2} \int_0^{+\infty} y e^{-y^2/(4t)} \delta(y - a) dy = \frac{a}{\sqrt{4\pi t^3}} e^{-a^2/(4t)}$.

(b) Since $e^{-a\sqrt{s}} = -\frac{d}{da}\frac{e^{-a\sqrt{s}}}{\sqrt{s}}$, we can obtain the inverse Laplace transform of $e^{-a\sqrt{s}}/\sqrt{s}$ by integrating the inverse Laplace transform of $e^{-a\sqrt{s}}$ with respect to the parameter a. From point (a) we find $\frac{1}{\sqrt{\pi t}} e^{-a^2/(4t)}$.

Exercise 2.6. By applying a Laplace transform to each term in the equation and using (2.14) and Example 2.1, we find $sX(s) - 1 + 3X(s) = \frac{1}{s-\alpha}$, where $X(s)$ is the Laplace transform of $x(t)$. Then $X(s) = \frac{1}{s+3} + \frac{1}{(s+3)(s-\alpha)}$. There are two methods for finding $x(t)$.

Method 1. Reduce $X(s)$ to the sum of Laplace transforms of elementary functions. For $\alpha \neq -3$ we can write $X(s) = \frac{1}{s+3} - \frac{1}{\alpha+3}(\frac{1}{s+3} - \frac{1}{s-\alpha})$. Using the elementary transforms of Exercise 2.3, we find $x(t) = e^{-3t} - \frac{1}{\alpha+3}(e^{-3t} - e^{\alpha t})$. For $\alpha = -3$, we can write $X(s) = \frac{1}{s+3} + \frac{1}{(s+3)^2} = \frac{1}{s+3} - \frac{d}{ds}\frac{1}{(s+3)}$. Using (2.16), we find $x(t) = e^{-3t} + te^{-3t}$. Remember to check that the initial condition, in this case $x(0) = 1$, is satisfied by the solution.

Method 2. Use the inversion formula (2.6) for the Laplace transform. The only contribution comes from $t > 0$ and we close the contour in the half-plane on the left of the integration line. For $\alpha \neq -3$, the integrand $X(s)e^{st}$ has a simple pole in $s = -3$ with residue $(1 - \frac{1}{\alpha+3})e^{-3t}$ and a simple pole in $s = \alpha$ with residue $\frac{1}{\alpha+3}e^{\alpha t}$. The result is again $x(t) = e^{-3t} - \frac{1}{\alpha+3}(e^{-3t} - e^{\alpha t})$. For $\alpha = -3$, the integrand $X(s)e^{st}$ has a double pole in $s = -3$. The result is $x(t) = \text{Res}[X(s)e^{st}, -3] = \frac{d}{ds}(X(s)e^{st}(s+3)^2)|_{s=-3} = e^{-3t} + te^{-3t}$.

Exercise 2.7. We apply a Laplace transform to the equations and, using (2.14), we find the system of algebraic equations $sY(s) - sX(s) + Y(s) + 2X(s) = \frac{1}{s-1}$, $sY(s) + sX(s) - 2 + X(s) = \frac{1}{s-2}$ where $X(s)$ and $Y(s)$ are the transforms of $x(t)$ and $y(t)$, respectively. The solution is $X(s) = \frac{2s^3 - 4s^2 + 3}{(s-2)(s-1)(2s^2+1)}$ and $Y(s) = \frac{2s^2 - 4s + 4}{(s-1)(2s^2+1)}$. To find $X(t)$ and $Y(t)$ we use the residue theorem to evaluate the inversion formula (2.6). For $t > 0$, we close the contour on the left of the integration line, where it contains the simple poles $s = 1, 2, \pm i/\sqrt{2}$. We find $x(t) = \frac{1}{3}e^{2t} - \frac{1}{3}e^t + \cos\frac{t}{\sqrt{2}} + \frac{2}{3}\sqrt{2}\sin\frac{t}{\sqrt{2}}$ and $y(t) = \frac{2}{3}e^t + \frac{1}{3}\cos\frac{t}{\sqrt{2}} - \frac{5}{3}\sqrt{2}\sin\frac{t}{\sqrt{2}}$.

Exercise 2.8. We apply a Laplace transform to the equation. Using the convolution property (2.17), Example 2.1 and Exercise 2.3, we find $\frac{1}{s^2+1} + \frac{F(s)}{s-1} = \frac{1}{s^2}$ where $F(s)$ is the transform of $f(x)$. Then $F(s) = \frac{1}{s} - \frac{1}{s^2} - \frac{s-1}{s^2+1}$. Using the elementary transforms given in Exercise 2.3, we find $f(x) = 1 - x - \cos x + \sin x$.

Exercise 2.9. By applying a Laplace transform in the variable t to the partial differential equation, we find $\partial_x^2 U(x,s) = s^2 U(x,s) - su(x,0) - \partial_t u(x,0) = s^2 U(x,s) - s\sin x$, where we used (2.15). We have found an ordinary linear differential equation in the variable x. A particular solution is easily found in the form $U(x,s) = C(s)\sin x$. Plugging it into the equation, we find $C(s) = s/(s^2+1)$. The general solution is obtained by adding the two linearly independent solutions of the homogeneous equation and we obtain $U(x,s) = \frac{s}{s^2+1}\sin x + Ae^{sx} + Be^{-sx}$ (see Example 4.4). The boundary conditions $u(0,t) = u(\pi,t) = 0$ imply $U(0,s) = U(\pi,s) = 0$ and therefore we choose $A = B = 0$. Using Exercise 2.3, we can invert the Laplace transform and find $u(x,t) = \cos t \sin x$.

Exercise 2.10. We can write $\mathcal{L}[L_n(t)](s) = \int_0^{+\infty} \frac{1}{n!}\frac{d^n}{dt^n}(t^n e^{-t})e^{-(s-1)t}dt = \frac{1}{n!}\mathcal{L}[\frac{d^n}{dt^n}(t^n e^{-t})](s-1)$. Using (2.15), we obtain $\mathcal{L}[L_n(t)](s) = \frac{1}{n!}(s-1)^n\mathcal{L}[(t^n e^{-t})](s-1)$, since all boundary terms vanish. Finally, we have $\mathcal{L}[L_n(t)](s) = \frac{1}{n!}(s-1)^n\mathcal{L}[t^n](s) = \frac{(s-1)^n}{s^{n+1}}$, where we used Exercises 2.2(a) and 2.3.

Chapter 3

Exercise 3.1. By repeatedly integrating by parts we find

$$\int_0^{+\infty} g(t)e^{-zt}dt = -g(t)\frac{e^{-zt}}{z}\Big|_0^{+\infty} + \frac{1}{z}\int_0^{+\infty} g'(t)e^{-zt}dt = \sum_{k=0}^{N}\frac{g^{(k)}(0)}{z^{k+1}} + R_N(z),$$

where the remainder is $R_N(z) = \frac{1}{z^{N+1}}\int_0^{+\infty} g^{(N+1)}(t)e^{-zt}dt$. One can show that, for every fixed N, the remainder is subleading with respect to the last term in the sum. The asymptotic series obtained in this way is precisely (3.19).

Exercise 3.2. We want to compare the integral $I(z) = \int_0^{+\infty} e^{-z\tau}\tau^k d\tau = k!/z^{k+1}$ with $I_b(z) = \int_b^{+\infty} e^{-z\tau}\tau^k d\tau$. By repeated integrations by parts we find

$$I_b(z) = e^{-bz}\sum_{n=0}^{k}\frac{k!}{(k-n)!}\frac{b^{k-n}}{z^{n+1}} \underset{z\to\infty}{\sim} e^{-bz}\frac{1}{z} \tag{B.2}$$

and therefore is subleading with respect to $I(z)$.

Exercise 3.3. Consider the integral (3.30) and assume, for simplicity, that the function $\phi(t)$ has a single maximum in $c \in (a,b)$. Split the integral as the sum of two integrals over the segments $[a,c]$ and $[c,b]$. On these segments, $\phi(t)$ is monotonic. By a change of variable $\tau = -\phi(t)$, the integrals reduce to the form (3.26) (with z replaced by x). For instance, the integral over the segment $[a, c]$ becomes $\int_{-\phi(c)}^{-\phi(a)} e^{-x\tau} \frac{g(t(\tau))}{\phi'(t(\tau))} d\tau$. As discussed in Section 3.2, this integral is dominated by the region around $\tau = -\phi(c)$, where the exponential has a maximum, and can be expanded using Watson's lemma (3.28) applied to the function $\frac{g(t(\tau))}{\phi'(t(\tau))}$. In the neighbourhood of $t = c$, $\tau = -\phi(t) = -\phi(c) - \phi''(c)(t - c)^2/2 + \cdots$ and $\phi'(t) = \phi''(c)(t - c) + \cdots$ with $\phi''(c) < 0$. We can use the expansions above to express $\phi'(t)$ in terms of τ. The integrand function then behaves as $g(c)e^{-x\tau}/\sqrt{2(\tau + \phi(c))|\phi''(c)|}$. Using Watson's lemma (3.28) with $a = -\phi(c)$, $c_0 = g(c)/\sqrt{2|\phi''(c)|}$ and $a_0 = -1/2$, we find $e^{x\phi(c)}\Gamma(1/2)g(c)/\sqrt{2|\phi''(c)|x}$. Since $\Gamma(1/2) = \sqrt{\pi}$ we find precisely half of the result (3.32). The other half comes from an analogous estimate of the contribution of the integral over $[c, b]$, which is again dominated by the neighbourhood of c.

Exercise 3.4. As in Example 3.8 we write the Gamma function as $\Gamma(x + 1) = x^{x+1} \int_0^{+\infty} e^{x\phi(t)} dt$ where $\phi(t) = -t + \ln t$. For large x, the integral is dominated by the region near the maximum of $\phi(t)$, $t = 1$. In order to evaluate higher orders in the asymptotic expansion we change variable to $\phi(t) - \phi(1) = -u^2$. We also extend the integral over u from $-\infty$ to $+\infty$ since the error is exponentially suppressed. We obtain $x^{x+1}e^{-x} \int_{\mathbb{R}} e^{-xu^2} \frac{dt}{du} du$. By inverting term by term $u^2 = (t - 1)^2/2 - (t-1)^3/3 + (t-1)^4/4 + O((t-1)^5)$, we find[1] $(t-1) = \sqrt{2}u + 2u^2/3 + \sqrt{2}u^3/18 + \cdots$. Thus $\Gamma(x+1) = x^{x+1}e^{-x} \int_{\mathbb{R}} e^{-xu^2}(\sqrt{2} + \sqrt{2}u^2/6 + \cdots) du$, since the odd terms in u vanish upon integration. The final result is $\Gamma(x+1) = \sqrt{\frac{2\pi}{x}} x^{x+1}e^{-x}(1 + \frac{1}{12}\frac{1}{x} + \cdots)$.

Exercise 3.5. (a) The relation can be proved by considering the product $\Gamma(p)\Gamma(q) = \int_0^{+\infty} du \int_0^{+\infty} dy\, e^{-(u+y)} u^{p-1} y^{q-1}$. We perform the change of variables $u = zt$ and $y = z(1 - t)$, with $t \in [0, 1]$ and $z \in [0, +\infty)$, and Jacobian $|\det \begin{pmatrix} t & z \\ 1-t & -z \end{pmatrix}| = z$. We obtain $\Gamma(p)\Gamma(q) = \int_0^{+\infty} dz\, z^{p+q-1}e^{-z} \int_0^1 dt\, t^{p-1}(1 - t)^{q-1} = \Gamma(p + q)B(p, q)$.

(b) We use Laplace's method. We have to expand the function $B(x + 1, x + 1) = \int_0^1 dt\, e^{x \ln[t(1-t)]}$ for $x \to +\infty$. On the real axis, the function $\phi(t) = \ln[t(1 - t)] = \ln(t) + \ln(1 - t)$ has a maximum at $t = \frac{1}{2}$ with value $\phi(1/2) = \ln(1/4)$. It is convenient to change variable to $\phi(t) - \phi(1/2) = -u^2$ or, equivalently, $4t - 4t^2 = e^{-u^2}$. The last relation can be inverted to $2t = 1 \pm \sqrt{1 - e^{-u^2}}$, where the two signs correspond to the right-hand and left-hand side of the maximum, respectively. To

---

[1] The same expansion can be obtained using Lagrange's formula (1.4).

obtain the first two terms of the asymptotic expansion, it is enough to expand the relation between t and u up to second order, $t = \frac{1}{2} + \frac{u}{2} - \frac{u^3}{8} + \cdots$, where we use a positive or negative u to parameterise the regions $t > 1/2$ and $t < 1/2$, respectively. We find $B(x+1, x+1) \sim \int_{-\infty}^{+\infty} du \, e^{-x[\ln 4 + u^2]}(\frac{1}{2} - \frac{3u^2}{8} + \cdots) = 2^{-(2x+1)}\sqrt{\frac{\pi}{x}}(1 - \frac{3}{8x} + \cdots)$.

We can give an alternative derivation of this result. Using Exercise 3.4, we easily find $[\Gamma(x+1)]^2 \sim 2\pi x^{2x+1} e^{-2x}(1 + \frac{1}{6x} + \cdots)$ and $\Gamma(2x+2) \sim \sqrt{2\pi}(2x)^{2x+\frac{3}{2}} e^{-2x}(1 + \frac{13}{24x} + \cdots)$. Then, using the relation proved in part (a), we obtain $B(x+1, x+1) = \frac{[\Gamma(x+1)]^2}{\Gamma(2x+2)} \sim 2^{-(2x+1)}\sqrt{\frac{\pi}{x}}(1 - \frac{3}{8x} + \cdots)$.

Exercise 3.6.   We use Laplace's method. In the interval $[0, \frac{\pi}{2}]$ the function $\phi(t) = \cos t$ has a maximum at $t = 0$. Making the change of variables $\phi(t) - \phi(0) = -\rho$ with ρ positive, namely $\cos t = 1 - \rho$, we obtain $I(x) = \int_0^1 e^{x(1-\rho)}(2\rho - \rho^2)d\rho$. The main contributions in the asymptotic regime come from a small interval $(0, \epsilon)$ around the origin, and extending the integral to $+\infty$ only introduces an exponentially small error. Therefore we can write $I(x) \sim e^x \int_0^{+\infty} e^{-x\rho}(2\rho - \rho^2)d\rho = \frac{2e^x}{x^2} - \frac{2e^x}{x^3}$. We can compute the integral exactly with the substitution $y = \cos t$. We find $I(x) = \int_0^1 dy e^{xy}(1 - y^2) = e^x(\frac{2}{x^2} - \frac{2}{x^3} + \frac{2e^{-x}}{x^3} - \frac{e^{-x}}{x})$. Comparing the two results, we see that the asymptotic expansion of the integral for $x \to +\infty$ captures only the first two terms, the other being exponentially suppressed.

Exercise 3.7.   In the interval $(-1, 1)$ the function $\phi(t) = t^3 - 3t$ has a maximum at $t = -1$. We perform the usual change of variables $\phi(t) - \phi(-1) = -\tau^2$. We need to find t as a function of τ. Plugging $t + 1 = c_1\tau + c_2\tau^2 + c_3\tau^3 + \cdots$ into the relation $\tau^2 = 3(t+1)^2 - (t+1)^3$, we can find the coefficients c_i iteratively. In order to calculate the first two terms of the asymptotic expansion, it is sufficient to solve for the first two coefficients. We find $t + 1 = \frac{1}{\sqrt{3}}\tau + \frac{1}{18}\tau^2 + \cdots$. Thus, the integral can be written as

$$\int_{-1}^1 \frac{e^{x(t^3-3t)}}{1+t^2}dt \sim \int_0^{+\infty} \frac{e^{x(2-\tau^2)}}{1+(-1+\frac{1}{\sqrt{3}}\tau+\cdots)^2}\left(\frac{1}{\sqrt{3}} + \frac{1}{9}\tau + \cdots\right)d\tau,$$

where, as usual, we have extended the upper limit of integration to infinity. Expanding the denominator in power series in τ and performing the Gaussian integrals, we find $I(x) \sim \frac{e^{2x}}{4\sqrt{3}}\sqrt{\frac{\pi}{x}}(1 + \frac{4}{3\sqrt{3\pi x}} + \cdots)$.

Exercise 3.8.   Observing that the integral can be written as $J_0(x) = \text{Re}\frac{2}{\pi}\int_0^{\frac{\pi}{2}} d\theta \, e^{ix\cos\theta}$, we can apply the stationary phase method. The stationary point is at $\theta = 0$. The leading term can be calculated by expanding $\cos\theta \sim 1 - \frac{\theta^2}{2}$. The result is

$$J_0(x) = \text{Re}\frac{2}{\pi}\int_0^{\frac{\pi}{2}} d\theta e^{ix\cos\theta} \sim \text{Re}\,\frac{2}{\pi}\int_0^{+\infty} d\theta e^{ix(1-\frac{\theta^2}{2})} = \sqrt{\frac{2}{\pi x}}\cos\left(x - \frac{\pi}{4}\right).$$

The same result is obtained by using (3.42) with $\phi(\theta) = \cos\theta$, $g(\theta) = 1$ and $c = 0$. Since the integration is only over the half-line on the right of the stationary point, we need to take half of the expression (3.42).

Exercise 3.9. The function $\phi(t) = t^2 - 2t$ has a stationary point at $t = 1$. The leading contribution to the integral comes from a small interval around $t = 1$, and can be obtained using (3.42) with $c = 1$ and $g(t) = (1+t^2)^{-1}$. We also need to divide by an extra factor of two, because the integration is only over the region $t \geq 1$ to the right of the stationary point. We thus find $I(x) \sim \frac{1}{4}\sqrt{\frac{\pi}{x}}e^{-ix+i\pi/4}$. We can perform the same computation using the saddle point method. The point $t = 1$ is a saddle point of $h(t) = i\phi(t)$. The paths with constant phase $t = 1$ $\text{Im}h(t) = \text{Im}h(1) = -1$ are the two lines of equations $v = \pm(u - 1)$, where $t = u + iv$. The steepest descent path is the line $v = u - 1$ where the real part of $h(t)$ presents a maximum. We can deform the original contour $[1, +\infty)$ to the half-line $v = u - 1$, with $u \geq 1$, without encountering any singularity. On the modified contour of integration we now perform the change of variable $h(t) - h(1) = i(t-1)^2 = -\tau^2$. We can thus write $t - 1 = e^{\frac{i\pi}{4}}\tau$, with τ real and positive, and we obtain

$$I(x) = e^{i\frac{\pi}{4}}\int_0^{+\infty} d\tau\frac{e^{-ix-x\tau^2}}{1+(1+e^{\frac{i\pi}{4}}\tau)^2} \sim \frac{e^{-ix+i\frac{\pi}{4}}}{2}\int_0^{+\infty} d\tau e^{-x\tau^2} = \frac{e^{-ix+i\frac{\pi}{4}}}{4}\sqrt{\frac{\pi}{x}},$$

where, at leading order, we approximated the denominator of the integrand with its value in $\tau = 0$. The result coincides with that obtained using the stationary phase method.

Exercise 3.10. We use the saddle point method. The integral can be written as $I = \int_0^1 e^{xh(t)}dt$, where $h(t) = it^3$ has a critical point of order two in $t = 0$. The curves of constant phase $\text{Im}(it^3) = 0$ passing through the saddle point $t = 0$ are three lines forming the angles $\pi/6, \pi/2$ and $5\pi/6$ with the real axis. The curve of constant phase passing through $t = 1$, $\text{Im}(it^3) = 1$, has equation $x^3 - 3xy^2 = 1$ with $t = x + iy$. Using Cauchy's theorem, we can deform the integral to become the combination of an integral I_1 over the half-line γ_1 at angle

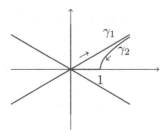

Fig. B.3.   Curves for Exercise 3.10.

$\pi/6$ and an integral I_2 along the curve γ_2 of constant phase passing through $t = 1$ as in Figure B.3. This can be done because there are no singularities in between. On γ_1 we can set $u = h(0) - h(t) = -it^3$, where $u > 0$ and the integral becomes $I_1 = \frac{1}{3}e^{i\pi/6}\int_0^{+\infty} e^{-xu}u^{-2/3}du = \frac{\Gamma(1/3)}{3}e^{i\pi/6}x^{-1/3}$, where we used (A.15). On γ_2, we set $u = h(1) - h(t) = -i(t^3 - 1)$ and we find $I_2 = -\frac{ie^{ix}}{3}\int_0^{+\infty} e^{-xu}(1 + iu)^{-2/3}du = -\frac{ie^{ix}}{3}\sum_{n=0}^{\infty}\frac{\Gamma(n+2/3)}{\Gamma(2/3)n!}\int_0^{+\infty} e^{-xu}(-iu)^n du = \frac{e^{ix}}{3}\sum_{n=0}^{\infty}\frac{\Gamma(n+2/3)}{\Gamma(2/3)(ix)^{n+1}}$, where we used (A.17). The sum $I_1 + I_2$ is the desired asymptotic expansion.

Exercise 3.11. We use the saddle point method. It is convenient to change variable to $z = iw$. We obtain $I(x) = i\int_{-\infty}^{+\infty} e^{-x(w^2+i\pi w)}(1 - i\sinh w)dw$. The critical point of $h(w) = -(w^2+i\pi w)$ is $w_0 = -\frac{i\pi}{2}$. Moreover, $h(w_0) = -\frac{\pi^2}{4}$, so the phase of $h(w)$ is zero at the saddle point. The paths with vanishing phase passing through w_0 are the lines $v = -\frac{\pi}{2}$ and $u = 0$, where we set $w = u + iv$. The real part of $h(w)$ has a maximum in w_0 on the line $v = -\frac{\pi}{2}$ and a minimum on the line $u = 0$. The path of steepest descent is then the line $v = -\frac{\pi}{2}$, parallel to the real axis. Since there are no singularities in the region between the path of steepest descent and the real axis, we can deform the integration contour to be the curve of steepest descent. We now perform the change of variable $h(w) - h(w_0) = -\tau^2$. On the path of steepest descent τ is real and the above relation is equivalent to $\tau = w + \frac{i\pi}{2}$. We obtain $I(x) = ie^{-x\frac{\pi^2}{4}}\int_{-\infty}^{+\infty} d\tau e^{-x\tau^2}(1 - \cosh\tau)$, where we used $\sinh(\tau - \frac{i\pi}{2}) = -i\cosh\tau$. Expanding $\cosh\tau$ in power series around $\tau = 0$, we find $I(x) = -ie^{-x\frac{\pi^2}{4}}\sum_{k=1}^{\infty}\frac{1}{(2k)!}\int_{-\infty}^{+\infty} d\tau e^{-x\tau^2}\tau^{2k} = -ie^{-x\frac{\pi^2}{4}}\sum_{k=1}^{\infty}\frac{\Gamma(k+1/2)}{(2k)!x^{k+1/2}}$, where we used (A.16). Using also $\Gamma(k + 1/2) = (2k)!\sqrt{\pi}/(4^k k!)$ we finally obtain

$$I(x) = -i\sqrt{\frac{\pi}{x}}e^{-x\frac{\pi^2}{4}}\sum_{k=1}^{\infty}\frac{1}{4^k k!}\frac{1}{x^k}. \tag{B.3}$$

Notice that the integral can be also evaluated exactly. Writing $\sinh w$ as a sum of exponentials and using repeatedly (A.14), we find $I(x) = i\sqrt{\frac{\pi}{x}} e^{-x} e^{\frac{\pi^2}{4}} (1 - e^{\frac{1}{4x}})$. By expanding the exponential in power series, we recover the expansion (B.3).

Exercise 3.12. The function $h(t) = \frac{i}{2}(\frac{t^2-1}{t})$ has two critical points at $t = \pm i$ with $\mathrm{Im}\, h(\pm i) = 0$. The paths with vanishing phase obey the equation $u(u^2 + v^2 - 1) = 0$, where $t = u + iv$. They correspond to the imaginary axis and the circle of radius 1 centred at the origin. We can deform the contour γ to be the unit circle. In a neighbourhood of the saddle points $t = \pm i$ we can write $h(t) = h(\pm i) \pm \frac{1}{2}(t \mp i)^2 + \cdots = \mp 1 \pm \frac{1}{2}(t \mp i)^2 + \cdots$. We see that, on the circle, the real part of $h(t)$ has a maximum at $t = -i$ and a minimum at $t = i$. The integral is then dominated by a neighbourhood of $t = -i$. We perform the usual change of variable $h(t) - h(-i) = -\frac{1}{2}(t + i)^2 + \cdots = -\tau^2$ with τ real. We then find the leading term

$$I(x) \sim \frac{e^x}{2\pi i} \int_{\gamma} \frac{e^{-\frac{x}{2}(t+i)^2}}{t^{n+1}} \mathrm{d}t \sim \frac{e^x}{2\pi i} \int_{-\infty}^{+\infty} \frac{e^{-x\tau^2}}{(-i)^{n+1}} \sqrt{2} \mathrm{d}\tau = \frac{e^x i^n}{\sqrt{2\pi x}}.$$

Exercise 3.13. The integral is on a line parallel to the imaginary axis in the s-plane. The function at the exponent, $h(s) = s - \sqrt{s}$, is multivalued. We choose the principal branch of the square root and the branch cut along the real negative axis, so that $\arg s \in (-\pi, \pi)$. The function $h(s)$ has a saddle point of order one in $s_0 = \frac{1}{4}$, where $h''(s_0) = 2 \neq 0$ and $h(s_0) = -\frac{1}{4}$. Using the parameterisation $s = u + iv = \rho e^{i\alpha}$, the paths with vanishing constant phase obey the equation $\sqrt{\rho} \sin(\alpha/2)[2\sqrt{\rho} \cos(\alpha/2) - 1] = 0$. They correspond to the real axis, where $\alpha = 0$, and to the locus where the second factor is zero. With some work, one can see that the curve passing through s_0 is the parabola of equation $u = \frac{1}{4} - v^2$. On the real axis, $\mathrm{Re}\, h(s) = \rho - \sqrt{\rho}$ has a minimum at $s = s_0$, while on the parabola $\mathrm{Re}\, h(s) = u - \frac{1}{2}$ has a maximum at the saddle point s_0. To find the asymptotic expansion, we have thus to deform the original path of integration to the parabola passing through the saddle point (see Figure B.4). Let us start with the case $g(s) = \frac{1}{s}$. By Cauchy's theorem, the original integral is equal to the integral on the parabola, since there are no singularities in the region between the two curves. On the parabola, we perform the usual change of variable $h(s) - h(s_0) = -\tau^2$, or $\sqrt{s} = \frac{1}{2} + i\tau$, where τ is real (the signs related to the imaginary part of $\sqrt{s}$ are taken into account by the signs of τ itself). The integral can thus be written as $I(x) = \frac{2e^{-x/4}}{\pi} \int_{-\infty}^{+\infty} \mathrm{d}\tau \frac{e^{-x\tau^2}}{1+2i\tau}$. The denominator can be expanded around $\tau = 0$ using the geometric series. Only even powers of τ contribute to the integral.

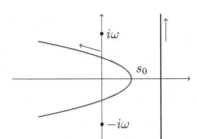

Fig. B.4.   Steepest descent curve and poles for Exercise 3.13 for $\omega > 1/2$.

Using (A.16), we obtain $I(x) = \frac{2e^{-x/4}}{\pi} \sum_{k=0}^{\infty} (-4)^k \frac{\Gamma(k+1/2)}{x^{k+1/2}}$. It is worth to notice that the leading term is $\frac{2}{\sqrt{\pi x}} e^{-x/4}$. Let us move to the case $g(s) = \frac{1}{s^2+\omega^2}$. The integrand has two simple poles at $s = \pm i\omega$. When $\omega < 1/2$, the two poles are outside the region between the original integration contour and the parabola. The asymptotic expansion can be calculated as in the previous case, and we leave the exercise to the reader. The leading term of the expansion is $I(x) \sim \frac{8}{\sqrt{\pi x}} e^{-x/4}/(1+16\omega^2)$. When $\omega > 1/2$, on the other hand, the two simple poles fall inside the region between the original integration contour and the parabola. Thus $I(x)$ does not coincide with the integral over the parabola. However, by the residue theorem, the difference between the two integrals is given by $2\pi i$ times the residues of the integrand at the two poles. By writing $f(s) = \frac{e^{xh(s)}}{s^2+\omega^2}$, the residues can be easily computed and give $\mathrm{Res}[f, i\omega] + \mathrm{Res}[f, -i\omega] = \frac{e^{-x\sqrt{\omega/2}}}{\omega} \sin[(\omega - \sqrt{\omega/2})x]$. As a consequence, since $\omega > 1/2$, the contribution of the residues is exponentially suppressed with respect to the asymptotic series coming from the integral over the parabola and can be neglected. The asymptotic expansion is then $I(x) \sim \frac{8}{\sqrt{\pi x}} e^{-x/4}/(1 + 16\omega^2)$.

Exercise 3.14.   Consider the Airy function (3.49). We find

$$y\,Ai(y) - Ai''(y) = \frac{1}{2\pi} \int_{\gamma_1} (w^2 + y) e^{i(yw + \frac{w^3}{3})} dw$$

$$= -\frac{i}{2\pi} \int_{\gamma_1} \frac{d}{dw} e^{i(yw + \frac{w^3}{3})} dw = 0,$$

because the exponential goes to zero at infinity along the contour γ_1.

Exercise 3.15. Using the results of Example 3.10, $Ai(x^{3/2}) \sim \frac{x^{1/3}}{2\pi} \int_C e^{xh(z)} dz$ where $h(z) = i(z + z^3/3)$ and C is the hyperbola $\operatorname{Im} h(z) = 0$ passing through $z = i$. The function $h(z)$ has a maximum along C at $z = i$. It is convenient to change variable to $u^2 = h(i) - h(z)$. Since $\operatorname{Im} h(z) = 0$, u is real along C so that this latter is mapped by the change of variables to the real axis. The integral then becomes $I = \int_C e^{xh(z)} dz = e^{xh(i)} \int_{\mathbb{R}} e^{-xu^2} \frac{dz}{du} du$. Since we are in the neighbourhood of $z = i$, it is useful to write u as a function of $z - i$. We have $u = (z - i)\sqrt{1 - i(z - i)/3}$. We need to invert this holomorphic function in the neighbourhood of $z = i$, $(z - i) = \sum_{n=0}^{\infty} b_n u^n$. The coefficients b_n can be evaluated using the Lagrange formula[2] (1.4): $b_0 = 0$ and $b_n = \frac{1}{n!} \frac{d^{n-1}}{dz^{n-1}} (1 - \frac{i}{3}(z - i))^{-n/2}|_{z=i}$. From the binomial expansion (A.17) we have $(1 - \frac{i}{3}(z - i))^{-n/2} = \sum_{k=0}^{\infty} \frac{\Gamma(k+n/2)}{\Gamma(n/2)} (\frac{i}{3})^k \frac{(z-i)^k}{k!}$ so that $b_n = \frac{\Gamma(3n/2-1)}{n!\Gamma(n/2)} (\frac{i}{3})^{n-1}$. We thus find

$$I = e^{-2x/3} \sum_{n=1}^{\infty} \frac{\Gamma(3n/2 - 1)}{(n - 1)!\Gamma(n/2)} \left(\frac{i}{3}\right)^{n-1} \int_{\mathbb{R}} e^{-xu^2} u^{n-1} du$$

$$= \frac{e^{-2x/3}}{\sqrt{x}} \sum_{k=0}^{\infty} \frac{\Gamma(3k + 1/2)}{(2k)!(-9x)^k},$$

where we used (A.16). Multiplying by the factor $\frac{x^{1/3}}{2\pi}$ and using $y = x^{2/3}$ we obtain (3.55).

Exercise 3.16. (a) The critical points of $h(z)$ are $z_{L,R} = \pm i e^{i\phi/2}$ and move in the plane when ϕ varies. Also the curves of steepest descent, which are the curve of constant phase and $\operatorname{Im}(h(z) - h(\pm i e^{i\phi/2})) = 0$, move in the plane. The case $\phi = 0$ is depicted in Figure 3.2 while $\phi = \pi/3, 2\pi/3$ and π are depicted in Figure B.5. All the steepest descent curves lie in regions of the plane where the integrals (3.49) and (3.57) are well-defined.

(b) Setting $y = x^{2/3} e^{i\phi}$ and $w = x^{1/3} z$ in (3.49), we find $Ai(x^{2/3} e^{i\phi}) = \frac{x^{1/3}}{2\pi} \int_{\gamma_1} e^{xh(z)} dz$, where $h(z) = i(z e^{i\phi} + \frac{z^3}{3})$. For $|\phi| < 2\pi/3$ the original contour of integration can be deformed into the steepest descent passing through the saddle point z_L, which therefore gives the leading contribution to the integral. For $|\phi| = 2\pi/3$ the two steepest descent curves meet since $\operatorname{Im} h(z_L) = \operatorname{Im} h(z_R)$.

---

[2] Set $f(z) = (z - i)\sqrt{1 - i(z - i)/3}$ and $z_0 = i$ in (1.4).

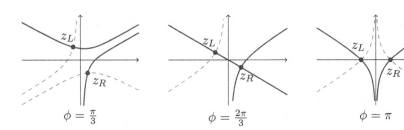

Fig. B.5. The solid (dashed) lines are the curves of steepest descent (ascent) for $h(z) = i(ze^{i\phi} + z^3/3)$. The dots are the critical points.

For $2\pi/3 < \phi < \pi$, the contour must be deformed into a combination of the two steepest descent curves and the integral receives contributions from both the saddle points. However, one can check that $\text{Re}(z_R) < \text{Re}(z_L)$, so the contribution of the saddle point z_L still dominates. The asymptotic expansion of $Ai(x^{2/3}e^{i\phi})$ with $0 \leq \phi < \pi$ can be evaluated as in Exercise 3.15. On the steepest descent passing through z_L we write $u^2 = h(z_L) - h(z)$. The relation $u = e^{i\phi/4}(z - z_L)\sqrt{1 - ie^{-i\phi/2}(z - z_L)/3}$ can be inverted using the Lagrange formula: $(z - z_L) = \sum_{n=0}^{\infty} b_n u^n$ with $b_0 = 0$ and $b_n = \frac{\Gamma(3n/2-1)}{n!\Gamma(n/2)}(\frac{i}{3})^{n-1}e^{-\frac{i(3n-2)\phi}{4}}$ for $n \geq 1$. Moreover $h(z_L) = -2e^{3i\phi/2}/3$. We thus find

$$Ai(y) = \frac{e^{-\frac{2}{3}xe^{\frac{3i\phi}{2}}}}{2\pi x^{1/6}} \sum_{k=0}^{\infty} \frac{\Gamma(3k + 1/2)}{(2k)!(-9x)^k} e^{-\frac{3}{2}ik\phi - i\frac{\phi}{4}} = \frac{e^{-\frac{2}{3}y^{3/2}}}{2\pi y^{1/4}} \sum_{k=0}^{\infty} \frac{\Gamma(3k + 1/2)}{(2k)!(-9y^{3/2})^k},$$

where we used $y = x^{2/3}e^{i\phi}$. We see that the asymptotic (3.55) is valid for all $0 \leq \arg y < \pi$. Consider now $\phi = \pi$. For this value of ϕ, $\text{Re}(z_R) = \text{Re}(z_L)$ and the two saddle points contribute equally.[3] We can now deform the original contour of integration to become the union of the two curves of steepest descent. The two integrals receive leading contributions from the neighbourhood of $z_L = -1$ and $z_R = 1$, respectively, and can be computed using Laplace's method. Near $z = \pm 1$ we set $-u^2 = h(z) - h(\pm 1) = \pm i(z \mp 1)^2 + O((z \mp 1)^3)$, where $u \in \mathbb{R}$, and we obtain $\int e^{xh(z)}dz = e^{xh(\pm 1)} \int_{-\infty}^{+\infty} e^{-xu^2}\frac{dz}{du}du \sim e^{\mp i2x/3} \int_{-\infty}^{+\infty} e^{-xu^2}e^{\pm i\pi/4}du = e^{\mp i(2x/3 - \pi/4)}\sqrt{\frac{\pi}{x}}$. Adding the two contributions, multiplying by the prefactor $\frac{x^{1/3}}{2\pi}$

---

[3] $\text{Re}(z_R) = \text{Re}(z_L)$ is also true for $\phi = \pm\pi/3$. However, z_R does not contribute to the asymptotic behaviour of the Airy function for $\phi = \pm\pi/3$ since it is not on the steepest descent path. It contributes instead to the behaviour of a generic solution of the Airy equation, as discussed in part (d).

and using the original variable $y = -x^{2/3}$ we find precisely (3.56). The analysis for $\phi \in [-\pi, 0]$ is similar.

(c) The Bairy function is given by the integral in (3.57). Consider first $\phi = 0$. The steepest descent curves are shown in Figure 3.2. γ_1 can be still deformed into the hyperbola C while γ_2 can be deformed into the union of the left half C, transversed from right to left, and the half-line $(-i\infty, i)$. Both saddle points $z_{L,R} = \pm i$ contribute to the asymptotic behaviour of the Bairy function. However, since $\operatorname{Re} h(-i) > \operatorname{Re} h(i)$, the leading contribution comes from the saddle point $z_R = -i$. On $(-i\infty, i)$, we set $-u^2 = h(z) - h(-i) = (z+i)^2 + O((z+i)^3)$, where u is real and we obtain $Bi(y) \sim \frac{-ix^{1/3}}{\pi} \int_{-i\infty}^{i} e^{xh(z)} dz \sim \frac{x^{1/3}}{\pi} e^{2x/3} \int_{\mathbb{R}} e^{-xu^2} du = \frac{1}{\sqrt{\pi}x^{1/6}} e^{2x/3}$. Setting $y = x^{2/3}$ we obtain (3.58). Consider next $\phi = \pi$. The steepest descent curves are given in Figure B.5. γ_1 can be still deformed into the union of the two steepest descent curves while γ_2 can be deformed into the left one. The contribution of these curves has been computed in part (b). By taking into account the orientation of the curves and the normalisation of the integrals, one easily finds (3.59).

(d) The Stokes lines are defined by $\operatorname{Im} h(z_L) = \operatorname{Im} h(z_R)$ and are $\phi = 0$ and $\phi = \pm 2\pi/3$. These are the values where the steepest descent curves merge. They also correspond to the directions where the contribution of one of the saddle points to a generic solution of the Airy equation is maximally suppressed. The anti-Stokes lines are defined by $\operatorname{Re} h(z_L) = \operatorname{Re} h(z_R)$ and are $\phi = \pi$ and $\phi = \pm \pi/3$. These are the directions where the contribution of the saddle points becomes equally important. In these directions there is a change in the asymptotic behaviour of a generic solution of the Airy equation. Notice that the solution $Ai(y)$ is special since there is no change at $\phi = \pm \pi/3$ but only at $\phi = \pi$. Instead, the linear combination $\alpha Ai(y) + \beta Bi(y)$ has a different asymptotic behaviour in the three regions $|\arg y| < \pi/3$, $\pi/3 < \arg y < \pi$ and $-\pi/3 < \arg y < -\pi$. This is another examples of Stokes phenomenon.

Exercise 3.17. (a) By changing variable $y \to \omega y$, we see that, if $u(y)$ solves the Airy equation, also $u(\omega y)$ does. One can check that $Ai(y)$ and $Ai(\omega y)$ are linearly independent, so that $u(y) = cAi(y) + dAi(\omega y)$ is the general solution of the Airy equation.

(b) From (3.55) we know that $Ai(y) \sim e^{-2y^{3/2}/3}/(2y^{1/4}\sqrt{\pi})$ for $|\arg y| < \pi$. We choose the branch cuts for powers on the negative real axis, so that $y^{\alpha} = \rho^{\alpha} e^{i\alpha\phi}$ if $y = \rho e^{i\phi}$ with $\phi \in (-\pi, \pi)$. The argument of ωy belongs to $(-\pi, \pi)$ for

$-\pi < \arg y < \pi/3$ so that $(\omega y)^\alpha = e^{2\alpha\pi i/3} y^\alpha$ in this sector. On the other hand, for $\pi/3 < \arg y < \pi$ the argument of ωy belongs to $(\pi, 2\pi)$ and we need to subtract 2π from the argument before taking the power, so that $(\omega y)^\alpha = e^{-4\alpha\pi i/3} y^\alpha$. We then find $Ai(\omega y) \sim e^{-i\pi/6} e^{2y^{3/2}/3} / (2y^{1/4}\sqrt{\pi})$ for $-\pi < \arg y < \pi/3$ and $Ai(\omega y) \sim e^{i\pi/3} e^{-2/3 y^{3/2}} / (2y^{1/4}\sqrt{\pi})$ for $\pi/3 < \arg y < \pi$. Now, $e^{2/3 y^{3/2}}$ is exponentially large compared to $e^{-2y^{3/2}/3}$ for $-\pi/3 < \arg y < \pi/3$ while the opposite is true for $\pi/3 < |\arg y| < \pi$. Combining all this information and keeping only the leading exponential, we find that $u(y) \sim c e^{-2/3 y^{3/2}} / (2y^{1/4}\sqrt{\pi})$ for $-\pi < \arg y < -\pi/3$, $u(y) \sim d e^{-i\pi/6} e^{2/3 y^{3/2}} / (2y^{1/4}\sqrt{\pi})$ for $-\pi/3 < \arg y < \pi/3$ and $u(y) \sim (c + d e^{i\pi/3}) e^{-2/3 y^{3/2}} / (2y^{1/4}\sqrt{\pi})$ for $\pi/3 < \arg y < \pi$. In the particular case where c or d are zero, the previous expressions should be modified in order to include the subleading terms. For example, the Airy function, $c = 1$ and $d = 0$, has the same leading asymptotic expansion (3.55) for all $\arg y \in (-\pi, \pi)$. We see that instead the generic solution $u(y)$ has anti-Stokes lines at $\arg y = \pm\pi/3, \pi$, as already discussed in Exercise 3.16. As we will discuss in Section 7.2.3, solutions of differential equations exhibit Stokes phenomenon near essential singular points.

(c) From (3.57) and (3.49) we have $Bi(y) = -iAi(y) - \frac{i}{\pi} \int_{\gamma_2} e^{i(yw + \frac{w^3}{3})} dw$. We now perform the change of variable $w = \omega\tilde{w}$ in the integral. The integrand remains of the same form with y replaced by ωy. The contour of integration for the variable $\tilde{w}$ is rotated by $2\pi/3$ clock-wise with respect to the original one. But γ_2 rotated by $2\pi/3$ can be continuously deformed into γ_1 without leaving the region of holomorphicity of the integrand and convergence of the integral. Therefore, using again (3.49), we find $Bi(y) = -iAi(y) - 2i\omega Ai(\omega y)$, so that $c = -i$ and $d = -2i\omega$. Using (b), we find the asymptotic behaviours of the Bairy function: $Bi(y) \sim -ie^{-2y^{3/2}/3} / (2y^{1/4}\sqrt{\pi})$ in the sector $\arg y \in (-\pi, -\pi/3)$, $Bi(y) \sim e^{2y^{3/2}/3} / (y^{1/4}\sqrt{\pi})$ in $(-\pi/3, \pi/3)$, in agreement with (3.58), and $Bi(y) \sim ie^{-2y^{3/2}/3} / (2y^{1/4}\sqrt{\pi})$ in $\arg y \in (\pi/3, \pi)$.

Chapter 4

Exercise 4.1. Both equations can be solved by separation of variables by writing them as $f(y)dy = g(x)dx$.

(a) Writing $y' = \frac{dy}{dx}$ we find $dy/(1 + y^2) = xdx$. Integrating both terms gives $\arctan y = x^2/2 + C$, where C is an arbitrary constant. Since $y(2) = 0$, we find $C = -2$ so that $y(x) = \tan(x^2/2 - 2)$.

(b) Integrating $dy/(y+2)^3 = dx/(x-1)^3$ we find $1/(y+2)^2 = 1/(x-1)^2 + C$, where C is an integration constant. From $y(0) = -1$ we find $C = 0$. We then obtain $y + 2 = \pm(x - 1)$. The solution satisfying the initial condition is $y = -x - 1$.

Exercise 4.2. We choose time as the independent variable, $y' = \frac{dy}{dt}$.

(a) The derivative of the energy vanishes on any solution, $E' = y'y'' + U'(y)y' = y'(y'' - f(y)) = 0$, since $y'' = f(y)$. Thus E is constant on any solution.

(b) To determine the general solution, we can use the conservation of energy, $E = E_0$. This is a first-order equation in y and it can be written as $y' = \pm\sqrt{2(E_0 - U(y))}$. By separating variables, we find the general solution in implicit form $\pm \int_{y_0}^{y} \frac{d\tau}{\sqrt{2(E_0 - U(\tau))}} = t - t_0$ with $y(t_0) = y_0$. Notice that the solution depends on two arbitrary constants E_0 and y_0.

(c) For the pendulum we just find $\pm \int_{y_0}^{y} \frac{d\tau}{\sqrt{2(E_0 + k \cos \tau)}} = t - t_0$.

Exercise 4.3. We can write the equation as $y' + (a_0(x)/a_1(x))y = g(x)/a_1(x)$. By introducing a function $p(x)$ that satisfies $p'(x) = a_0(x)/a_1(x)$, we can also write $(ye^{p(x)})' = e^{p(x)}g(x)/a_1(x)$. This has solution $y(x)e^{p(x)} = \int_{x_0}^{x} e^{p(w)} \frac{g(w)}{a_1(w)} dw + const$. We can also write the solution as $y(x) = e^{-p(x)}(\int_{x_0}^{x} e^{p(w)} \frac{g(w)}{a_1(w)} dw + y_0 e^{p(x_0)})$, where we chose the integration constants in such a way that $y(x_0) = y_0$.

Exercise 4.4. Consider first (4.26). If the characteristic equation (4.28) has two coincident roots $\alpha_1 = \alpha_2$, one solution will be given by $e^{\alpha_1 x}$. The second linearly independent one is $xe^{\alpha_1 x}$. Indeed, by plugging it into the equation, we find $(a_2\alpha_1^2 + a_1\alpha_1 + a_0)xe^{\alpha_1 x} + (2a_2\alpha_1 + a_1)e^{\alpha_1 x}$. The term proportional to $xe^{\alpha_1 x}$ is zero because of (4.28). The one proportional to $e^{\alpha_1 x}$ is also zero because $\alpha_1 = -a_1/(2a_2)$. Indeed we know that, as for any equation of second degree, a_1 is minus the sum of the two roots of (4.28) multiplied by a_2. In this case, the two roots coincide and we obtain $a_1 = -2a_2\alpha_1$. The case of (4.29) is completely analogous.

Exercise 4.5. Substituting the ansatz (4.35) into (4.34) and using $a_2y_i'' + a_1y_i' + a_0y_i = 0$, with $i = 1, 2$, we find $a_1(c_1'y_1 + c_2'y_2) + a_2(c_1''y_1 + c_2''y_2 + 2c_1'y_1' + 2c_2'y_2') = g$. If we impose $c_1'y_1 + c_2'y_2 = 0$, the term proportional to a_1 cancels. Taking a derivative of this condition we also find $c_1''y_1 + c_2''y_2 + c_1'y_1' + c_2'y_2' = 0$, which can be used to obtain $a_2(c_1'y_1' + c_2'y_2') = g$. We see that, if we can simultaneously satisfy the two conditions $c_1'y_1 + c_2'y_2 = 0$ and $c_1'y_1' + c_2'y_2' = g/a_2$, the original differential equation is solved. We can explicitly determine c_1' and c_2': $c_1' = -gy_2/(a_2W)$ and $c_2' = gy_1/(a_2W)$, where $W = y_1y_2' - y_2y_1'$ is the Wronskian. c_1 and c_2 are then found by integration. Notice that we can solve the system only if y_i are linearly independent so that $W \neq 0$. The general solution of the equation is obtained by adding the particular solution just found to the general solution of the homogeneous equation. The general solution is then $y(x) = \int_{x_0}^x \frac{g(w)}{a_2(w)W(w)}(y_1(w)y_2(x) - y_2(w)y_1(x))dw + C_1 y_1(x) + C_2 y_2(x)$, where C_i are arbitrary constants. The parameter x_0 is not an independent constant, since its variation can be reabsorbed in a variation of C_1 and C_2. Indeed, we can also write the solution in the form $y(x) = -y_1(x)\int_{x_1}^x \frac{g(w)}{a_2(w)W(w)}y_2(w)dw + y_2(x)\int_{x_2}^x \frac{g(w)}{a_2(w)W(w)}y_1(w)dw$ where x_1 and x_2 are two arbitrary constants.

Exercise 4.6.   Using (4.37) we set $y = Y/x$ obtaining the equation $Y'' - 2Y/(1 + x)^2 = 2/(1 + x)$. Setting also $t = x + 1$ we obtain the inhomogeneous Euler equation $t^2 Y'' - 2Y = 2t$, where, from now on, the prime indicates the derivative with respect to t. We use the method of variation of arbitrary constants to solve it. The two linearly independent solutions Y_i of the homogeneous equation are obtained as in Example 4.5: $Y_1 = t^2$ and $Y_2 = 1/t$. A particular solution of the inhomogeneous equation is $c_1 Y_1 + c_2 Y_2$ where c_1 and c_2 are determined as in Exercise 4.5. We find $c_1' = 2/(3t^2)$ and $c_2' = -2t/3$ and therefore $c_1 = -2/(3t) + C_1$ and $c_2 = -t^2/3 + C_2$ where C_1 and C_2 are two constants. The solution is then $Y = c_1 Y_1 + c_2 Y_2 = -t + C_1 t^2 + C_2/t$. This is also the general solution of the equation since the freedom of adding linear combinations of solutions of the homogeneous equation is already incorporated in C_1 and C_2. In terms of the original variables we have $y = x^{-1}(-x - 1 + C_1(x + 1)^2 + C_2/(x + 1))$.

Exercise 4.7.   (a) Plugging $y = u^{1/(1-n)}$ and $y' = u'u^{n/(1-n)}/(1 - n)$ into the Bernoulli equation gives $u' + (1 - n)Au = (1 - n)B$, which can be solved using the method of Exercise 4.3.

(b) Using part (a) with $n = 2$, we set $y = 1/u$ and find $u' - u/x = -x^3$. Writing the equation as $(u/x)' = -x^2$, we can find the solution by integration, $u = -x^4/3 + Cx$, where C is an arbitrary constant. In terms of the variable y the solution is $y = 3/(3Cx - x^4)$.

(c) Plugging $y = y_1 + u$ in the equation and using $y_1' = p + qy_1 + ty_1^2$, we find $u' - (q + 2ty_1)u = tu^2$, which is a Bernoulli equation.

Exercise 4.8.   Applying a Laplace transform to (4.48) and using (2.14), we find $sU(x, s) - u_0(x) = \kappa U''(x, s)$, where $U(x, s) = \int_0^{+\infty} u(x, t)e^{-st}dt$ and, from now on, the derivatives are assumed to be with respect to x. This is a second-order linear differential equation in the variable x for $U(x, s)$, and can be solved using the method of variation of arbitrary constants (see Exercise 4.5). The two solutions of the homogeneous equation are $U_1 = e^{x\sqrt{s/\kappa}}$ and $U_2 = e^{-x\sqrt{s/\kappa}}$. In the notation of Exercise 4.5 we have $y_1 = U_1, y_2 = U_2, g = -u_0, a_2 = \kappa$ and $W = U_1 U_2' - U_2 U_1' = -2\sqrt{s/\kappa}$. Then the general solution is $U(x,s) = -e^{x\sqrt{s/\kappa}} \int_{x_1}^x \frac{u_0(w)}{2\sqrt{\kappa s}} e^{-w\sqrt{s/\kappa}} dw + e^{-x\sqrt{s/\kappa}} \int_{x_2}^x \frac{u_0(w)}{2\sqrt{\kappa s}} e^{w\sqrt{s/\kappa}} dw$. The requirement that $u(x,t)$ vanishes for large $|x|$ can be satisfied if we choose $x_1 = +\infty$ and $x_2 = -\infty$. With this choice we can write the final result in compact form: $U(x,s) = \frac{1}{2\sqrt{\kappa s}} \int_{-\infty}^{+\infty} e^{-|x-w|\sqrt{s/\kappa}} u_0(w) dw$. The function $u(x,t)$ can be obtained with an inverse Laplace transform. Since $\mathcal{L}^{-1}(e^{-a\sqrt{s}}/\sqrt{s}) = e^{-a^2/(4t)}/\sqrt{\pi t}$ (see part (b) of Exercise 2.5), we find $u(x,t) = \frac{1}{\sqrt{4\pi\kappa t}} \int_{-\infty}^{+\infty} e^{-|x-w|^2/(4\kappa t)} u_0(w) dw$, in agreement with the results of Example 4.6.

Exercise 4.9. Applying a Fourier transform to (4.54) and using Appendix A, we find $\partial_t^2 U(p,t) + p^2 U(p,t) = 0$ where $U(p,t) = \frac{1}{\sqrt{2\pi}} \int_{-\infty}^{+\infty} u(x,t) e^{-ipx} dx$. This is a second-order linear differential equation in the variable t for $U(p,t)$, whose solution is $U(p,t) = c_+(p) e^{ipt} + c_-(p) e^{-ipt}$. Notice that the two integration constants $c_\pm$ can depend on p. By inverting the Fourier transform, we find $u(x,t) = \frac{1}{\sqrt{2\pi}} \int_{-\infty}^{+\infty} U(p,t) e^{ipx} dp = f_+(x+t) + f_-(x-t)$, where the functions $f_\pm(x)$ are the inverse Fourier transforms of $c_\pm(p)$. Since $c_\pm$ are arbitrary functions of p, $f_\pm$ are arbitrary functions of x. We have thus found the general solution of the two-dimensional wave equation (4.57).

Exercise 4.10. Applying a Laplace transform to (4.48) using (2.14) and the initial condition $u(x,0) = 0$, we find $sU(x,s) = \kappa \partial_x^2 U(s,x)$, where $U(x,s) = \int_0^{+\infty} u(x,t) e^{-st} dt$. This is a second-order linear differential equation in the variable x for $U(x,s)$, with solution $U(x,s) = c_1(s) e^{x\sqrt{s/\kappa}} + c_2(s) e^{-x\sqrt{s/\kappa}}$. Notice that the two integration constants c_i can depend on s. Since u must be bounded at infinity, we set $c_1(s) = 0$. The condition $u(0,t) = g(t)$ implies that $c_2(s) = G(s)$, where $G(s)$ is the Laplace transform of $g(t)$. Using the convolution theorem (2.17), we find $U(x,t) = \int_0^t g(\tau) L(t-\tau) d\tau$ where L is the inverse Laplace transform of $e^{-x\sqrt{s/\kappa}}$, $L(t) = \frac{x}{\sqrt{4\pi\kappa t^3}} e^{-x^2/(4\kappa t)}$ (see Exercise 2.5).

Chapter 5

Exercise 5.1. This follows from an explicit computation. Plugging the expressions for p, q and ω in (5.12) we immediately find (5.11). Notice that, if $a_2(x) > 0$, $p(x) > 0$ and $\omega(x) > 0$.

Exercise 5.2.   (a) We proceed as in Section 5.1.1. Integrating by parts twice, we find $(f, Lg) = -(\bar{f}g' - \bar{f}'g)|_0^\pi + (Lf, g)$. Suppose that $g \in \mathcal{D}(L)$, so that $g(0) = \alpha g(\pi)$ and $g'(\pi) = \alpha g'(0)$. The boundary terms read $-g'(0)(\alpha \bar{f}(\pi) - \bar{f}(0)) + g(0)(\bar{f}'(\pi)/\alpha - \bar{f}'(0))$ and vanish if and only if $f \in \mathcal{D}(L)$. We conclude that L is self-adjoint.

(b) The equation $y'' = -(\lambda - 4)y$ is an equation with constant coefficients. The characteristic equation (4.28) is $\alpha^2 + \lambda - 4 = 0$ so that the general solution of the equation is $y(x) = Ce^{i\sqrt{\lambda - 4}x} + De^{-i\sqrt{\lambda - 4}x}$, where C and D are arbitrary constants. We can also write the exponential in terms of trigonometric function, obtaining $y(x) = A\sin\sqrt{\lambda - 4}x + B\cos\sqrt{\lambda - 4}x$, where A and B are new constants. Imposing the boundary conditions $y(0) = \alpha y(\pi)$ and $y'(\pi) = \alpha y'(0)$ we find a linear system of algebraic equations for A and B: $\alpha A\sin\sqrt{\lambda - 4}\pi + B(\alpha\cos\sqrt{\lambda - 4}\pi - 1) = 0$ and $(\cos\sqrt{\lambda - 4}\pi - \alpha)A - B\sin\sqrt{\lambda - 4}\pi = 0$. The system has non-trivial solutions only if the determinant of the coefficients is zero. We thus find $\cos\sqrt{\lambda - 4}\pi = \frac{2\alpha}{1+\alpha^2}$. This equation determines the eigenvalues $\lambda_n = 4 + (\pm\phi_0 + 2\pi n)^2/\pi^2$ where $\phi_0 = \arccos\frac{2\alpha}{1+\alpha^2}$ and $n \in \mathbb{Z}$. Using $\sin\sqrt{\lambda_n - 4}\pi = \frac{1-\alpha^2}{1+\alpha^2}$ we easily find $B = \alpha A$. The eigenvectors are then $y_n = A_n(\sin\sqrt{\lambda_n - 4}x + \alpha\cos\sqrt{\lambda_n - 4}x)$, where A_n are arbitrary constants. Since L is self-adjoint, the eigenvectors are mutually orthogonal and A_n can be chosen such that $(y_n, y_m) = \delta_{nm}$.

(c) For any α, the eigenvalues of L are different from zero and L is invertible. The equation $Ly = f$ has then a unique solution for all f. We can find it as follows. Since the Sturm–Liouville problem is regular, the eigenvectors y_n are a Hilbert basis in $L^2[0, \pi]$ so that we can expand $y = \sum_n c_n y_n$ and $f = \sum_n d_n y_n$. Plugging these expansions into the equation $Ly = f$ we find $\sum_n(c_n\lambda_n - d_n)y_n = 0$, which is solved by $c_n = d_n/\lambda_n$.

Exercise 5.3.   We showed in Section 4.2 that, for an equation of the form (4.17), the Wronskian satisfies the differential equation $W' = -a_1 W/a_2$. For a Sturm–Liouville equation, $a_2 = p$ and $a_1 = p'$. Therefore we find $W' = -p'W/p$, or $(Wp)' = 0$. This proves that the product pW is constant.

Exercise 5.4.   (a) Consider two eigenfunctions $y_1(x)$ and $y_2(x)$ of the problem (5.12) associated to the same eigenvalue λ. We know from Exercise 5.3 that the quantity $p(x)W(x)$, where $W(x) = y_1(x)y_2'(x) - y_2(x)y_1'(x)$, is actually independent of x. We can then evaluate this quantity in $x = a$ and we find $p(x)W(x) = p(a)(y_1(a)y_2'(a) - y_2(a)y_1'(a)) = 0$, since $y_1(x)$ and $y_2(x)$ satisfy the boundary conditions (5.15). Then $p(x)W(x) = 0$ for all $x \in [a, b]$. Since for a regular Sturm–Liouville problem $p(x) \neq 0$ on $[a, b]$, we conclude that $W(x) = 0$ and $y_1(x)$ and $y_2(x)$ are linearly dependent.

(b) If $y(x)$ is an eigenfunction of the problem (5.12) associated with the eigenvalue λ, $\overline{y(x)}$ is also. Indeed, since $p(x), q(x), \omega(x), \alpha_i, \beta_i$ and the eigenvalues of a Sturm–Liouville problem are real, if $y(x)$ satisfies (5.12) and (5.15) also $\overline{y(x)}$ does. By part (a), $y(x)$ and $\overline{y(x)}$ are linearly dependent: $\overline{y(x)} = cy(x)$ for some constant c. Taking the complex conjugate of this relation and using the relation again, we obtain $y(x) = \overline{cy(x)} = |c|^2 y(x)$, so that $|c| = 1$. Then $u(x) = e^{i\phi/2} y(x)$, where $c = e^{i\phi}$, is still an eigenfunction and it is real.

(c) Consider two eigenfunctions $y_1(x)$ and $y_2(x)$ of the problem (5.12) associated with the eigenvalues λ_1 and λ_2, with $\lambda_2 > \lambda_1$. Multiplying the equations $py_1'' + p'y_1' + qy_1 = -\lambda_1 \omega y_1$ and $py_2'' + p'y_2' + qy_2 = -\lambda_2 \omega y_2$ by y_2 and y_1, respectively, and subtracting them, we obtain $(p(y_1'y_2 - y_2'y_1))' = (\lambda_2 - \lambda_1)\omega y_1 y_2$. Integrating the previous relation on the interval $[x_1, x_2]$ where x_1 and x_2 are consecutive zeros of $y_1(x)$, we find $p(x_2)y_1'(x_2)y_2(x_2) - p(x_1)y_1'(x_1)y_2(x_1) = (\lambda_2 - \lambda_1) \int_{x_1}^{x_2} y_1(t)y_2(t)\omega(t)\mathrm{d}t$. Since x_1 and x_2 are consecutive zeros, $y_1(x)$ maintains the same sign on the interval (x_1, x_2). Suppose that y_1 is positive (the case where it is negative is analogous). Then $y_1'(x_1) > 0$ and $y_1'(x_2) < 0$. Assuming that $y_2(x)$ also has the same sign on the whole interval (x_1, x_2), we obtain a contradiction. Indeed the left-hand side of the previous relation has sign opposite to y_2 while the right-hand side has the same sign as y_2. Hence, $y_2(x)$ has at least one zero in the interval (x_1, x_2).

Notice that these properties are not valid in the case of periodic boundary conditions as Example 5.3 shows. It is easy to see that they are instead valid for the singular Sturm-Liouville problem $-y''(x) = Q(x)y(x)$ with $x \in (-\infty, +\infty)$ and boundary conditions where y and y' vanish at infinity. This includes the case of the one-dimensional Schrödinger equation for a particle in a potential $V(x)$ discussed in Section 9.3 (see also Exercise 9.2).

Exercise 5.5. We look for a solution of the form $u(x, y) = X(x)Y(y)$. Plugging it into the equation we find $YX' = 3Y'X + 2XY$. Diving by XY and reorganising the terms we find $\frac{X'(x)}{X(x)} = 3\frac{Y'(y)}{Y(y)} + 2$. Since the left-hand side of the last equation only depends on x and the right-hand side only on y, they both must be equal to a constant λ. We find two ordinary differential equations, $X' = \lambda X$ and $Y' = (\lambda - 2)Y/3$. The solutions are $X(x) = e^{\lambda x}$ and $Y(y) = e^{(\lambda-2)y/3}$. The general solution of the partial differential equation is given by a linear combination $u(x, y) =$

$\sum_\lambda c_\lambda e^{\lambda x} e^{(\lambda-2)y/3}$. The boundary condition $u(x,0) = (e^x + e^{-x})/2$ selects the two values $\lambda = \pm 1$ with $c_{\pm 1} = 1/2$. The solution is then $u = (e^x e^{-y/3} + e^{-x} e^{-y})/2$.

Exercise 5.6. We use the method of separation of variables and look for a solution of the form $u(x,y) = X(x)Y(y)$. Plugging it into the equation and dividing by XY we obtain $-X''(x)/X(x) = Y''(y)/Y(y)$. Since the left-hand side of the last equation only depends on x and the right-hand side only on y, they both must be equal to a constant λ. We obtain the two ordinary differential equations $X''(x) = -\lambda X(x)$ and $Y''(y) = \lambda Y(y)$, whose general solutions are $X(x) = A \sin \sqrt{\lambda} x + B \cos \sqrt{\lambda} x$ and $Y(y) = C \sinh \sqrt{\lambda} y + D \cosh \sqrt{\lambda} y$. The boundary condition $u(x,0) = X(x)Y(0) = 0$ should be satisfied for all x. This implies $Y(0) = 0$ so that $D = 0$. The boundary conditions $u(0,y) = u(L,y) = 0$ similarly imply $X(0) = 0$ and $X(L) = 0$. We then obtain $B = 0$ and $A \sin \sqrt{\lambda} L = 0$. The second condition gives $\sqrt{\lambda} = n\pi/L$ where $n \in \mathbb{Z}$. Since $n = 0$ gives a vanishing result and $-n$ gives the same solution up to a sign, we restrict to $n = 1, 2, \ldots$. We have found a set of solutions of the form $u_n(x,y) = \sin \frac{\pi n x}{L} \sinh \frac{\pi n y}{L}$. The general solution is the linear combination: $u(x,y) = \sum_{n=1}^{\infty} c_n \sin \frac{\pi n x}{L} \sinh \frac{\pi n y}{L}$. The last boundary condition, $u(x,L) = 1$, imposes $1 = \sum_{n=1}^{\infty} c_n \sin \frac{\pi n x}{L} \sinh \pi n$. Notice that the functions $s_n = \sqrt{\frac{2}{L}} \sin \frac{\pi n x}{L}$ are a complete orthonormal system in $L^2[0,L]$ (see footnote 8 in Chapter 5). We find therefore $c_n = \frac{1}{\sinh \pi n} \frac{2}{L} \int_0^L \sin \frac{\pi n x}{L} dx$.

Exercise 5.7. We use the method of separation of variables in polar coordinates and look for a solution of the form $u(\rho,\phi) = R(\rho)\Phi(\phi)$. The computation proceeds as in Example 5.7 and we obtain equations (5.41) and (5.42). The conditions that $u(\rho,0) = u(\rho,\pi) = 0$ for all ρ imply $\Phi(0) = \Phi(\pi) = 0$. The solutions of (5.42) satisfying these conditions are $\Phi(\phi) = a_m \sin m\phi$ with $m = 1, 2, \ldots$. As in Example 5.7 the solutions of (5.41) are $R(r) = c_m \rho^m + d_m \rho^{-m}$, where we kept also the term ρ^{-m} since it is regular for $1 \leq \rho \leq 2$. The boundary condition $u(2,\phi) = 0$ implies $R(2) = 0$ so that $d_m = -c_m 4^m$. The general solution is then $u(\rho,\phi) = \sum_{m=1}^{\infty} b_m((\rho/2)^m - (2/\rho)^m) \sin m\phi$, where we reabsorbed all constants into the arbitrary constant b_m. We finally impose $u(1,\phi) = \sin 2\phi$ obtaining $\sin 2\phi = \sum_{m=1}^{\infty} b_m(2^{-m} - 2^m) \sin m\phi$. We see that $b_m = 0$ for $m \neq 2$ and $b_2 = -4/15$. The solution is then $u(\rho,\phi) = -\frac{4}{15}\left(\frac{\rho^4 - 16}{4\rho^2}\right) \sin 2\phi$.

Exercise 5.8. The general solution of the Laplace equation in spherical coordinates is (5.74). To have a regular solution of the Dirichlet problem for the region inside the unit sphere we set $d_{lm} = 0$. The boundary condition at $r = 1$ implies $\sum_{lm} c_{lm} Y_{lm}(\theta,\phi) = 1 + \cos\theta$. By looking at the first spherical harmonics we see that $1 = \sqrt{4\pi} Y_{00}$ and $\cos\theta = \sqrt{\frac{4\pi}{3}} Y_{10}$. We find therefore that the only non-zero coefficients are $c_{00} = \sqrt{4\pi}$ and $c_{10} = \sqrt{4\pi/3}$. The solution is then $u = \sqrt{4\pi}Y_{00} + \sqrt{\frac{4\pi}{3}}Y_{10}r = 1 + r\cos\theta$. To solve the problem in the region outside the sphere, we require that u is bounded at infinity. This requires that all c_{lm} vanish. The boundary condition at $r = 1$ now implies $\sum_{lm} d_{lm}Y_{lm}(\theta, \phi) = 1 + \cos\theta$. This is solved as before. The solution is $u = \sqrt{4\pi}Y_{00} + \sqrt{\frac{4\pi}{3}}\frac{Y_{10}}{r^2} = 1 + \frac{1}{r^2}\cos\theta$.

Exercise 5.9. (a) We use $|\mathbf{x} - \mathbf{x}'| = (\sum_{k=1}^{3}(x_k - x'_k)^2)^{1/2}$. By explicit derivation we find $\partial_{x_i}^2 |\mathbf{x} - \mathbf{x}'|^{-1} = (3(x_i - x'_i)^2 - \sum_{k=1}^{3}(x_k - x'_k)^2)/|\mathbf{x} - \mathbf{x}'|^5$. Summing over $i = 1, 2, 3$ we find $\Delta|\mathbf{x} - \mathbf{x}'|^{-1} = \sum_{i=1}^{3}\partial_{x_i}^2|\mathbf{x} - \mathbf{x}'|^{-1} = 0$.

(b) We choose spherical coordinates with the z-axis along the vector $\mathbf{x}'$. With this choice, θ is equal to the angle γ between $\mathbf{x}$ and $\mathbf{x}'$. We also call $r = |\mathbf{x}|$ and $r' = |\mathbf{x}'|$. Since $|\mathbf{x} - \mathbf{x}'|^{-1}$ solves the Laplace equation, we can write it in the form (5.74). The function $|\mathbf{x} - \mathbf{x}'|^{-1}$ is invariant under rotations around the direction of $\mathbf{x}'$, and therefore is independent of the angle ϕ. Its expansion (5.74) should also be independent of ϕ and therefore only the terms proportional to $Y_{l0} \sim P_l$ can appear. We thus have an expansion of the form $|\mathbf{x} - \mathbf{x}'|^{-1} = \sum_{l=0}^{\infty}(a_l r^l + b_l/r^{l+1})P_l(\cos\gamma)$. In order to find the coefficients a_l and b_l, we can evaluate the previous expression for $\mathbf{x}$ along the z-axis and obtain $|r - r'|^{-1} = \sum_{l=0}^{\infty}(a_l r^l + b_l/r^{l+1})$, since $P_l(1) = 1$. Moreover, as $r > r'$, using the geometric series, we can write $|r - r'|^{-1} = \frac{1}{r}\sum_{l=0}^{\infty}(\frac{r'}{r})^l$. Comparing the two expressions we can determine a_l and b_l. The final result is $|\mathbf{x} - \mathbf{x}'|^{-1} = \frac{1}{r}\sum_{l=0}^{\infty}(\frac{r'}{r})^l P_l(\cos\gamma)$.

(c) Using $|\mathbf{x} - \mathbf{x}'|^2 = |\mathbf{x}|^2 + |\mathbf{x}'|^2 - 2(\mathbf{x}, \mathbf{x}') = r^2 + r'^2 - 2rr'\cos\gamma$ and the results of point (b), we can write $(r^2 + r'^2 - 2rr'\cos\gamma)^{-1/2} = \frac{1}{r}\sum_{l=0}^{\infty}(\frac{r'}{r})^l P_l(\cos\gamma)$. Setting $r = 1$, $r' = t < 1$ and $\cos\gamma = x$, we find $(1 + t^2 - 2tx)^{-1/2} = \sum_{l=0}^{\infty}t^l P_l(x)$.

Exercise 5.10. We proceed as in Exercise 5.6. We find $u(x, y) = X(x)Y(y)$ with $X(x) = A\sin\sqrt{\lambda}x + B\cos\sqrt{\lambda}x$ and $Y(y) = C\sinh\sqrt{\lambda}y + D\cosh\sqrt{\lambda}y$. Imposing $u(x, 0) = 0$ we find $D = 0$. The general solution is then of the form $u(x, y) = \sum_{\lambda}(a_\lambda \sin\sqrt{\lambda}x + b_\lambda \cos\sqrt{\lambda}x)\sinh\sqrt{\lambda}y$. Imposing $\partial_y u(x, 0) = \epsilon\sin(x/\epsilon)$ we find that only $\lambda = 1/\epsilon^2$ contributes, and, for this particular value, $b_\lambda = 0$, and $a_\lambda = \epsilon^2$. The solution is then $u(x, y) = \epsilon^2 \sin(x/\epsilon)\sinh(y/\epsilon)$.

Exercise 5.11. We would like to reduce the problem to the one discussed in Example 5.8 with vanishing boundary conditions at $x = 0$ and $x = 1$, which we know has solutions of the form (5.51). This can be done by looking for a solution of the form $\tilde{u}(x, t) + v(x, t)$ where $\tilde{u}$ satisfies boundary conditions $\tilde{u}(0, t) = 1$ and $\tilde{u}(1, t) = 2$ and v satisfies the boundary conditions of Example 5.8. We easily find a time-independent solution for $\tilde{u}$, namely $\tilde{u}(x, t) = x + 1$. Then $v(x, y)$ must solve the heat equation with boundary conditions $v(0, t) = 0$, $v(1, t) = 0$

and $v(x,0) = x^2 - x$. The general solution for v is given in (5.51), $v(x,t) = \sum_{n=1}^{\infty} c_n e^{-\kappa \pi^2 n^2 t} \sin(\pi n x)$. The coefficients c_n are given by (5.52) and read $c_n = 2 \int_0^1 (x^2 - x) \sin(n\pi x) \mathrm{d}x = 4((-1)^n - 1)/(n\pi)^3$.

Exercise 5.12.   We proceed as in Example 5.8. Setting $u(x,t) = X(x)T(t)$ we obtain equation (5.49). Imposing the boundary conditions, we obtain a regular Sturm–Liouville problem for X: $\kappa X''(x) = \lambda X(x)$ with $X(0) = 0$ and $X'(L) = -cX(L)$. The solution is $X(x) = A \sin \sqrt{-\lambda/\kappa} x + B \cos \sqrt{-\lambda/\kappa} x$. The boundary conditions give $B = 0$ and $\cot \sqrt{-\lambda/\kappa} L = -c\sqrt{-\kappa/\lambda}$. This equation can only be solved graphically. We find an infinite number of eigenvalues λ_n and corresponding eigenvectors $X_n(x) = A_n \sin \sqrt{-\lambda_n/\kappa} x$. The eigenvectors are mutually orthogonal and we can choose the normalisation constants A_n such that $(X_n, X_m) = \delta_{nm}$. Since the Sturm–Liouville problem is regular, the X_n are an orthonormal basis in $L^2[0,L]$. The solutions of the equation for T are the exponentials $T_n(t) = e^{\lambda_n t}$. The general solution is $u(x,t) = \sum_{n=1}^{\infty} c_n e^{\lambda_n t} X_n(x)$. The last boundary condition requires $u(x,0) = g(x)$ and therefore $\sum_{n=1}^{\infty} c_n X_n(x) = g(x)$. The coefficient c_n are computed as $c_n = (X_n, g)$.

Exercise 5.13.   We proceed as in Example 5.9. Setting $u(x,t) = X(x)T(t)$ we obtain equation (5.54). The boundary conditions at $x = 0$ and $x = L$ imply $X(0) = X(L) = 0$ and the equation for X is solved by $\lambda_n = -n^2 \pi^2 v^2/L^2$ and $X_n = \sin n\pi x/L$. The condition $u(x,0) = 0$ implies $T(0) = 0$. The equation for T with $T(0) = 0$ has the solution $T(t) = \sin n\pi vt/L$. Then the general solution is $u(x,t) = \sum_{n=1}^{\infty} c_n \sin \frac{n\pi vt}{L} \sin \frac{\pi n x}{L}$. Imposing $\partial_t u(x,0) = \sin(3\pi x/L)$, we see that $c_3 = L/(3\pi v)$ and all other c_n are zero. Therefore the solution of the boundary problem is $u(x,t) = \frac{L}{3\pi v} \sin \frac{3\pi vt}{L} \sin \frac{3\pi x}{L}$.

Exercise 5.14.   We need to find the solutions of $\Delta u = \lambda u$ that satisfy $u(x,0) = u(x,L) = u(0,y) = u(L,y) = 0$. Using the method of separation of variables we set $u(x,y) = X(x)Y(y)$. Plugging it into the equation and dividing by XY we obtain $X''(x)/X(x) + Y''(y)/Y(y) = \lambda$. The two terms of the right-hand side must be separately constant and we obtain $X''(x) = \lambda_1 X(x)$ and $Y''(y) = \lambda_2 Y(y)$ with $\lambda_1 + \lambda_2 = \lambda$. The boundary conditions require $X(0) = X(L) = 0$ and $Y(0) = Y(L) = 0$. The solutions of these Sturm–Liouville problems can be obtained as in Exercise 5.6. They read $\lambda_{1\,n} = -\pi^2 n^2/L^2$ with $X_n = \sin(n\pi x/L)$ and $\lambda_{2\,m} = -\pi^2 m^2/L^2$ with $Y_m = \sin(m\pi y/L)$, where $n, m = 1, 2, \ldots$. The eigenvalues $\lambda_{nm} = -\frac{\pi^2}{L^2}(n^2 + m^2)$ of the Laplace equation depend on two integers and the corresponding eigenfunctions are $u_{nm} = \frac{2}{L} \sin \frac{n\pi x}{L} \sin \frac{m\pi y}{L}$. One can show that the u_{nm} are a Hilbert basis in $L^2([0,L] \times [0,L])$.

(a) Separation of variables, $v(x, y, t) = u(x, y)T(t)$, leads to $T'(t) = \kappa\lambda T(t)$ and $\Delta u = \lambda u$. The boundary conditions for v imply that u is one of the eigenvector of the Laplacian with vanishing boundary conditions that we found above. The equation for T has solutions $T_n(t) = e^{\lambda_{nm}\kappa t}$ and the general solution is then $v(x, y, t) = \sum_{nm} c_{nm} \sin\frac{n\pi x}{L} \sin\frac{m\pi y}{L} e^{\lambda_{nm}\kappa t}$. The last boundary condition, $v(x, y, 0) = g(x, y)$, gives $g(x, y) = \sum_{n,m=1}^{\infty} c_{nm} \sin\frac{n\pi x}{L} \sin\frac{m\pi y}{L}$. The coefficients can be found as usual as $c_{nm} = \frac{4}{L^2} \int_0^L dx \int_0^L dy g(x, y) \sin\frac{n\pi x}{L} \sin\frac{m\pi y}{L}$.

(b) We expand both $u(x, y) = \sum_{n,m=1}^{\infty} a_{nm} u_{nm}(x, y)$ and $f(x, y) = \sum_{n,m=1}^{\infty} b_{nm} u_{nm}(x, y)$ in eigenfunctions of the Laplacian. The Poisson equation becomes $-\sum_{n,m=1}^{\infty} \frac{\pi^2}{L^2}(n^2 + m^2) a_{nm} u_{nm} = \sum_{n,m=1}^{\infty} b_{nm} u_{nm}$. By equating the coefficients of the same u_{nm} we find $a_{nm} = -\frac{L^2 b_{nm}}{\pi^2(n^2+m^2)}$. Notice that the equation has a solution for any function f. This is due to the fact that the Laplacian Δ with vanishing boundary condition is an invertible operator, since it has no zero eigenvalue.

Exercise 5.15. (a) V_0 only contains the polynomial 1 and it has complex dimension one. V_1 is the vector space spanned by the three monomials n_1, n_2, n_3 and has complex dimension three. V_2 is the vector space of quadratic polynomials $\sum_{ij} a_{ij} n_i n_j$ with complex symmetric coefficients $a_{ij} = a_{ji}$ with vanishing trace $\sum_i a_{ii} = 0$. A generic symmetric rank two tensor has six independent components. Since the trace condition removes one, V_2 has complex dimension 5.

(b) The constant harmonic $Y_{00} = 1/\sqrt{4\pi}$ is obviously a basis for V_0. Consider now the three harmonics with $l = 1$. We can write them as $Y_{10} = \sqrt{\frac{3}{4\pi}} n_3$ and $Y_{1,\pm 1} = \mp\sqrt{\frac{3}{8\pi}}(n_1 \pm in_2)$. We see that the complex vector space spanned by Y_{1m}, $m = -1, 0, 1$ is the same as the complex vector space V_1 spanned by n_i with $i = 1, 2, 3$. Consider now the five harmonics with $l = 2$. All of them are quadratic polynomials in the n_i. Indeed, it is easy to check that $Y_{20} = \sqrt{\frac{5}{16\pi}}(2n_3^2 - n_1^2 - n_2^2)$, $Y_{2,\pm 1} = \mp\sqrt{\frac{15}{8\pi}} n_3(n_1 \pm in_2)$ and $Y_{2,\pm 2} = \sqrt{\frac{15}{32\pi}}(n_1 \pm in_2)^2$. The corresponding matrices a_{ij} are

$$Y_{20} : \sqrt{\frac{5}{16\pi}} \begin{pmatrix} -1 & 0 & 0 \\ 0 & -1 & 0 \\ 0 & 0 & 2 \end{pmatrix}, \quad Y_{2,\pm 1} : \sqrt{\frac{15}{32\pi}} \begin{pmatrix} 0 & 0 & \mp 1 \\ 0 & 0 & -i \\ \mp 1 & -i & 0 \end{pmatrix},$$

$$Y_{2,\pm 2} : \sqrt{\frac{15}{32\pi}} \begin{pmatrix} 1 & \pm i & 0 \\ \pm i & -1 & 0 \\ 0 & 0 & 0 \end{pmatrix}.$$

As all the matrices a_{ij} associated to $Y_{2,m}$ are traceless, $Y_{2,m} \in V_2$. Moreover, since the five matrices are linearly independent, they form a basis for the vector space V_2.

Chapter 6

Exercise 6.1. We saw in Exercise 4.5 that the general solution for a second-order linear equation of the form (4.34) is $y(x) = \int_{x_0}^{x} \frac{g(t)}{a_2(t)W(t)}(y_1(t)y_2(x) - y_2(t)y_1(x))dt + c_1 y_1(x) + c_2 y_2(x)$, where $y_i(x)$ are the linearly independent solutions of the homogeneous equation and W is their Wronskian. For equation (6.23), $a_2(x) = 1$, $y_1(x) = e^{i\omega x}$, $y_2(x) = e^{-i\omega x}$ and $W = y_1 y_2' - y_2 y_1' = -2i\omega$. If we choose $x_0 = -\infty$, we find $y(x) = \int_{-\infty}^{+\infty} \theta(x - t)\frac{\sin \omega(x-t)}{\omega}g(t)dt + c_1 e^{i\omega x} + c_2 e^{-i\omega x}$, which is the same as the result in Example 6.3.

Exercise 6.2. If the self-adjoint operator L has discrete spectrum, its eigenvectors u_n, $Lu_n = \lambda_n u_n$, form a Hilbert basis. We can expand both f and h in this basis: $f = \sum_n f_n u_n$ and $h = \sum_n h_n u_n$. Plugging these into the equation $Lf = h$ we find $\sum_n f_n \lambda_n u_n = \sum_n h_n u_n$. A vector in a Hilbert space is zero if and only if all its Fourier coefficients are. We obtain therefore the infinite set of equations $f_n \lambda_n = h_n$.

(a) If $\ker L = \{0\}$, all $\lambda_n \neq 0$ and we find $f_n = h_n/\lambda_n$. The solution is unique.

(b) If $\ker L \neq \{0\}$ at least one λ_n vanishes, and the corresponding equation $0 = h_n$ can be solved only if $h_n = 0$. Therefore $Lf = h$ can be solved only if $h \in (\ker L)^\perp$. In this case, the coefficients f_n associated with the eigenvalues $\lambda_n \neq 0$ are given as before by $f_n = h_n/\lambda_n$, while those associated with the eigenvalues $\lambda_n = 0$ remain arbitrary. The solution is not unique. This just corresponds to the fact that, if f is a solution, so is $f + g$, where $g \in \ker L$.

Exercise 6.3. It is convenient to bring the problem to the canonical form (4.36). To this aim, we make the substitution (4.37), $f(x) = \varphi(x)/x$. The differential equation becomes $\varphi'' + \varphi = 1$ with boundary conditions $\varphi(\pi/2) = \varphi(\pi) = 0$. The general integral of the homogeneous equation is $\varphi_h(x) = A \cos x + B \sin x$ and it is easy to see that there is no choice of A and B that satisfies the boundary conditions. Then the Green function of the problem exists and is unique: it is given by (6.29), where $u(x) = \cos x$ satisfies the boundary condition in $x = \pi/2$, $\tilde{u}(x) = \sin x$ the one in $x = \pi$ and $W(x) = a_2(x) = 1$. We find

$$G(x,t) = \begin{cases} \cos x \sin t, & \dfrac{\pi}{2} < x < t, \\[2mm] \sin x \cos t, & t < x < \pi. \end{cases}$$

We then have $\varphi(x) = \int_{\frac{\pi}{2}}^{\pi} dt\, G(x,t) = \cos x \int_x^{\pi} \sin t\, dt + \sin x \int_{\frac{\pi}{2}}^x \cos t\, dt = 1 + \cos x - \sin x$. The solution of the original problem is $f(x) = \frac{1}{x}(1 + \cos x - \sin x)$. Now we move to the problem with boundary conditions $f(\pi) = 0$ and $f(\pi/2) + \frac{\pi}{2} f'(\pi/2) = 0$. In the new variable φ, these are equivalent to $\varphi(\pi) = 0$ and $\varphi'(\pi/2) = 0$. In this case, there is a solution of the homogeneous boundary problem, $\varphi_0(x) = \sin x$. Therefore, the self-adjoint differential operator $L = \frac{d^2}{dx^2} + 1$ is not invertible and the Green function does not exist. As discussed in Exercise 6.2, a necessary condition for the existence of a solution of the inhomogeneous boundary problem is that the source term $f(x) = 1$ is orthogonal to the kernel of the operator. Since $f(x) = 1$ and $\varphi_0(x) = \sin x$ are not orthogonal in $L^2[\pi/2, \pi]$, the problem has no solutions.

Exercise 6.4. The operator can be put in canonical form $L^c = \frac{d^2}{dt^2} - 4$ by the change of variables (4.37), $u(t) = \frac{y(t)}{1+t^2}$. We still impose boundary conditions $u(\pm\infty) = 0$ so that L^c is an extension of L. The linearly independent solutions of the equation $L^c u = 0$ are $u = e^{2t}$ and $\tilde{u} = e^{-2t}$. They satisfy the boundary conditions at $t = -\infty$ and $t = +\infty$, respectively, and their Wronskian is $W = -4$. Thus, from (6.29) we have $G(t,y) = -\frac{1}{4}(e^{2(t-y)}\theta(y-t) + e^{-2(t-y)}\theta(t-y))$.

Exercise 6.5. The homogeneous equation $y' + 5x^4 y = 0$ can be solved by separation of variables. From $\frac{y'}{y} = \frac{d\ln(y)}{dy} = -5x^4$ we find $y_h = ce^{-x^5}$. To find a particular solution of the inhomogeneous equation we use the method of variation of arbitrary constants. Plugging $y_p = c(x)e^{-x^5}$ into the inhomogeneous equation, we obtain $c(x) = \frac{1}{5}e^{x^5}$ and $y_p = \frac{1}{5}$. Thus, $y(x) = y_p + y_h = \frac{1}{5} + ce^{-x^5}$. Imposing the boundary condition $y(0) = 0$ we find $c = -\frac{1}{5}$, so that $y(x) = \frac{1}{5}(1 - e^{-x^5})$. We can also find the solution using the Green function method. This is a Cauchy problem with homogeneous boundary condition. We proceed as at the end of Section 6.2.2 with the only difference that the equation is now first order. We want to write the solution as $y(x) = \int_0^{+\infty} G(x,y)f(y)dy$ where G satisfies $\frac{dG(x,y)}{dx} + 5x^4 G(x,y) = \delta(x-y)$. To satisfy $y(0) = 0$ we use a retarded Green function with $G(x,y) = 0$ for $x < y$. The Green function can thus be written as $G(x,y) = c(y)y_h(x)\theta(x-y)$. Near $x = y$ we find $\delta(x-y) = \frac{dG(x,y)}{dx} + 5x^4 G(x,y) \sim \delta(x-y)c(x)y_h(x)$. This implies $c(y) = 1/y_h(y)$. Thus we find $G(x,y) = \exp(-x^5 + y^5)\theta(x-y)$ and the solution is $y(x) = \int_0^{+\infty} \exp(-x^5 + y^5)\theta(x-y)y^4 dy = \int_0^x \exp(-x^5 + y^5)y^4 dy = \frac{1}{5}(1 - e^{-x^5})$, as found previously.

Exercise 6.6. The homogeneous equation is of Euler type and has linearly independent solutions $u_1 = x$ and $u_2 = x^3$ (see Example 4.5). It is easy to see that a particular solution of the inhomogeneous equation is $u_p = -2x^2$. Imposing the boundary conditions on $f = u_p + Au_1 + Bu_2$, we find the solution $f(x) = x - 2x^2 + x^3$. We now derive the same result using the Green function method. As discussed at the end of Section 6.2.2, the appropriate Green function is the retarded one, given in (6.34). Using $u_1(x) = x, u_2(x) = x^3, a_2(x) = x^2$ and $W(x) = 2x^3$, we find $G(x,y) = -\theta(x - y)(xy^3 - x^3y)/(2y^5)$ and $f(x) = \int_1^x 2G(x,y)y^2dy = x - 2x^2 + x^3$, as expected.

Exercise 6.7.   We need to show that the equality $xT = 1$ holds in $\mathcal{D}'(\mathbb{R})$. To this purpose we apply the equation to a test function. This gives $\langle xT(x) - 1, \phi(x)\rangle = P\int \frac{1}{x}x\phi(x)dx + c\int \delta(x)x\phi(x)dx - \int \phi(x)dx = 0$, where we used the fact that the integral with δ gives zero because of x and the integral in principal part is the same as the ordinary integral because the integrand is regular in $x = 0$.

Exercise 6.8.   We need to find solutions of (6.64).

(a) Consider $\mathbf{x}, \mathbf{y} \in \Omega$. We have $\Delta_y G_L(\mathbf{y} - \mathbf{x}') = 0$ since, if $\mathbf{x} \in \Omega$ then $\mathbf{x}' \notin \Omega$ and $\mathbf{x}' \neq \mathbf{y}$. Therefore $\Delta_y G(\mathbf{x} - \mathbf{y}) = \Delta_y G_L(\mathbf{y} - \mathbf{x}) = \delta(\mathbf{y} - \mathbf{x})$. Notice also that $G_L(\mathbf{y} - \mathbf{x})$ in (6.52) is a function of the modulus $|\mathbf{y} - \mathbf{x}|$ only. If $\mathbf{y} \in \partial\Omega$, $y_d = 0$ and then $|\mathbf{y} - \mathbf{x}| = |\mathbf{y} - \mathbf{x}'|$. Therefore $G(\mathbf{y}, \mathbf{x}) = G_L(\mathbf{y} - \mathbf{x}) - G_L(\mathbf{y} - \mathbf{x}') = 0$ for all $\mathbf{y} \in \partial\Omega$.

(b) Consider $\mathbf{x}, \mathbf{y} \in \Omega$. As in part (a), we have $\Delta_y G_L(\mathbf{y} - \mathbf{x}) = \delta(\mathbf{y} - \mathbf{x})$ and $\Delta_y G_L(|\mathbf{x}|(\mathbf{y} - \mathbf{x}')) = 0$ since $\mathbf{x}' \notin \Omega$ and then $\mathbf{x}' \neq \mathbf{y}$. Therefore $\Delta_y G(\mathbf{y} - \mathbf{x}) = \delta(\mathbf{y} - \mathbf{x})$. We can easily see that, if $\mathbf{y} \in \partial\Omega$ is on the unit sphere, we have $|\mathbf{y} - \mathbf{x}| = |\mathbf{x}||\mathbf{y} - \mathbf{x}'|$. Indeed, $|\mathbf{x}|^2|\mathbf{y} - \mathbf{x}'|^2 = |\mathbf{x}|^2(|\mathbf{y}|^2 - 2(\mathbf{x},\mathbf{y})/|\mathbf{x}|^2 + 1/|\mathbf{x}|^2)$ is equal to $|\mathbf{y} - \mathbf{x}|^2 = |\mathbf{y}|^2 + |\mathbf{x}|^2 - 2(\mathbf{x},\mathbf{y})$ if $|\mathbf{y}| = 1$. Therefore $G(\mathbf{y}, \mathbf{x}) = G_L(\mathbf{y} - \mathbf{x}) - G_L(|\mathbf{x}|(\mathbf{y} - \mathbf{x}')) = 0$ for all $\mathbf{y} \in \partial\Omega$, since $G_L(\mathbf{x})$ is a function of the length of the argument only.

Exercise 6.9.   (a) It is immediate to see that the vector $\mathbf{u}(t) = U(t)\mathbf{u}_0$ satisfies the equation $\mathbf{u}' + A\mathbf{u} = 0$ and the initial condition $\mathbf{u}(0) = \mathbf{u}_0$.

(b) $\mathbf{u}(t) = \int_0^t U(t - \tau)\mathbf{f}(\tau)d\tau + U(t)\mathbf{u}_0$ satisfies $\mathbf{u}(0) = U(0)\mathbf{u}_0 = \mathbf{u}_0$. Moreover, $\mathbf{u}'(t) + A\mathbf{u}(t) = U(0)\mathbf{f}(t) + \int_0^t (U'(t - \tau) + AU(t - \tau))\mathbf{f}(\tau)d\tau + (U'(t) + AU(t))\mathbf{u}_0 = \mathbf{f}(t)$.

(c) Take $\mathcal{H} = L^2(\mathbb{R}^d)$ and $A = -\kappa\Delta$. The generalisation of Example 4.6 to $\mathbb{R}^d$ shows that the evolution operator $U(t)$ is given by $U(t)\phi(\mathbf{x}) = \int_{\mathbb{R}^d} h(t, \mathbf{x} - \mathbf{y})\phi(\mathbf{y})d\mathbf{y}$. The result then follows from part (b).

(d) Using the notations of part (c), we can write $G(t, \mathbf{x}) = \theta(t)h(t, \mathbf{x}) = \theta(t)U(t)\delta(\mathbf{x})$. It follows that $(\partial_t - \kappa\Delta)G(t, \mathbf{x}) = \delta(t)U(t)\delta(\mathbf{x}) + \theta(t)(U'(t) + AU(t))\delta(\mathbf{x}) = \delta(t)\delta(\mathbf{x})$ since $U(0) = \mathbb{I}$ and $U'(t) + AU(t) = 0$.

Exercise 6.10. We use the same argument as in Example 6.5. We first prove that $(\Delta + m^2)G_\pm = 0$ for $\mathbf{x} \neq 0$. This follows from a simple computation. Since $r = (\sum_i x_i^2)^{1/2}$, we have $\partial_{x_i} r = x_i r^{-1}$. Therefore, $\partial_{x_i}(e^{\pm imr}r^{-1}) = x_i(-1 \pm imr)e^{\pm imr}r^{-3}$ and $\partial_{x_i}^2(e^{\pm imr}r^{-1}) = (-1 \pm imr)e^{\pm imr}r^{-3} + x_i^2(3 \mp 3imr - m^2r^2)e^{\pm imr}r^{-5}$. Summing over $i = 1, 2, 3$ we find $(\Delta + m^2)(e^{\pm imr}r^{-1}) = 0$. $(\Delta + m^2)G_\pm$ is then a distribution with support in $\mathbf{x} = 0$. To see that it is equal to $\delta(\mathbf{x})$ we can show that the integral of $(\Delta + m^2)G_\pm$ over a ball of infinitesimal radius $R = \epsilon$ is one. Indeed, for $\epsilon \to 0$, we can approximate $G_\pm \sim -1/(4\pi r)$ so that $\int_{r \leq \epsilon} \Delta G_\pm = 1$ as proved in Example 6.5. Moreover, in three dimensions, $G_\pm$ is integrable near $\mathbf{x} = 0$ and therefore $\int_{r \leq \epsilon} m^2 G_\pm \to 0$ for $\epsilon \to 0$. More rigorously, one can prove, as in Example 6.5, that $\langle(\Delta + m^2)G_\pm, \phi\rangle = \phi(0)$ where ϕ is a test function.

Chapter 7

Exercise 7.1. It is convenient to change variable from x to $t = e^{-x}$. The equation becomes $t\ddot{y} + \dot{y} + (E - 4t)y = 0$, where the dot indicates the derivative with respect to t. In the new variable, we have to determine the values of E such that the solution is bounded at $t = 0$ and $t = +\infty$. To find the behaviour of the solution at $t = +\infty$, we approximate the equation by neglecting terms that are subleading for large t: $\ddot{y} - 4y = 0$.[4] Then for large t the solution goes as $y \sim e^{\pm 2t}$. Only the exponential with negative sign is bounded for $t \to \infty$. To analyse the behaviour of the solution around the origin, it is useful to isolate the behaviour at infinity and to make the ansatz $y = e^{-2t}\sum_{n=0}^{\infty} c_n t^n$. Inserting it in the original equation we obtain the recursion relation $c_n(E - 4n - 2) + c_{n+1}(n + 1)^2 = 0$ for the coefficients c_n, which is solved by $c_n = \frac{4n-2-E}{n^2}c_{n-1}$, $n = 1, 2, \ldots$. If n is very large, the ratio c_n/c_{n-1} goes like $4/n$, the same behaviour as the coefficients of the Taylor expansion of e^{4t}. This means that the solution grows like $e^{-2t}e^{4t} \sim e^{2t}$ and is not bounded. This can be avoided only if the series truncates to a polynomial. If $E - 4\hat{n} + 2 = 0$ for some integer $\hat{n}$, the series reduces to a polynomial of order $\hat{n}$. We thus find $E = 4\hat{n} - 2$ with $\hat{n} = 1, 2, \ldots$.

Exercise 7.2. The equation is a hypergeometric equation of parameters $a = 1$, $b = -1/2$ and $c = 1$ and exhibits three Fuchsian points in $z_1 = 0$, $z_2 = 1$

---

[4]Notice that we have neglected $\dot{y}$ compared to $\ddot{y}$. This is justified a posteriori by the form of the solution.

and $z_3 = \infty$ (see Example 7.4). The solutions near $z = 1$ are given in (7.37). The regular solution is $y_1^{(1)}(z) = F(1, -1/2, 1, 1 - z)$. Using the hypergeometric expansion (7.32) we find $u(z) = 1 + \frac{1}{2}(z - 1) - \frac{1}{8}(z - 1)^2 + \cdots$.

Exercise 7.3. The equation has two Fuchsian points at $z = 0$ and $z = 1$. The point at infinity is regular, as it can be easily verified. By expanding the coefficient of the equation around $z = 0$, we find that the indicial equation (7.19) is $\alpha^2 + \alpha - 2 = 0$, with solutions $\alpha_1^{(0)} = 1$ and $\alpha_2^{(0)} = -2$. For $\alpha_1^{(0)} = 1$, solving the first few terms of the recurrence relations (7.18), one realises that $c_n = c_0$ for any n and the solution (7.20) becomes $z \sum_{n=0}^{+\infty} z^n = z/(1 - z)$. Since $\alpha_1^{(0)} - \alpha_2^{(0)} = 3$, we should expect a second solution with a logarithm term as in (7.22). However, in this case the coefficient C of the logarithm is zero. Indeed, it is not difficult to check that for $\alpha_2^{(0)} = -2$ and $n = 3$ (7.18) vanishes identically, so that c_3 is arbitrary. We then find $c_1 = -2c_0$, $c_2 = c_0$ and $c_n = c_3$ for $n > 3$. Setting $c_3 = 0$, the second solution is $(1 - 1/z)^2$. Setting $c_3 \neq 0$ would simply add another copy of the solution for $\alpha_1^{(0)} = 1$. The general solution of the equation is then $u(z) = a_1 z/(1 - z) + b_1(1 - 1/z)^2$, and is defined on the whole $\mathbb{C} - \{0, 1\}$. The expansion around the point $z = 1$, with indicial exponents $\alpha_1^{(1)} = 2$ and $\alpha_2^{(1)} = -1$, obviously gives the same result. The equation can also be solved exactly. With the change of variable $w = \frac{z}{z-1}$ it reduces to an Euler's equation (see Example 4.5), $w^2 u''(w) + 2wu'(w) - 2u(w) = 0$, whose general solution is $u(w) = a_1 w + b_1 w^{-2}$. In terms of the original variable we recover the solutions obtained by power series.

Exercise 7.4. $z = 0$ is a Fuchsian point whose indicial exponents are $\alpha_\pm = (1 \pm \sqrt{1 + 4\lambda})/2$. $z = -1/2$ and $z = \infty$ are Fuchsian points both with vanishing indicial exponents. The existence of a regular and single-valued solution around $z = 0$ requires that α_+ is a positive integer. This happens for $\lambda = n(n - 1)$, with n integer and $n \geq 0$. For $\lambda = 0$ the equation reduces to $u''(z) + \frac{2}{1+2z}u' = 0$ and can be solved explicitly: $u(z) = A + B \ln(1 + 2z)$.[5]

Exercise 7.5. $F(a, b, c; z)$ reduces to a polynomial of degree $-b$ if b is a negative integer. $F(l + 1, -l, 1; (1 - x)/2)$ is thus a polynomial of degree l. The first three polynomials, $l = 0, 1, 2$, are easily obtained writing the first few terms of the expansion of the hypergeometric functions (7.32). We find $F(1, 0, 1, (1-x)/2) = 1$,

---

[5] Notice that the solution is holomorphic in $z = 0$. Although the difference of the indicial indices around the origin is $\alpha_+ - \alpha_- = 1$, the coefficient C of the term $\ln z$ predicted by (7.22) turns out to be zero.

$F(2, -1, 1, (1 - x)/2) = x$ and $F(3, -2, 1, (1 - x)/2) = (3x^2 - 1)/2$. We see that they coincide with the first three Legendre polynomials (5.34).

Exercise 7.6. (a) Using (7.32) and $\Gamma(n + 1) = n!$, we find $F(1, 1, 2; -z) = \sum_{n=0}^{\infty}(-1)^n \frac{z^n}{n+1} = \ln(1 + z)/z$.

(b) Using (7.32), $\Gamma(n+1) = n!$ and $\Gamma(x+1) = x\Gamma(x)$, we find $F(1/2, 1, 3/2; -z^2) = \sum_{n=0}^{\infty}(-1)^n \frac{z^{2n}}{2n+1} = \arctan(z)/z$.

(c) Using (7.49), we find $\Phi(a, a; z) = \sum_{n=0}^{\infty} \frac{z^n}{n!} = e^z$.

(d) Using (7.49) and $\Gamma(n + 1) = n!$, we find $\Phi(1, 2; z) = \sum_{n=0}^{\infty} \frac{z^n}{(n+1)!} = \frac{e^z - 1}{z}$.

(e) Using (7.49) and $\Gamma(n+1) = n!$, we find $\Phi(2, 1; z) = \sum_{n=0}^{\infty}(n+1)\frac{z^n}{n!} = (z+1)e^z$.

Exercise 7.7. (a) We replace $z \to z/b$ in (7.26). The derivatives transform as $y'' \to b^2 y''$ and $y' \to by'$ and the equation becomes $bz(1 - z/b)y'' + [bc - (a + b + 1)z]y' - aby = 0$. Dividing by b and taking the limit $b \to \infty$ we obtain the confluent hypergeometric equation $zy'' + (c - z)y' - ay = 0$.

(b) Replacing $z \to z/b$ in (7.32), we find $\frac{\Gamma(c)}{\Gamma(a)\Gamma(b)} \sum_{n=0}^{\infty} \frac{\Gamma(a+n)\Gamma(b+n)}{\Gamma(c+n)} \frac{z^n}{n!b^n}$. We then take the limit $b \to \infty$. Using $\Gamma(z + 1) = z\Gamma(z)$ we can write $\Gamma(n + b)/\Gamma(b) = (n + b - 1)(n + b - 2) \cdots b \sim b^n$. We see that $\lim_{b \to \infty} F(a, b, c; z/b) = \frac{\Gamma(c)}{\Gamma(a)} \sum_{n=0}^{\infty} \frac{\Gamma(a+n)}{\Gamma(c+n)} \frac{z^n}{n!} = \Phi(a, c; z)$ (see (7.49)). By the same logic the second solution (7.35) is sent into the solution (7.50) multiplied by b^{c-1}.

(c) The indicial equation (7.19) for the confluent hypergeometric equation (7.48) near $z = 0$ reads $\alpha(\alpha - 1) + c\alpha = 0$ and it is solved by $\alpha_1^{(0)} = 0$ and $\alpha_2^{(0)} = 1 - c$. Consider first $\alpha_1^{(0)} = 0$. The solution is of the form $y = \sum_{n=0}^{\infty} c_n z^n$ where the coefficients satisfy the recurrence relation (7.18), $n(n-1+c)c_n = (n-1+a)c_{n-1}$. This relation is solved by $c_n = \frac{c_0}{n!} \frac{\Gamma(a+n)}{\Gamma(a)} \frac{\Gamma(c)}{\Gamma(c+n)}$, where we used (7.31). The solution with $c_0 = 1$ is then the confluent hypergeometric function (7.49). Similarly, for $\alpha_2^{(0)} = 1 - c$ (c not integer), the solution is of the form $y = z^{1-c} \sum_{n=0}^{\infty} c_n z^n$. Equation (7.18) now reads $n(n + 1 - c)c_n = (n - c + a)c_{n-1}$ and it is obtained from the previous one by the replacement $a \to a - c + 1, c \to 2 - c$. We thus obtain (7.50).

Exercise 7.8. (a) We expand the integrand of $I = \int_0^1 dt\, t^{a-1}(1 - t)^{c-a-1}(1 - zt)^{-b}$ in power series in z using the binomial expansion (A.17) and the integral in part (a) of Exercise 3.5

$$\sum_{n=0}^{\infty} \frac{z^n \Gamma(n + b)}{n!\Gamma(b)} \int_0^1 dt\, t^{a-1+n}(1 - t)^{c-a-1} = \frac{\Gamma(c - a)}{\Gamma(b)} \sum_{n=0}^{\infty} \frac{z^n}{n!} \frac{\Gamma(a + n)\Gamma(b + n)}{\Gamma(c + n)}.$$

Thus $\frac{\Gamma(c)}{\Gamma(a)\Gamma(c-a)}I = \frac{\Gamma(c)}{\Gamma(a)\Gamma(b)} \sum_{n=0}^{\infty} \frac{\Gamma(a+n)\Gamma(b+n)}{\Gamma(c+n)} \frac{z^n}{n!} = F(a, b, c; z)$ (see (7.32)).

(b) Using $\lim_{b\to\infty}(1 - zt/b)^{-b} = e^{zt}$ in the integral representation of part (a), we find $\Phi(a, c; z) = \lim_{b\to\infty} F(a, b, c; z/b) = \frac{\Gamma(c)}{\Gamma(a)\Gamma(c-a)} \int_0^1 dt\, t^{a-1}(1 - t)^{c-a-1}e^{zt}$. This agrees with (7.55).

Exercise 7.9.   (a) With the change of variable $y(z) = e^{z/2}z^{-\nu}u(z)$ the confluent hypergeometric equation (7.48) becomes

$$zu''(z) + (c - 2\nu)u'(z) + \left(\frac{\nu^2 + \nu - c\nu}{z} - \frac{z}{4} - a + \frac{c}{2}\right)u(z) = 0. \qquad (B.4)$$

Then setting $c = 2a = 2\nu + 1$ and sending $z \to 2iz$ we arrive at the Bessel equation (7.57). We also recover (7.65), up to an overall constant.

(b) For n integer we have

$$J_{-n}(z) = \sum_{k=0}^{\infty} \frac{(-1)^k}{k!\Gamma(k - n + 1)}\left(\frac{z}{2}\right)^{2k-n} = \sum_{k=n}^{\infty} \frac{(-1)^k}{k!\Gamma(k - n + 1)}\left(\frac{z}{2}\right)^{2k-n}. \qquad (B.5)$$

To derive the last equality, we used the fact that, since the gamma functions $\Gamma(k - n + 1)$ have simple poles for $k \le n - 1$, the corresponding coefficients in $J_{-n}(z)$ vanish. Redefining the summation index as $k' = k - n$ and using the fact that $\Gamma(n + 1) = n!$, we find

$$J_{-n}(z) = (-1)^n \sum_{k'=0}^{\infty} \frac{(-1)^{k'}}{k'!\Gamma(k' + n + 1)}\left(\frac{z}{2}\right)^{2k'+n} = (-1)^n J_n(z). \qquad (B.6)$$

Exercise 7.10.   We need to compute $\lim_{\nu\to n} Y_\nu(z)$, where $Y_\nu = [\cos\nu\pi J_\nu(z) - J_{-\nu}(z)]/\sin\nu\pi$ and the Bessel functions of first kind $J_{\pm\nu}$ are given in (7.60). We apply de l'Hôpital rule. Keeping the terms that are non zero in the limit, we find

$$\lim_{\nu\to n} \frac{\cos\nu\pi J_\nu(z) - J_{-\nu}(z)}{\sin\nu\pi} = \lim_{\nu\to n} \frac{1}{\pi}\left[\frac{dJ_\nu(z)}{d\nu} - \frac{1}{\cos\nu\pi}\frac{dJ_{-\nu}(z)}{d\nu}\right]. \qquad (B.7)$$

We find $\frac{dJ_{\pm\nu}(z)}{d\nu} = \pm\ln\frac{z}{2}J_{\pm\nu}(z) \mp \sum_{k=0}^{\infty} \frac{(-1)^k}{k!}\frac{\Gamma'(\pm\nu+k+1)}{\Gamma^2(\pm\nu+k+1)}\left(\frac{z}{2}\right)^{\pm\nu+2k}$. For $\nu = n > 0$, we have $\Gamma(\nu + k + 1) = (n + k)!$ and $\Gamma'(k + n + 1)/\Gamma(k + n + 1) = \psi(k + n + 1) = \sum_{p=0}^{k+n-1} \frac{1}{1+p} - \gamma = h(k + n) - \gamma$ (see Appendix A). Then

$$\frac{dJ_\nu(z)}{d\nu}\bigg|_{\nu=n} = (\ln\frac{z}{2} + \gamma)J_n(z) - \sum_{k=0}^{\infty} \frac{(-1)^k}{k!(n + k)!}h(k + n)\left(\frac{z}{2}\right)^{n+2k}.$$

Consider now $J_{-\nu}$. When $k \le n - 1$, the function $\Gamma(-\nu + k + 1)$ has simple poles in $\nu = n$: $\Gamma(-\nu + k + 1) \sim (-1)^{-n+k+1}[(n - k - 1)!(-\nu + n)]^{-1}$ and

$\Gamma'(-\nu+k+1) \sim (-1)^{-n+k}[(n-k-1)!(-\nu+n)^2]^{-1}$. Splitting the series in terms with $k \leq n-1$ and terms with $k \geq n$, we can write

$$\frac{dJ_{-\nu}(z)}{d\nu}\bigg|_{\nu=n} = -\log(z/2)J_{-n}(z) + \sum_{k=0}^{n-1} \frac{(-1)^n(n-k-1)!}{k!}\left(\frac{z}{2}\right)^{-n+2k}$$

$$+ \sum_{k=0}^{\infty} \frac{(-1)^{k+n}h(k)}{k!(n+k)!}\left(\frac{z}{2}\right)^{n+2k} - (-1)^n\gamma J_n(z),$$

where in the second sum we rescaled $k \to k+n$ and we used the definition of $J_n(z)$. Plugging the previous expressions in (B.7) we find (7.64). We also used part (b) of Exercise 7.9.

Exercise 7.11. We need to write (7.7) in terms of the variable $w = 1/z$. We obtain $\frac{d^2y}{dw^2} + \frac{2w-p}{w^2}\frac{dy}{dw} + \frac{q}{w^4}y = 0$ where now y, p and q are considered functions of w: $p = p_0 + p_1 w + \cdots$ and $q = q_0 + q_1 w + \cdots$. Equation (7.7) has an essential singular point at infinity since, if p_0 and q_0 are non-vanishing, the coefficient of dy/dw has a pole of order greater than one and the coefficient of y has a pole of order greater than two. $z = \infty$ becomes a Fuchsian point if $p_0 = q_0 = q_1 = 0$.

Exercise 7.12. We need to solve (7.7) with $p(z) = p_0 + p_1/z$ and $q(z) = q_2/z^2$ near infinity. From (7.70) and (7.71) we find two pairs of exponents $\lambda = \alpha = 0$ and $\lambda = -p_0, \alpha = -p_1$. The two linearly independent solutions have then the expansion $y_1(z) = \sum_{n=0}^{\infty} c_n/z^n$ and $y_2(z) = z^{-p_1}e^{-p_0 z}\sum_{n=0}^{\infty} c_n/z^n$. Substituting $y_1(z)$ into (7.7) and collecting the terms of order $1/z^{n+2}$ we obtain the recursion relation $\frac{c_{n+1}}{c_n} = \frac{n(n+1)-np_1+q_2}{p_0(n+1)}$. The radius of convergence of the series in y_1 is given by $\frac{1}{|z|} < R$ with $R = \lim_{n\to\infty}|c_n/c_{n+1}| = 0$. The series does not converge for any finite z. Substituting $y_2(z)$ into (7.7) and collecting the terms proportional to $z^{-p_1}e^{-p_0 z}/z^{n+2}$ we obtain the recursion relation $\frac{c_{n+1}}{c_n} = -\frac{n^2+n(1+p_1)+p_1+q_2}{p_0(n+1)}$. As before the series diverges since $R = \lim_{n\to\infty}|c_n/c_{n+1}| = 0$.

Exercise 7.13. As discussed in Example 7.10, around $z = \infty$ the Bessel equation (7.57) is of the form (7.67) with non-zero coefficients $p_1 = 1$, $q_0 = 1$ and $q_2 = -\nu^2$. In this case the first two recurrence relations in (7.69) become $\lambda^2 + 1 = 0$ and $2\alpha + 1 = 0$, which have solutions $\lambda = \pm i$ and $\alpha = -1/2$. The other relations give $c_n = -c_{n-1}[4\nu^2 - (2n-1)^2]/(8\lambda n)$, which can be solved as

$$c_k^1 = \frac{i^k}{8^k}\frac{1}{k!}\prod_{j=1}^k [4\nu^2 - (2j-1)^2]c_0^1, \qquad c_k^2 = \frac{(-i)^k}{8^k}\frac{1}{k!}\prod_{j=1}^k [4\nu^2 - (2j-1)^2]c_0^2,$$

where c_k^1 and c_k^2 correspond to $\lambda = i$ and $\lambda = -i$, respectively. Setting $c_0^1 = a_1/\sqrt{2\pi}$ and $c_0^2 = a_2/\sqrt{2\pi}$ we recover (7.73).

Exercise 7.14.   We work in polar coordinates and use the method of separation of variables as in Example 5.7. By setting $u(r, \phi, t) = R(r)\Phi(\phi)T(t)$ we find three equations

$$r^2 R'' + r R' + (\lambda^2 r^2 - \mu^2)R = 0, \qquad \Phi'' + \mu^2 \Phi = 0, \qquad T'' + v^2 \lambda^2 T = 0,$$

where λ and μ are two arbitrary constants. We immediately find $\Phi(\phi) = A\cos\mu\phi + B\sin\mu\phi$. Requiring periodicity of Φ and its derivative, we also find $\mu = m = 0, 1, 2, \dots$. Similarly we have $T(t) = C\cos v\lambda t + D\sin v\lambda t$. The equation for R is a Bessel equation with general solution $R(r) = EJ_m(\lambda r) + FY_m(\lambda r)$ (see Example 7.9). Since we look for solutions that are regular at the centre of the membrane and the Bessel function $Y_m(\lambda r)$ is singular for $r = 0$, we set $F = 0$. Imposing the condition that the boundary of the membrane is fixed, $u(a, \phi, t) = 0$, we find $R(a) = 0$. Therefore $J_m(\lambda a) = 0$. The Bessel function $J_m(x)$ oscillates and has an infinite number of zeros x_{nm}, $n = 1, 2, 3, \dots$. We find $\lambda = \lambda_{nm} = x_{nm}/a$. The general solution is therefore $u(r, \phi, t) = \sum_{nm} J_m(\lambda_{nm}r)(a_{nm}\cos m\phi + b_{nm}\sin m\phi)\cos v\lambda_{nm}t + \sum_{nm} J_m(\lambda_{nm}r)(a'_{nm}\cos m\phi + b'_{nm}\sin m\phi)\sin v\lambda_{nm}t$. The various terms in the previous expression are the normal modes of the membrane.

Exercise 7.15.   Define a new independent variable $x = \frac{2}{3}(-z)^{3/2}$ and a new unknown $\tilde{y}(x) = y(z)/(-z)^{-1/2}$. With these redefinitions the Airy equation becomes a Bessel equation of order $1/3$, $x^2\tilde{y}'' + x\tilde{y}' + (x^2 - 1/9)\tilde{y} = 0$.

(a) From the asymptotic expansion of the solution of the Bessel equation, (7.73), we find that, for large x, $\tilde{y}(x) \sim a_1 x^{-1/2}e^{ix}\sum_{n=0}^{\infty}\frac{\tilde{c}_n}{x^n} + a_2 x^{-1/2}e^{-ix}\sum_{n=0}^{\infty}(-1)^n\frac{\tilde{c}_n}{x^n}$. Substituting $x = \frac{2}{3}(-z)^{3/2}$ and $\tilde{y}(x) = y(z)/(-z)^{1/2}$, and absorbing numerical factors into the arbitrary constants, we find the required asymptotic expansion for the Airy function.

(b) Consider first $Ai(z)$. By using (3.54), we see that, for $|\arg z| < \pi$, $a_1 = 1/(2\sqrt{\pi})$ and $a_2 = 0$. On the other hand, using (3.56) with $|y| = -z$, for large negative z, we find $a_1 = 1/(2\sqrt{\pi})$ and $a_2 = i/(2\sqrt{\pi})$ (one can show that this expansion can be extended to the sector $\pi/3 < \arg z < 5\pi/3$). The coefficients are different in different sectors. This is another instance of the Stokes phenomenon. Notice that the different asymptotic expansions have overlapping domains of validity. Consider next $Bi(z)$. Using (3.58) we see that, for large positive z, $a_2 = 1/\sqrt{\pi}$ and

a_1 can be arbitrary since it is subleading. This expansion is valid for $|\arg z| < \pi/3$. On the other hand, using (3.59) for large negative z, we find $a_1 = i/(2\sqrt{\pi})$ and $a_2 = 1/(2\sqrt{\pi})$. This expansion is valid for $\pi/3 < \arg z < 5\pi/3$. For a generic solution of the Airy equation, Stokes phenomenon occurs at $z = \pm\pi/3$ and $z = \pi$.
(c) As the Airy equation reduces to the Bessel equation by part (a), $Ai(-z)$ and $Bi(-z)$ must be linear combinations of the two independent solutions of the Bessel equation: $Ai(-z) = \sqrt{z}(a_1 J_{1/3}(\frac{2}{3}z^{3/2}) + a_2 J_{-1/3}(\frac{2}{3}z^{3/2}))$ and $Bi(-z) = \sqrt{z}(b_1 J_{1/3}(\frac{2}{3}z^{3/2}) + b_2 J_{-1/3}(\frac{2}{3}z^{3/2}))$. The coefficients can be determined by comparing the asymptotic behaviours. Using (3.56), (3.59), (7.75) and the identities $\cos(x \pm y) = \cos x \cos y \mp \sin x \sin y$, we find $a_1 = a_2 = 1/3$ and $b_1 = -b_2 = -1/\sqrt{3}$.

Chapter 8

Exercise 8.1. Consider a sequence of compact operators A_k that converges to A in the norm of $\mathcal{B}(\mathcal{H})$, $||A_k - A|| \to 0$ for $k \to \infty$. Let $\{\mathbf{x}_n\}$ be a bounded sequence. We have to prove that $\{A\mathbf{x}_n\}$ contains a convergent subsequence. Since A_1 is compact, $\{A_1\mathbf{x}_n\}$ contains a convergent subsequence, which we denote with $\{A_1\mathbf{x}_n^{(1)}\}$. The sequence $\{x_n^{(1)}\}$ is contained in the original sequence $\{\mathbf{x}_n\}$ and therefore is bounded. Since A_2 is compact, $\{A_2\mathbf{x}_n^{(1)}\}$ contains a convergent subsequence, which we denote with $\{A_2\mathbf{x}_n^{(2)}\}$. Proceeding in this way, we construct a sequence of subsequences $\{\mathbf{x}_n^{(k)}\}$ of $\{\mathbf{x}_n\}$, such that $\{A_k\mathbf{x}_n^{(k)}\}$ converges and $\{\mathbf{x}_n^{(k)}\} \subset \{\mathbf{x}_n^{(k-1)}\}$. Consider now the diagonal sequence with elements $\mathbf{z}_n = \mathbf{x}_n^{(n)}$. We can write, for all k, $||A\mathbf{z}_i - A\mathbf{z}_j|| \leq ||A\mathbf{z}_i - A_k\mathbf{z}_i|| + ||A_k\mathbf{z}_i - A_k\mathbf{z}_j|| + ||A_k\mathbf{z}_j - A\mathbf{z}_j||$. We will now show that all three terms on the right-hand side can be made smaller than any positive ϵ for i and j large enough. As a consequence $\{A\mathbf{z}_i\}$ converges, which proves that A is compact, since $\{\mathbf{z}_i\}$ is a subsequence of $\{\mathbf{x}_n\}$. To start with, consider the first and third term of the previous inequality. Since $\{\mathbf{z}_n\}$ is bounded and the sequence $||A_k - A||$ tends to zero, for all $\epsilon > 0$ and all i we can find k large enough so that $||A\mathbf{z}_i - A_k\mathbf{z}_i|| \leq ||A - A_k|| \, ||\mathbf{z}_i|| < \epsilon/3$. Consider now the second term. Being a subsequence of $\{A_k\mathbf{x}_i^{(k)}\}$, which converges, $\{A_k\mathbf{z}_i\}_{i>k}$ is a Cauchy sequence and therefore $||A_k\mathbf{z}_i - A_k\mathbf{z}_j|| < \epsilon/3$ for i and j large enough. This completes the proof.

Exercise 8.2. The unit sphere $S = \{\mathbf{u} \in \mathcal{H} \mid ||\mathbf{u}|| = 1\}$ is obviously bounded. We can easily see that it is also closed. Indeed, if $\mathbf{u}_n \to \mathbf{u}$ for $n \to \infty$ and $||\mathbf{u}_n|| = 1$, for the continuity of the norm we have $||\mathbf{u}|| = ||\lim_{n\to\infty} \mathbf{u}_n|| = \lim_{n\to\infty} ||\mathbf{u}_n|| = 1$.

Thus $\mathbf{u} \in S$. However, S is not compact. In a metric space, a set is compact if and only if every sequence contains a convergent subsequence. This is not the case for the sequence $\{\mathbf{e}_i\}$ in S, where $\mathbf{e}_i$ is a Hilbert basis. No subsequence of $\{\mathbf{e}_i\}$ can be Cauchy, since $||\mathbf{e}_i - \mathbf{e}_j||^2 = 1 + 1 = 2$ and, therefore, no subsequence of $\{\mathbf{e}_i\}$ converges.

Exercise 8.3.   Consider the operator (8.11) where A is a compact operator, $A\mathbf{u}_n = \lambda_n \mathbf{u}_n$ and $f(\lambda)$ is bounded in the neighbourhood of $\lambda = 0$.

(i) It is clear from the hypothesis that $f(\lambda)$ is bounded on $\sigma(A)$ and therefore there exists a constant k such that $|f(\lambda_n)| \leq k$ for any n. Then $||f(A)\mathbf{v}||^2 = \sum_n |f(\lambda_n)c_n|^2 \leq k^2 \sum_n |c_n|^2 = k^2 ||\mathbf{v}||^2$ and $||f(A)\mathbf{v}|| \leq k||\mathbf{v}||$ for all $\mathbf{v}$. $f(A)$ is then bounded.

(ii) We have $f(A)\mathbf{u}_n = f(\lambda_n)\mathbf{u}_n$. Therefore $\mathbf{u}_n$ is an eigenvector of $f(A)$ with eigenvalue $f(\lambda_n)$.

(iii) Any $\mathbf{v}, \mathbf{w} \in \mathcal{H}$ can be expanded as $\mathbf{v} = \sum_n c_n \mathbf{u}_n$ and $\mathbf{w} = \sum_n d_n \mathbf{u}_n$. If $f(\lambda_n)$ is real, we have $(\mathbf{w}, f(A)\mathbf{v}) = \sum_n \bar{d}_n f(\lambda_n)c_n = (f(A)\mathbf{w}, \mathbf{v})$. Then $f(A)$ is self-adjoint.

(iv) We can write $f(A)$ as the limit of the sequence of operators F_k defined as $F_k\mathbf{v} = \sum_{n \leq k} f(\lambda_n)c_n\mathbf{u}_n$. The F_k are operators of finite rank. Indeed the image of F_k is the direct sum of a finite number of eigenspaces of A. By the properties of compact operators, those corresponding to $\lambda_n \neq 0$ are of finite dimension, while that corresponding to $\lambda = 0$ does not contribute since $f(0) = 0$. Being a limit of operators of finite rank, $f(A)$ is compact.

Exercise 8.4.   It is useful to think about A as an infinite matrix made of 2×2 blocks. Indeed, we can decompose $l^2 = \bigoplus_n \mathcal{H}_n$, $n = 1, 2, \ldots$, where $\mathcal{H}_n$ is the two-dimensional space spanned by the elements $\mathbf{e}_{2n-1}$ and $\mathbf{e}_{2n}$ of the canonical basis, and we see that A sends $\mathcal{H}_n$ into itself. The restriction of A to $\mathcal{H}_n$ is the two-dimensional matrix σ_2/n (see Example 9.1).

(a) Since

$$||A\mathbf{x}||^2 = \sum_{k=1}^{\infty} |(A\mathbf{x})_k|^2 = \sum_{n=1}^{\infty} \frac{|x_{2n}|^2 + |x_{2n-1}|^2}{n^2} \leq \sum_{k=1}^{\infty} |x_k|^2 = ||\mathbf{x}||^2,$$

we see that A is bounded. From this inequality we also see that $||A|| \leq 1$. We can actually find a vector that saturates the inequality, $||A\mathbf{e}_1|| = ||i\mathbf{e}_2|| = ||\mathbf{e}_2|| = ||\mathbf{e}_1||$, and therefore $||A|| = 1$. To prove that A is self-adjoint it is enough to prove that each of its blocks is. Since σ_2 is self-adjoint and n is real, A is self-adjoint. To prove that A is compact we show that it is the limit of operators of finite rank. Consider the operators A_k obtained by restricting A to the finite-dimensional space $\bigoplus_{n=1}^{k} \mathcal{H}_n$. One can easily see that $||(A - A_k)\mathbf{v}||^2 \leq ||\mathbf{v}||^2/k^2$ so that $||A - A_k|| \leq 1/k$. Therefore $A_k \to A$ in $\mathcal{B}(\mathcal{H})$.

(b) It is enough to work block by block. Since the eigenvalues of σ_2 are ± 1 with eigenvectors $(1, \pm i)/\sqrt{2}$, we see that the eigenvalues of A are $\pm 1/n$, $n = 1, 2, \cdots$ with eigenvectors $(\mathbf{e}_{2n-1} \pm i\mathbf{e}_{2n})/\sqrt{2}$.

(c) A is self-adjoint. Therefore we know that the residual spectrum is zero and the spectrum is real (see Appendix A). We already computed the discrete spectrum of A and it consists of the points $\pm 1/n$. Since the spectrum $\sigma(A)$ is closed, it must contains the point $\lambda = 0$. But $\lambda = 0$ is not an eigenvalue, and so it belongs to the continuous spectrum. The general theory of compact operators states that $\lambda = 0$ is the only point in the continuous spectrum. Let us verify it explicitly. Recall that the continuous spectrum consists of those λ such that $(A - \lambda\mathbb{I})^{-1}$ exists, is densely defined but is not bounded. We can take $\lambda \in \mathbb{R}$ since we know that the spectrum is real. For all $\lambda \neq \pm 1/n$, $A - \lambda\mathbb{I}$ can be inverted. Since $A - \lambda\mathbb{I}$ is block diagonal, the inverse also is and it is given by the inverse of each block. Using $\sigma_2^2 = \mathbb{I}$ we find that $(\sigma_2/n - \lambda\mathbb{I})^{-1} = (\sigma_2/n + \lambda\mathbb{I})/(-\lambda^2 + 1/n^2)$. It is not difficult to see that for $\lambda \neq 0$, $(A - \lambda\mathbb{I})^{-1}$ is bounded. Indeed, we can write $(A - \lambda\mathbb{I})^{-1} = B_k + \bar{B}_k$, where B_k contains the first k rows and columns and it is zero otherwise and $\bar{B}_k$ contains the rest. B_k is an operator from a finite-dimensional space to itself and therefore is bounded. By taking k sufficiently large we see that $\bar{B}_k$ approaches the diagonal matrix with the first k entries equal to zero and the rest of the entries equal to $-1/\lambda$, and this operator is also bounded. We conclude that $\lambda \neq 0$ cannot be an element of the continuous spectrum. However, for $\lambda = 0$, A^{-1} restricts to $n\sigma_2$ in the block $\mathcal{H}_n$. Since $||A^{-1}\mathbf{e}_{2n}|| = n$ is not a bounded function of n, A^{-1} is not bounded. However, A^{-1} is at least densely defined since it is defined on the elements of the canonical basis $\mathbf{e}_n$, which span l^2. Since A^{-1} is densely defined but not bounded, $\lambda = 0$ is in the continuous spectrum.

Exercise 8.5. Set $\omega(x) = 1$ for simplicity. For a regular Sturm–Liouville problem $p(x)$ and $q(x)$ are continuous and $p(x) \neq 0$ on the closed interval $[a, b]$. The solutions of the differential equation $Ly = \lambda y$ are then also continuous on $[a, b]$ for all $\lambda \in \mathbb{C}$.

(a) Since L is self-adjoint, we know that the spectrum is real and the residual spectrum is empty. The eigenvalues λ_n form a discrete set.[6] We want to show that

---

[6]The corresponding eigenvectors are mutually orthogonal and, in a separable Hilbert space, a set of orthogonal vectors is at most countable.

the continuous spectrum is empty. Consider a $\lambda \in \mathbb{R}$ that is not an eigenvalue. The resolvent $(L - \lambda\mathbb{I})^{-1}$ is an integral operator constructed with the Green function $G(x, y; \lambda)$ of the Sturm–Liouville problem, $(L - \lambda\mathbb{I})^{-1}f(x) = \int G(x, y; \lambda)f(y)dy$ (see Section 6.2.2). $G(x, y; \lambda)$ is given by (6.29),

$$G(x, y; \lambda) = \begin{cases} C(\lambda)\tilde{u}_\lambda(y)u_\lambda(x), & a \leq x < y, \\ C(\lambda)u_\lambda(y)\tilde{u}_\lambda(x), & y < x \leq b, \end{cases} \tag{B.8}$$

where $u_\lambda(x)$ and $\tilde{u}_\lambda(x)$ are the solutions of the differential equation $Lu = \lambda u$ that satisfy the boundary conditions (5.15) at $x = a$ and $x = b$, respectively. Moreover, $1/C(\lambda) = -p(x)W(x) = -p(x)(\tilde{u}'_\lambda(x)u_\lambda(x) - \tilde{u}_\lambda(x)u'_\lambda(x))$ is independent of x by Exercise 5.3. Since λ is not an eigenvalue, u_λ and $\tilde{u}_\lambda$ are linearly independent solutions and $C(\lambda)$ is finite. Then $G(x, y; \lambda)$ is continuous on $[a, b] \times [a, b]$ and hence square-integrable. It follows that $(L - \lambda\mathbb{I})^{-1}$ is compact and therefore bounded. Hence λ cannot belong to the continuous spectrum, which is then empty.

(b) By shifting L if necessary, we can assume that $\lambda = 0$ is not an eigenvalue. L is then invertible and L^{-1} compact. Any eigenvalue λ_n of L, $Lu_n = \lambda_n u_n$, is also an eigenvalue of L^{-1}, $L^{-1}u_n = u_n/\lambda_n$. Since the eigenspaces of the compact operator L^{-1} are finite dimensional, and the u_n are a Hilbert basis, there must exist an infinite number of different λ_n. The spectral theorem for compact operators tells us that $\lim_{n\to\infty} 1/\lambda_n = 0$.

Exercise 8.6. If $\mathbf{x} \in \ker(A^\dagger - \bar{\lambda}\mathbb{I})$, $A^\dagger\mathbf{x} = \bar{\lambda}\mathbf{x}$. Then $0 = (A^\dagger\mathbf{x} - \bar{\lambda}\mathbf{x}, \mathbf{y}) = (\mathbf{x}, A\mathbf{y} - \lambda\mathbf{y})$ for all $\mathbf{y} \in \mathcal{H}$ and therefore $\mathbf{x}$ is orthogonal to all vectors in $\mathrm{im}(A - \lambda\mathbb{I})$. It follows that $\mathbf{x} \in \mathrm{im}(A - \lambda\mathbb{I})^\perp$. The argument can also be run in the opposite direction. We conclude that $\mathrm{im}(A - \lambda\mathbb{I})^\perp = \ker(A^\dagger - \bar{\lambda}\mathbb{I})$.

Exercise 8.7. The kernel $k(x, y) = x - y$ is degenerate and it can be written as in (8.31) with $P_1(x) = x, P_2(x) = -1$ and $Q_1(x) = 1, Q_2(x) = x$. From (8.34) we find $a_{11} = -a_{22} = 1/2, a_{12} = -1$ and $a_{21} = 1/3$. The eigenvalues of K are the eigenvalues $\pm/(2i\sqrt{3})$ of the matrix a, in agreement with Example 8.4. We now want to solve (8.28). We first compute $b_1 = 1$, $b_2 = 1/2$ from (8.34). Then (8.33) gives: $\zeta_1 = -12\lambda/(1 + 12\lambda^2)$ and $\zeta_2 = -(1 + 6\lambda)/(1 + 12\lambda^2)$. The solution of (8.28) is then given by (8.35): $\phi(x) = 6(1 - 2x - 2\lambda)/(1 + 2\lambda^2)$ as found in Example 8.4.

Exercise 8.8. The equation is written in the traditional notation for Fredholm equations of second type (see footnote 5). To compare with Section 8.2 we just need to set $\mu = 1/\lambda$, $f(x) = -x/\mu$ and $k(x, y) = 1/x + 1/y$. The kernel is degenerate with $P_1(x) = 1, P_2(x) = 1/x$ and $Q_1(y) = 1/y, Q_2(y) = 1$. The solution can be written as in (8.35): $\phi(x) = x + \mu\zeta_1 + \frac{\mu}{x}\zeta_2$, where $\zeta_1 = \int_1^2 dy \frac{1}{y}\phi(y)$ and $\zeta_2 = \int_1^2 dy\ \phi(y)$. Using (8.34) the linear system (8.33) for ζ_i becomes: $[1 - \mu \ln 2]\zeta_1 - \frac{1}{2}\mu\zeta_2 = 1$, $-\mu\zeta_1 + [1 - \mu \ln 2]\zeta_2 = \frac{3}{2}$. There is a unique solution when the determinant of the matrix of coefficients, $D = (1 - \mu \ln 2)^2 - \mu^2/2$, is different from zero, namely for $\mu \neq \mu_\pm = \frac{\sqrt{2}}{\sqrt{2}\ln 2 \pm 1}$. In this case $\zeta_1 = D^{-1}(1 - \mu \ln 2 + \frac{3}{4}\mu)$ and $\zeta_2 = (2D)^{-1}(3 + 2\mu - 3\mu \ln 2)$. It is easy to see that for $\mu = \mu_+$ and $\mu = \mu_-$ there are no solutions.

Exercise 8.9. To find the spectrum, we have to solve the homogeneous equation $K\phi(x) = \int_0^1 (6x^2 - 12xy)\phi(y)dy = \lambda\phi(x)$. The kernel is degenerate and we choose $P_1(x) = 6x^2, P_2(x) = -12x$ and $Q_1(y) = 1, Q_2(y) = y$. From (8.34) we easily find $a = \begin{pmatrix} 2 & -6 \\ 3/2 & -4 \end{pmatrix}$ with eigenvalue $\lambda = -1$ of multiplicity two and eigenvector $(\zeta_1, \zeta_2) = (2, 1)$. Thus the only eigenvalue of the Fredholm operator is $\lambda = -1$. The corresponding eigenvector is given by (8.37) and, up to normalisation, is $\phi_{\lambda=-1} = x - x^2$. We need now to solve the Fredholm equation. We use the notation of Section 8.2 and we set $\lambda = 1/\mu$ and $f(x) = -1/\mu$. The system (8.33) is $[1 - 2\mu]\zeta_1 + 6\mu\zeta_2 = 1$ and $-3\mu\zeta_1 + [2 + 8\mu]\zeta_2 = 1$. For $\mu \neq -1$, the solution is $\zeta_1 = (1 + \mu)^{-1}$ and $\zeta_2 = (2 + 2\mu)^{-1}$, so that $\phi(x) = 1 + \frac{6\mu}{1+\mu}(x^2 - x)$ is the unique solution of the Fredholm equation. In the case $\mu = -1$, the system reduces to $3\zeta_1 = 6\zeta_2 + 1$, and the equation has a one-parameter family of solutions $\phi(x) = 1 + 12\zeta_2(x - x^2) - 2x^2$. The Fredholm alternative is at work here: $\lambda = -1$ is an eigenvalue and the arbitrariness corresponds to adding to the solution an arbitrary multiple of the eigenvector.

Exercise 8.10. The differential equation $y'' = 1 - xy$ can be transformed into an integral one by integrating both sides with respect to x and using $y(0) = y'(0) = 0$. We obtain $y'(x) = \int_0^x (1 - ty(t))dt = x - \int_0^x ty(t)dt$. Integrating again we obtain $y(x) = x^2/2 - \int_0^x ds \int_0^s ty(t)dt$. Exchanging the integrals over t and s (we are integrating over a triangle), we find $y(x) = x^2/2 - \int_0^x dt \int_t^x ds[ty(t)] = x^2/2 - \int_0^x (x - t)ty(t)dt$ which is the required Volterra equation. Vice versa, if $y(x)$ is solution of the Volterra equation, by differentiating it, we find $y'(x) = x - \int_0^x ty(t)dt$. Differentiating again we find $y'' = 1 - xy$. Moreover, from the integral equation and the equation for $y'(x)$, we immediately read $y'(0) = y(0) = 0$.

Exercise 8.11. We can write the Volterra equation as $\phi(x) = f(x) + \mu \int_0^{+\infty} k(x, t)\phi(t)dt$ with $k(x, t) = e^{x-t}\theta(x - t)$. Using the method of iterations discussed in Section 8.2.1, we find $\phi(x) = \sum_{n=0}^{\infty} \mu^n K^n f$ where K is the integral operator on $L^2[0, \infty)$ with kernel $k(x, t)$. The kernel of K^2 is $k^{(2)}(x, z) = \int_0^{+\infty} k(x, y)k(y, z)dy = \int_0^{+\infty} \theta(x - y)\theta(y - z)e^{x-z}dy = \theta(x - z)(x - z)e^{x-z}$. By iteration, we find that the kernel of K^n is $k^{(n)}(x, z) = \theta(x - z)(x - z)^n e^{x-z}/n!$. Therefore $\phi(x) = \sum_{n=0}^{\infty} \mu^n \int_0^x \frac{(x-z)^n}{n!} e^{x-z} f(z) dz = \int_0^x e^{(\mu+1)(x-z)} f(z) dz$.

Chapter 9

Exercise 9.1. We first prove that $(\mathbf{S} \cdot \mathbf{n})^2 = \mathbb{I}/4$. Using the properties of the Pauli matrices given in Example 9.1, we have $(\mathbf{S} \cdot \mathbf{n})^2 = \sum_{i,j=1}^3 n_i n_j \sigma_i \sigma_j/4 = \sum_{i=1}^3 n_i^2 \sigma_i^2/4 + i \sum_{i \neq j \neq k}^3 \epsilon_{ijk} n_i n_j \sigma_k/4 = \sum_{i=1}^3 n_i^2 \mathbb{I}/4 = \mathbb{I}/4$. The term with ϵ_{ijk} vanishes because it is multiplied by the quantity $n_i n_j$ which is symmetric in i and j. It follows from $(\mathbf{S} \cdot \mathbf{n})^2 = \mathbb{I}/4$ that the eigenvalues of $\mathbf{S} \cdot \mathbf{n}$ can only be $\pm 1/2$. The two eigenvalues cannot be equal because otherwise $\mathbf{S} \cdot \mathbf{n}$ would be proportional to the identity matrix. The explicit eigenvectors of $\mathbf{S} \cdot \mathbf{n}$ are given in (9.67).

Exercise 9.2. (a) Suppose that there are two different square integrable solutions u_1 and u_2 of (9.36) corresponding to the same E. We need to prove that u_1 and u_2 are linearly dependent. They satisfy $u_1'' = 2m(V(x) - E)u_1$ and $u_2'' = 2m(V(x) - E)u_2$. Therefore $u_1''/u_1 = u_2''/u_2$ or, equivalently, $u_2 u_1'' - u_1 u_2'' = (u_2 u_1' - u_1 u_2')' = 0$, and $u_2 u_1' - u_1 u_2' = const$. The constant is actually zero because u_i, being square-integrable, vanish at infinity. Therefore $u_1'/u_1 = u_2'/u_2$. Integrating this equation, we find $u_1 = cu_2$ with c constant. u_1 and u_2 are then linearly dependent.

(b) Let u be a square-integrable solution of (9.36). Since E is real, $\bar{u}$ also satisfies (9.36). By part (a), $\bar{u} = cu$. Now $u = \overline{(\bar{u})} = \bar{c}\bar{u} = |c|^2 u$ so that c is a complex number of modulus one. Since wave functions are defined up to a phase factor, we can always set $c = 1$ and choose u real.

(c) Let u be a square-integrable solution of (9.36). If $V(x)$ is even, $u(-x)$ also satisfies (9.36). By part (a), $u(x) = cu(-x)$, where c is a constant. Now $u(x) = cu(-x) = c^2 u(x)$ so that $c = \pm 1$. We see that $u(-x) = \pm u(x)$ or, in other words, $u(x)$ is either even or odd.

Exercise 9.3. The potential is singular in $x = \pm a$ and we need to understand how to treat the differential equation. As in Example 9.8, we solve the equation in the three regions $x < -a$ (region I), $|x| < a$ (region II), $x > a$ (region III), and we patch the solutions using appropriate conditions. To find the patching conditions we integrate the equation $u'' = 2m(V(x) - E)u$ over the small intervals

$[\pm a - \epsilon, \pm a + \epsilon]$ around the singularities $x = \pm a$ of $V(x)$. The result is $u'(\pm a + \epsilon) - u'(\pm a - \epsilon) = 2m \int_{\pm a - \epsilon}^{\pm a + \epsilon}(V(x) - E)u(x)dx$. We now send ϵ to zero. If $V(x)$ and $u(x)$ are bounded, the right-hand side of the previous equation goes to zero for $\epsilon \to 0$ and we obtain that u' is continuous in $x = \pm a$. Then also u is continuous. Since we are interested in the discrete spectrum, we consider the values of E for which the classical motion is bounded, $-V_0 < E < 0$. The solutions of the Schrödinger equation are $u_I = A_1 e^{\rho x} + A_1' e^{-\rho x}$ in region I, $u_{II} = A_2 e^{ipx} + A_2' e^{-ipx}$ in region II and $u_{III} = A_3 e^{\rho x} + A_3' e^{-\rho x}$ in region III, where $\rho = \sqrt{-2mE}$ and $p = \sqrt{2m(E + V_0)}$. Square-integrability at $x = \pm\infty$ requires $A_1' = A_3 = 0$. Continuity of u and u' at $x = \pm a$ gives a homogeneous linear system for the remaining four constants A_1, A_2, A_2', A_3', which has non-trivial solutions only for specific values of E. From part (c) of Exercise 9.2 we can assume that the solutions are either even or odd. Even solutions have $A_1 = A_3'$ and $A_2 = A_2'$. Given the symmetry of u, it is enough to impose continuity of u and u' at $x = a$. We find $2A_2 \cos pa = A_3' e^{-\rho a}$ and $-2A_2 p \sin pa = -\rho A_3' e^{-\rho a}$. Taking the ratio of the two equations we find $\tan pa = \sqrt{p_0^2 - p^2}/p$, where $p_0 = \sqrt{2mV_0}$. Odd solutions have $A_1 = -A_3'$ and $A_2 = -A_2'$. Imposing continuity of u and u' in $x = a$ we find $2A_2 i \sin pa = A_3' e^{-\rho a}$ and $2A_2 ip \cos pa = -\rho A_3' e^{-\rho a}$, which imply $\cot pa = -\sqrt{p_0^2 - p^2}/p$. The two equations $\tan pa = \sqrt{p_0^2 - p^2}/p$ and $-\cot pa = \sqrt{p_0^2 - p^2}/p$ have a finite number of solutions p_n corresponding to the energy levels $E_n = p_n^2/2m$ (see Figure B.6). In all there are n energy levels for $\frac{(n-1)^2 \pi^2}{8ma^2} < V_0 < \frac{n^2 \pi^2}{8ma^2}$.

Exercise 9.4. (a) It follows from $[Q, P] = i\mathbb{I}$ that $[a, a^\dagger] = (-i[Q, P] + i[P, Q])/2 = \mathbb{I}$. Moreover, $a^\dagger a = m\omega Q^2/2 + P^2/(2m\omega) + i(QP - PQ)/2 = H/\omega - \mathbb{I}/2$, so that $H = \omega(N + \mathbb{I}/2)$.

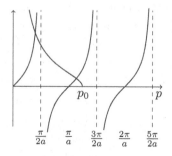 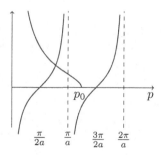

Fig. B.6. The three intersections of $\sqrt{p_0^2 - p^2}/p$ with $\tan pa$ and $-\cot pa$ for $\frac{\pi^2}{2ma^2} < V_0 < \frac{9\pi^2}{8ma^2}$.

(b) We use $[AB, C] = A[B, C] + [A, C]B$, which is easily proved by comparing the two sides. We find $[N, a] = [a^\dagger a, a] = [a^\dagger, a]a = -a$ since $[a, a^\dagger] = \mathbb{I}$ by part (a). $[N, a^\dagger] = a^\dagger$ is proved similarly.

(c) Using $N|n\rangle = n|n\rangle$ and (9.98), we find $0 \le ||a|n\rangle||^2 = \langle n|a^\dagger a|n\rangle = \langle n|N|n\rangle = n$. Then $n \ge 0$, and $n = 0$ if and only if $||a|n\rangle|| = 0$, or equivalently $a|n\rangle = 0$.

(d) Using part (b), we have $N(a^\dagger|n\rangle) = a^\dagger(N|n\rangle) + [N, a^\dagger]|n\rangle = (n+1)a^\dagger|n\rangle$ so that $a^\dagger|n\rangle$ is an eigenvector of N with eigenvalue $n+1$. Therefore $a^\dagger|n\rangle = c|n+1\rangle$ where c is a constant. Taking the norm of the two sides of the previous equation, we find $|c|^2 = ||a^\dagger|n\rangle||^2 = \langle n|aa^\dagger|n\rangle = \langle n|N + 1|n\rangle = (n+1)\langle n|n\rangle = n+1$. The phase of c can be reabsorbed in the definition of the eigenstates $|n\rangle$ without spoiling the normalization condition $\langle n'|n\rangle = \delta_{nn'}$. We can then take c real and we find $c = \sqrt{n+1}$ and $a^\dagger|n\rangle = \sqrt{n+1}|n+1\rangle$. The results for a are proved similarly.

(e) Consider an eigenvector $|n\rangle$ of N with eigenvalue n. Since N is self-adjoint, n is real. By repeatedly applying a we find an infinite tower of eigenstates $|n'\rangle$ with $n' = n - k$, where k is a positive integer by part (d). However, we proved in part (c), that n' must be positive. The tower should then truncate. This is only possible if there is an n' such that $a|n'\rangle = 0$. This implies $n' = 0$ by part (c) and therefore n must be an integer. We find a tower of eigenstates $|0\rangle, |1\rangle, |2\rangle, \ldots$. These can be obtained acting on $|0\rangle$ with $a^\dagger$. Indeed, since $|n+1\rangle = a^\dagger|n\rangle/\sqrt{n+1}$, we easily find $|n\rangle = (a^\dagger)^n|0\rangle/\sqrt{n!}$.

(f) Since the eigenvalues of N are $n \in \mathbb{N}$, the result follows from part (a).

Exercise 9.5. In analogy with Example 9.8 we expect discrete levels for $E < 0$. In the three regions, $x < -l$ (region I), $-l < x < l$ (region II) and $x > l$ (region III), the potential is zero and the solutions of the time-independent Schrödinger equation are $u_I = ae^{\rho x}$, $u_{II} = c_1 e^{\rho x} + c_2 e^{-\rho x}$ and $u_{III} = be^{-\rho x}$ where $\rho = \sqrt{-2mE}$. In region I and III we kept only the square-integrable solution. From part (c) of Exercise 9.2 we can assume that the solutions are either even or odd. Even solutions have $a = b$ and $c_1 = c_2$. As in Example 9.8 we require the continuity of u in $x = \pm l$ and a jump of the first derivative $u'(\pm l_+) - u'(\pm l_-) = -\mu u(\pm l)$. This implies $c_1(e^{2l\rho} + 1) = b$ and $\rho c_1(e^{2l\rho} - 1) = (\mu - \rho)b$. We then find $e^{-2l\rho} = -1 + 2\rho/\mu$. We can see graphically that this equation has always a solution for ρ where the exponential $e^{-2l\rho}$ intersects the line $-1 + 2\rho/\mu$. Odd solutions have $-a = b$ and $c_1 = -c_2$. The continuity of u and the jump of u' in $x = l$ give $c_1(e^{2l\rho} - 1) = b$ and $\rho c_1(e^{2l\rho} + 1) = (\mu - \rho)b$. We then find $e^{-2l\rho} = 1 - 2\rho/\mu$. This equation has a solution only if the modulus of the derivative of $e^{-2l\rho}$ in the origin is greater than $2/\mu$, that is $l > 1/\mu$. In total we find one energy level for $l < 1/\mu$ and two for $l > 1/\mu$.

Exercise 9.6.   (a) The angular momentum operators $L_i = \sum_{j,k=1}^{3} \epsilon_{ijk} Q_j P_k$ are obtained by replacing the position x_i and momentum p_i with the operators Q_i and P_i in the classical expression of the angular momentum $L_i = \sum_{j,k=1}^{3} \epsilon_{ijk} x_j p_k$. Since j and k in the definition of L_i are necessarily different and Q_j and P_k commute for $j \neq k$, there is no ordering problem in the definition of L_i. As Q_j and P_k are self-adjoint and commute, also L_i is self-adjoint.

(b) We will use repeatedly the identity $[AB, C] = A[B, C] + [A, C]B$ and the commutators $[Q_i, Q_j] = [P_i, P_j] = 0$ and $[Q_i, P_j] = i\delta_{ij}$. Consider first $[L_1, L_2]$. We have $[L_1, L_2] = [Q_2 P_3 - Q_3 P_2, Q_3 P_1 - Q_1 P_3] = Q_2[P_3, Q_3]P_1 + Q_1[Q_3, P_3]P_2 = i(Q_1 P_2 - Q_2 P_1) = iL_3$. The other identities are proved similarly.

(c) Using part (b), we have $[\mathbf{L}^2, L_i] = \sum_{k=1}^{3}[L_k^2, L_i] = \sum_{k=1}^{3} L_k[L_k, L_i] + \sum_{k=1}^{3}[L_k, L_i]L_k = \sum_{k,p=1}^{3} i\epsilon_{kip}(L_k L_p + L_p L_k) = \sum_{k,p=1}^{3} i\epsilon_{kip}(L_k L_p - L_k L_p) = 0$, where in the last step we just renamed indices and used the antisymmetry of the tensor ϵ_{ijk}.

(d) We have $[L_i, Q_j] = \sum_{k=1}^{3} \epsilon_{ikp}[Q_k P_p, Q_j] = \sum_{k=1}^{3} \epsilon_{ikp} Q_k[P_p, Q_j] = -i\sum_{k=1}^{3} \epsilon_{ikj} Q_k = i\sum_{k=1}^{3} \epsilon_{ijk} Q_k$. The identity $[L_i, P_j] = i\sum_{k=1}^{3} \epsilon_{ijk} P_k$ is proved similarly.

Exercise 9.7.   It is convenient to write the angular momentum operators in spherical coordinates. After some algebra, one finds

$$L_3 = -i\frac{\partial}{\partial \phi}, \qquad \mathbf{L}^2 = -\frac{1}{\sin\theta}\frac{\partial}{\partial\theta}\left(\sin\theta \frac{\partial}{\partial\theta}\right) - \frac{1}{\sin^2\theta}\frac{\partial^2}{\partial\phi^2}. \qquad (\text{B.9})$$

We see that $\mathbf{L}^2 = -\hat{\Delta}$, where $\hat{\Delta}$ is the Laplace operator on the unit sphere defined in (5.59). The eigenvectors of $\hat{\Delta}$ are the spherical harmonics (5.69). Since the functions (5.69) are also eigenvectors of $-i\frac{\partial}{\partial\phi}$, we conclude that the spherical harmonics Y_{lm} are a common basis of eigenvectors of $\mathbf{L}^2$ and L_3: $\mathbf{L}^2 Y_{lm} = l(l + 1)Y_{lm}$ and $L_3 Y_{lm} = mY_{lm}$, with $l = 0, 1, 2, \ldots$ and $m = -l, \ldots, l$. We see that the spectrum of the angular momentum operators is discrete. The modulus square of the angular momentum can assume only the values $l(l + 1)$ with $l = 0, 1, 2, \ldots$, and, for every fixed l, its projection along the z-axis can assume only the values $m = -l, \ldots, l$. We were able to find a common basis of eigenvectors for $\mathbf{L}^2$ and L_3 because they commute, as proved in part (c) of Exercise 9.6.

Exercise 9.8. (a) The Fourier transform of the Gaussian $\psi(x)$ is again a Gaussian $\hat\psi(p) = \pi^{-1/4}a^{-1/2}e^{-p^2/(2a^2)}$. The more the Gaussian is peaked around zero the more its transform is spread. It is easy to check that ψ and $\hat\psi$ are correctly normalized, $\int_{-\infty}^{+\infty}|\psi(x)|^2\mathrm{d}x = \int_{-\infty}^{+\infty}|\hat\psi(p)|^2\mathrm{d}p = 1$. As ψ and $\hat\psi$ are even, the expectation values of P and Q vanish, $\langle Q\rangle = \langle P\rangle = 0$, and the standard deviations are $(\Delta Q)^2 = 1/(2\,a^2)$ and $(\Delta P)^2 = a^2/2$, where we used (A.16). We see that the Gaussian wave function saturates the inequality (9.72), $\Delta Q\Delta P = 1/2$.

(b) Using the residue theorem, where we close the contour in the upper half-plane for $p < 0$ and in the lower half-plane for $p > 0$, we easily compute $\hat\psi(p) = \sqrt{a}e^{-a|p|}$. The more ψ is peaked around zero the more $\hat\psi$ is spread and vice versa. We find again $\langle Q\rangle = \langle P\rangle = 0$, while $\Delta Q = a$ and $\Delta P = 1/(a\sqrt2)$. We see that the Heisenberg principle (9.72) is satisfied, $\Delta Q\Delta P = 1/\sqrt2 > 1/2$.

(c) The Fourier transform of a plane wave is a delta function, $\hat\psi(p) = \frac{1}{2\pi}\int_{-\infty}^{+\infty}e^{-i(p-p_0)x}\mathrm{d}x = \delta(p - p_0)$. We cannot literally apply (9.72) since ψ is a distribution and the integrals are not well-defined. However we see that, since $|\psi(x)|^2 = 1$, the distribution $\psi(x)$ is totally uniform, while the distribution of momenta is localized at p_0.

Finally, from (9.71), we see that (9.72) is saturated if and only if $R\psi = 0$. Replacing A with Q and B with P we find the differential equation $Q\psi + i\lambda P\psi = (x + \lambda\,d/dx)\psi = 0$, which is solved by a Gaussian $\psi(x) = ce^{-x^2/(2\lambda)}$.

Exercise 9.9. By expanding the exponential in power series, we find $U(t)^\dagger = (\mathbb{I} - iHt - H^2\frac{t^2}{2} + \cdots)^\dagger = \mathbb{I} + iHt - H^2\frac{t^2}{2} + \cdots = e^{iHt}$ so that $U(t)^\dagger U(t) = e^{iHt}e^{-iHt} = \mathbb{I}$ and similarly $U(t)U^\dagger(t) = \mathbb{I}$.

Exercise 9.10. Since the Hamiltonian is time independent, we can use (9.75) to evaluate $\psi(t)$. In the notations of Example 9.4, the eigenstates of the Hamiltonian are $e_\pm^{(z)}$ with eigenvalues $E = \pm B$. We write $\psi(0) = (e_+^{(z)} + e_-^{(z)})/\sqrt2$. Using (9.76) we find $\psi(t) = (e_+^{(z)}e^{-iBt} + e_-^{(z)}e^{iBt})/\sqrt2 = (e^{-iBt}, e^{iBt})/\sqrt2$. The probability that a measure of S_3 at time t gives $\pm1/2$ is $|(e_\pm^{(z)}, \psi(t))|^2 = 1/2$. The probability that a measure of S_1 at time t gives $1/2$ is $|(e_+^{(x)}, \psi(t))|^2 = \cos^2 Bt$ and the probability that it gives $-1/2$ is $|(e_-^{(x)}, \psi(t))|^2 = \sin^2 Bt$. The probability that a measure of S_2 at time t gives $1/2$ is $|(e_+^{(y)}, \psi(t))|^2 = (\cos Bt + \sin Bt)^2/2$ and the probability that it gives $-1/2$ is $|(e_-^{(y)}, \psi(t))|^2 = (\cos Bt - \sin Bt)^2/2$.

Exercise 9.11. (a) Replace L_i with J_i in Exercise 9.6.

(b) We find $[J_3, J_\pm] = [J_3, J_1 \pm iJ_2] = iJ_2 \pm J_1 = \pm J_\pm$. Moreover $J_+ J_- = (J_1 + iJ_2)(J_1 - iJ_2) = J_1^2 + J_2^2 - i(J_1 J_2 - J_2 J_1) = \mathbf{J}^2 - J_3^2 + J_3$ and then $\mathbf{J}^2 = J_+ J_- + J_3^2 - J_3$. The other identity is obtained by considering $J_- J_+$.

(c) Since the J_i are self-adjoint we have $(J_+)^\dagger = J_-$. Then, using part (b), $||J_+|jm\rangle||^2 = \langle jm|J_- J_+|jm\rangle = \langle jm|\mathbf{J}^2 - J_3^2 - J_3|jm\rangle = j(j+1) - m^2 - m = (j-m)(j+m+1)$. Similarly $||J_-|jm\rangle||^2 = j(j+1) - m^2 + m = (j+m)(j-m+1)$. Since these norms are positive numbers we learn that $-(j+1) \le m \le j$ and $-j \le m \le j+1$.[7] The intersection of the two inequalities implies $-j \le m \le j$.

(d) Using $[J_3, J_+] = J_+$ we find $J_3(J_+|jm\rangle) = (J_+ J_3 + [J_3, J_+])|jm\rangle = (m+1) J_+|jm\rangle$ so that $J_+|jm\rangle$ is an eigenvector of J_3 with eigenvalue $m+1$. Similarly, $J_-|jm\rangle$ is an eigenvector of J_3 with eigenvalue $m-1$. Since, by hypothesis, the eigenvectors are non-degenerate we have $J_\pm|jm\rangle = c_\pm|j, m\pm 1\rangle$. Using the norms of $J_\pm|jm\rangle$ computed in part (c), we find $|c_\pm|^2 = (j \mp m)(j \pm m + 1)$. A phase in the constants $c_\pm$ can be reabsorbed in the definition of the eigenvectors (the orthonormality condition will not change) and we can assume that the $c_\pm$ are real. We then find $J_\pm|jm\rangle = \sqrt{(j \mp m)(j \pm m + 1)}|j, m \pm 1\rangle$.

(e) By repeatedly applying J_- and J_+ to an eigenstate $|jm\rangle$ we find an infinite tower of eigenstates $|jm'\rangle$ with $m' - m \in \mathbb{Z}$, by part (d). However, by part (c), we must have $-j \le m' \le j$. This means that the tower $|jm'\rangle$ should truncate. This is only possible if there are m_1' and m_2' such that $J_-|jm_1'\rangle = 0$ and $J_+|jm_2'\rangle = 0$. Since $||J_\pm|jm'\rangle||^2 = (j \mp m')(j \pm m' + 1)$ we learn that $m_1' = -j$ and $m_2' = j$. The tower consists then of the $2j + 1$ states $|j, -j\rangle, |j, -j+1\rangle, \ldots, |j, j\rangle$ with $m = -j, -j+1, \ldots, j$. Since $2j+1$ must be an integer, we also find $j = 0, 1/2, 1, 3/2, \ldots$.

Exercise 9.12. (a) We have $y'(x) = \sum_{n=0}^\infty \hbar^{n-1} S_n'(x)y(x)$ and $y''(x) = [(\sum_{n=0}^\infty \hbar^{n-1} S_n'(x))^2 + \sum_{n=0}^\infty \hbar^{n-1} S_n''(x)]y(x)$. By equating the terms with the same power of $\hbar$, we find $(S_0')^2 = v$, $2S_1' S_0' + S_0'' = 0$ and $2S_0' S_n' + S_{n-1}'' + \sum_{k=1}^{n-1} S_k' S_{n-k}' = 0$ for $n \ge 2$.

(b) The Airy equation is obtained by setting $\hbar = 1$ and $v(x) = x$. Solving the first three equations in this case, we find $S_0 = \pm\frac{2}{3}x^{3/2}$, $S_1 = -\frac{1}{4}\ln x$

---

[7] We take $j \ge 0$. Indeed $j(j+1) \ge 0$ since it is the eigenvalue of $\mathbf{J}^2$. As j and $-1-j$ correspond to the same eigenvalue, we can take j positive.

and $S_2 = \pm\frac{5}{48}x^{-3/2}$, which give $y(x) \sim x^{-1/4}e^{\pm 2x^{3/2}/3}(1 \pm \frac{5}{48}x^{-3/2} + \cdots)$, in agreement with (3.55) and Exercise 7.15. The reason for this agreement is that the WKB expansion is valid if $\hbar S_n$ is much smaller than S_{n-1}. In our case, $S_n(x) = o(S_{n-1}(x))$ for x large, so the relation is valid also for $\hbar = 1$ if x is sufficiently large. The WKB expansion is a standard trick to find the asymptotic behaviour of the solutions of various differential equations.

Bibliography

Andrews, G., Askey, R. and Roy, R. (1999). *Special Functions*, Encyclopedia of Mathematics and its Applications (Cambridge University Press).

Bak, J. and Newman, D. (2010). *Complex Analysis*, Undergraduate Texts in Mathematics (Springer, New York).

Bateman, H., Erdélyi, A., Bateman Manuscript Project and U.S. Office of Naval Research (1953). *Higher Transcendental Functions*, Vol. 1 (McGraw-Hill). By Permission of the Copyright Owner, Scanned Copies are Publicly Available.

Bender, C. and Orszag, S. (1978). *Advanced Mathematical Methods for Scientists and Engineers I: Asymptotic Methods and Perturbation Theory*, Advanced Mathematical Methods for Scientists and Engineers (Springer).

Cohen-Tannoudji, C., Diu, B. and Laloe, F. (1992). *Quantum Mechanics, 2 Volume Set* (Wiley).

Erdélyi, A. (2012). *Asymptotic Expansions*, Dover Books on Mathematics (Dover Publications).

Gelfand, I. and Vilenkin, N. (1964). *Generalized Functions: Applications of Harmonic Analysis*, Applications of Harmonic Analysis, Vol. 4 (Academic Press).

Griffiths, D. (2016). *Introduction to Quantum Mechanics* (Cambridge University Press).

Kolmogorov, A. and Fomin, S. (1999). *Elements of the Theory of Functions and Functional Analysis*, Dover Books on Mathematics, Vol. 1 (Dover Publications).

Landau, L. and Lifshitz, E. (2013). *Statistical Physics* (Elsevier Science).

Petrini, M., Pradisi, G. and Zaffaroni, A. (2017). *A Guide to Mathematical Methods for Physicists* (World Scientific).

Pradisi, G. (2012). *Lezioni di Metodi Matematici Della Fisica*, Appunti/Scuola Normale Superiore (Scuola Normale Superiore).

Reed, M. and Simon, B. (1981). *I: Functional Analysis*, Methods of Modern Mathematical Physics (Elsevier Science).

Rudin, W. (1987). *Real and Complex Analysis*, Mathematics Series (McGraw-Hill).

Rudin, W. (1991). *Functional Analysis*, International Series in Pure and Applied Mathematics (McGraw-Hill).

Tricomi, F. (2012). *Integral Equations*, Dover Books on Mathematics (Dover Publications).

Index

Printed in the United States
By Bookmasters